Analytical Technology in Electromagnetic Field Theory

in RF, Wireless and Optical fiber communications

For Electric Engineering
For Physics
For Applied Mathematics

Han Xiong Lian
Professor in Beijing University of Posts and Telecommunications, Beijing, China
Visiting Professor in Simon Fraser University,Vancouver, Canada
Visiting Scholar in Colorado University, Boulder,USA

iUniverse, Inc.
New York Bloomington

Analytical Technology in Electromagnetic Field Theory in RF, Wireless and Optical Fiber Communications

Copyright © 2009 by Han Xiong Lian

All rights reserved. No part of this book may be used or reproduced by any means, graphic, electronic, or mechanical, including photocopying, recording, taping or by any information storage retrieval system without the written permission of the publisher except in the case of brief quotations embodied in critical articles and reviews.

iUniverse books may be ordered through booksellers or by contacting:

iUniverse
1663 Liberty Drive
Bloomington, IN 47403
www.iuniverse.com
1-800-Authors (1-800-288-4677)

Because of the dynamic nature of the Internet, any Web addresses or links contained in this book may have changed since publication and may no longer be valid. The views expressed in this work are solely those of the author and do not necessarily reflect the views of the publisher, and the publisher hereby disclaims any responsibility for them.

ISBN: 978-1-4401-4782-1 (pbk)
ISBN: 978-1-4401-4781-4 (ebk)

Printed in the United States of America

Preface

As my knowledge, the students in the departments of the Physics and Electrical Engineering in the universities of USA and Canada still learn traditional analytical technology from physics for their theoretical analysis in electromagnetic field. However, the traditional analysis can only be used to analyze the *time harmonic* electromagnetic field in a *closed* region filled with a *homogeneous medium* such as a metal wave-guides etc.

This book is written for the students, teachers and the scientists in the Electric Engineering, Physics, and Applied Mathematics.

The writing of this book was prompted by the numerous and basic developments in the theory of electromagnetic field that have taken place since the development of the Information Technology (IT) such as the RF and the Wireless communication, the Satellite communication and the Optical fiber communication.

For instance, when we analyze the RF and wireless wave propagation in the city, the field is in open space and the buildings should be viewed as a discrete medium in the air. When we analyze the microwave propagation in the forest, the forest should be viewed as a random discrete medium in an open space since the scatters (such as the leaves, the branches and the tree trunks) are randomly positioned and randomly oriented.

When we analyze the optical fiber communications system, we have to analyze the non-linearity such as the Stimulated Raman Scattering, the Stimulated Brillouin Scattering, the Self-Focus Phenomenon, the Kerr Effect, the Four-Photon Mixture , the Nonlinear Absorption and the Soliton Transmission in a single-mode fiber. In this case, the single-mode optical fiber should be viewed as a non-linear medium. From the mathematic point of view, we have to discuss *the characteristic method, the perturbation approach, the function transformation method* and *the inverse scattering transformation method* to transform the nonlinear equations into the linear equations so as to get the solutions.

When we analyze the electromagnetic field in a plasma, the plasma should be viewed as a non-homogeneous dynamic medium.

When we analyze the electro-optical devices and the magnetic-optical devices, the devices above are anisotropic medium.

When we analyze the electromagnetic field in some optical fiber sensor, we have to analyze the field in a non-orthogonal coordinators system.

To sum up, we have to analyze the electromagnetic field in an open space filled with a discrete medium, an anisotropic medium, a dynamic medium, a random medium, or a non-linear medium as well as the electromagnetic field in a non-orthogonal coordinates system.

The author is a professor in microwave and optical fiber communications. He has been so lucky to be involved in most of the subjects mentioned above. When he was an associated professor, as a visiting scholar, he has been doing research on *the wireless propagation in the forest* and *the dielectric waveguides* in the University of Colorado, Boulder, USA under the supervising of his supervisors Dr. David C. Cheng (IEEE Fellow)[1] [2][3][13] and Professor Leonard Lewin (IEEE Fellow) [29][30] from 1981 to 1984, and then when he was a full professor, he has been teaching "Applied Mathematics" for the PhD students and doing research on *the nonlinearity in the single-mode optical fiber* [63] [64] [66][65] in Beijing University of Posts and Telecommunications in China from 1985 to 1989, and published a book named "Mathematic Method in the Electromagnetic Field Theory" [6] in China, and then when he was a visiting professor, he has been doing research on *the electro-optical devices* in Simon Fraser University in Burnaby, Canada with Dr. Andrew H. Rawicz [44], Dr. S. Stapleton [43] and Dr. Jamal Deen from 1989 to 1991.

Now the author tries to sum up the analytical technology in the electromagnetic field theory that has taken place in these several ten years, which includes the papers and materials not only from author but also from other authors. All of those have been listed in the bibliography and have been mentioned in each chapter.

Obviously, the problems mentioned above are so complicated which can not be analyzed by means of the traditional analysis, but to set up a new system to analyze the electromagnetic field based on the functional analysis.

Nowadays, the functional analysis has been displaced everywhere on the high level magazine to analyze the complicated boundary value problem in the theory of electromagnetic field. However, "the functional analysis" and "the electromagnetic field theory" are still separated into two causes in USA and Canada.

The functional analysis has been successfully used in the Quantum mechanic [20] [21], those, we would like to say, are the complementary of the applied mathematics and the functional analysis.

In this book, the author

1. Tries to introduce *the principle of the functional analysis* (in Chapter 4),

2. Then to analyze the complicated boundary value problems of the electromagnetic field *based on the functional analysis.* Namely, making use of *the vector spaces and the operator theory* to analyze the boundary value problem of the electromagnetic field

(a) in *an inhomogeneous dynamic medium* (in Chapter 5),

(b) in *a discrete medium* and *anisotropic medium* (in Chapter 6),

(c) in *a nonlinear medium* (in Chapter 7), respectively,

3. Then to analyze the electromagnetic field in *a non-orthogonal coordinates system* (in chapter 8), and

4. To analyze the electromagnetic field in *an anisotropic multi-waveguides system* (in chapter 9),

5. Then following the complex function to explain the conformal transformation (in Chapter 2) and the Wiener-Hopf technology (in Chapter 3) to analyze the field in the micro-strip waveguide and the dielectric waveguide.

Especially, in Chapter 7, we have discussed the nonlinear phenomena in single mode fiber such as the Stimulated Raman Scattering, the Stimulated Brillouin Scattering, the Self-Focus Phenomenon, the Kerr Effect, the Four-Photon Mixture , the Nonlinear Absorption and the Soliton Transmission in a single-mode fiber. From the mathematic point of view, we have discussed *the characteristic method, the perturbation approach, the function transformation method* and *the inverse scattering transformation method* to transform the nonlinear equations into the linear equations to get the solutions.

Now, the numerical method has been widely used in the microwave devices and optical devices to do the calculation by computer. However, the analytical technology still is a powerful tool in the computer aided deign. For instance, a very powerful computer aided deign program named "EESOF" from EESOF Inc. and the "Advanced Design System" (ADS) from Agilent Inc. are mostly based on the analytical solutions. Now this software has been used widely in the world to settle down a big project with a small amount of computer time compared with that from the purely numerical method. Meanwhile, the analytical technology can provide the medelling, analytical technology and the physical meaning for the analytical solution. Which encourages us to learn more analytical technology.

Finally, the author wish to express his gratefully acknowledge for his supervisors Dr. David C. Chang (IEEE Fellow) and professor Leonard Lewin (IEEE Fellow). When both of them were in the University of Colorado, Boulder, USA the supervising from both of them led him to connect the great projects and made him to get great progress. In addition, he wish to special thank Professor Ye, Peida (IEEE Fellow), he gave him a change to teach the applied mathematics for PhD students in Beijing University of Posts and Telecommunications. Meanwhile, he wish to express his gratefully acknowledge for the helpful comments, criticism of early version of this book to the follows: Professor Renhai Lou, Professor Jingren Qian, Professor Zhengming Hu, Professor Kejin Rao, Professor Quanrang Yang, and Professor Zesong Quan in China.

Finally, the author wish to express his gratefully acknowledge for his publisher iUniverse Inc., for their continued support of this book.

Han Xiong Lian
2008.09.01

Contents

Part I

Application of complex function

In Part I, we will mainly discuss the complicated boundary value problem for the time harmonic electromagnetic field in a linear homogenous medium. In which, the boundary may be the one between a dielectric medium and a metal or a dielectric medium and another dielectric medium in an open space.

Conformal Transformation: As well known, the complex functions still is a powerful tool for the boundary value problem of the *time harmonic* electromagnetic field in a *linear homogenous medium* since the wave function of the time harmonic electromagnetic field in a linear homogenous medium satisfies the two-dimensional Laplace equation. Meanwhile, both the real part and the imaginary part of an analytic function also satisfy the two-dimensional Laplace equation. Therefore, the real part or the imaginary part of the analytic function can be used to represent the wave function of the time harmonic electromagnetic field in a linear homogenous medium. In addition, the coordinates transformation of the analytic function is a conformal transformation. That means that the wave function represented by an analytic function also satisfies the conformal transformation. In this case, the complicated boundary value problem of wave function can be transformed to a simple boundary value problem by means of the conformal transformation. This is the so call the *Conformal Transformation*, where, the Conformal Transformation is an angle-preserving transformation.

The *Schwartz transformation* is also a coordinates transformation. However the *Schwartz transformation* is not an angle-preserving transformation but, instead a coordinates transformation from a polygon region to a upper plane in a complex plane.

Here, we are going to spend little bit time to explain the Conformal Transformation and the Schwartz transformation in this chapter, then to spend more time to introduce the Wiener-Hopf technology deduced from the integral transformation in Part I. The Wiener-Hopf technology is a powerful tool to deal with the half space boundary value problem in open region such as the microstrip and the dielectric waveguide.

As a preliminary knowledge, we will briefly discuss some basic concepts [7] and some formulas, which will be used frequently after then.

Chapter 1

From closed region to open space

1.1 Introduction

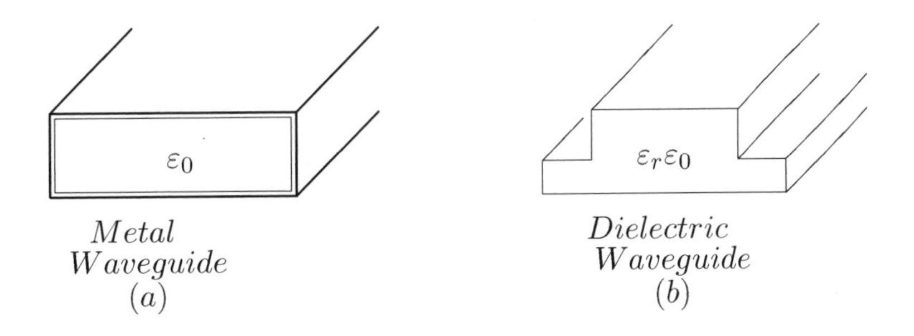

Figure 1-1 From closed region to open space

Recently, the rapid development of the wireless communications, satellite communications and optical fiber communication encourage scientists to analyze various dielectric waveguides used for the optical devices, lasers and microwave devices. In this case, the study of the guidewave turns to study the electromagnetic field in an open space such the microstrips and the dielectric waveguides as indicated in Figure 1-1 (b) other then the usual study in a closed region such as the metal waveguide as indicated in Figure 1-1 (a).

As well known, the fields in a closed metal waveguide in consist of the normal modes. From the spectral point of view, the spectral of the electromagnetic field in a closed region such as the metal waveguide is consist of a discrete spectral. However, in open space such as a dielectric waveguide, the spectral of the electromagnetic field is consist of not only the discrete spectral but also the continuous spectral.

As well known, the fields in a closed region such as in a metal waveguide can be represented by a series of normal modes, which are so called the discrete spectral. In which, one normal model in the metal waveguide is a guided mode and the others are non-guide modes, which are so called the cutoff modes and the cutoff modes stop inside the metal waveguide. In this case, the state in waveguide is called the single mode propagation. Which can be down by the waveguide design, namely, the size of the cross section in waveguide can be controlled to get the single mode propagation condition.

Physically, if the size of the waveguide is getting larger and larger, the difference between two neighboring eigenvalues will be getting smaller and smaller, and finally, the eigenvalues will become continuous spectrum when the size of the metal waveguide becomes infinity, which is just the case of an open space.

In fact, in open space such as a dielectric waveguide, the spectral of the electromagnetic field is consist of not only the discrete spectral but also the continuous spectral. And the single mode propagation con be realized by the geometric design in dielectric waveguide. Which is the same as the metal waveguide, the guided mode is propagation inside the dielectric waveguide. However, the cutoff modes become the radiation modes leaking to outside the dielectric waveguide.

1.2 The mode representative in an open space

Now we will discuss the mode representative of the field in an open space.

1. For instance, suppose that $\phi(x, z)$ is an electric field component or a magnetic field component in a two dimensional open space such that

$$\left(\frac{\partial^2}{\partial x^2} + \frac{\partial^2}{\partial z^2} + k^2\right) \phi(x, z) = 0, \tag{1.1}$$

After make a Fourier Transformation of (1.1), we have the solution in form of

$$\phi(x, z) = \frac{1}{2\pi} \int_{-\infty}^{\infty} \Phi(\alpha) \exp(-i\alpha z) \exp(-\Upsilon x)\, d\alpha, \quad x > 0 \tag{1.2}$$

Where the solution has been inverse Fourier Transformation already. In which, α is the eigenvalue of the normal mode, $\Phi(\alpha)$ is a continuous spectrum, and

$$\Upsilon = (\alpha^2 - k^2)^{1/2} \tag{1.3}$$

From mathematic of view, the α in (1.3) can be $-\infty < \alpha < \infty$, and then γ is a complex in general. Meanwhile, γ is a multiple function of α. However, from physical point of view, γ should be uniquely determined, which can be down by the *radiation condition*, namely, since $x > 0$, the radiation condition require that

$$Re\,\Upsilon > 0 \tag{1.4}$$

$$Im\,\Upsilon \leq 0 \tag{1.5}$$

The first one is to satisfies the radiation condition, namely, the field

$$\phi(x,z) \to 0 \quad when\ x \to \infty \tag{1.6}$$

and the second one is to match the time factor $\exp(-i\omega t)$ in this chapter. To meet the requirement in (1.4) and (1.5), we have to discuss the *branch cut* of γ and Reimann surfaces after then.

2. Now we would like to introduce a dielectric wavguide as indicated in Figure 1-2.

This dielectric waveguide will be discussed in detail in Chapter 3. In which, when an electric field E_y^i is incident obliquely on to the interface between the grounded dielectric waveguide and the pure dielectric waveguide, the scattering field $E_y^s(x,y)$ may be obtained by the Maxwell equations and the boundary conditions as follows:

$$\begin{aligned} E_y^s(x,y) &= \int_{-\infty}^{\infty} E_y^s(\lambda,y)e^{-ik_0\lambda x}d\lambda \\ &= \int_{-\infty}^{\infty} -\Big[\frac{(u_n+\varepsilon_r u_0 \tanh u_n k_0 2d)\cosh u_n k_0(y-2d)}{u_n(\varepsilon_r u_0 + u_n \tanh u_n k_0 2d)} \\ &\quad + \frac{1}{u_0}\sinh u_n k_0(y-2d)\Big]F_-(\lambda)e^{-ik_0\lambda x}d\lambda \end{aligned} \tag{1.7}$$

Where

$$u_0 = (\lambda^2+\alpha^2-1)^{1/2} \tag{1.8}$$

$$u_n = (\lambda^2+\alpha^2-n^2)^{1/2} \tag{1.9}$$

and λ is the eigenvalue of the normal mode.

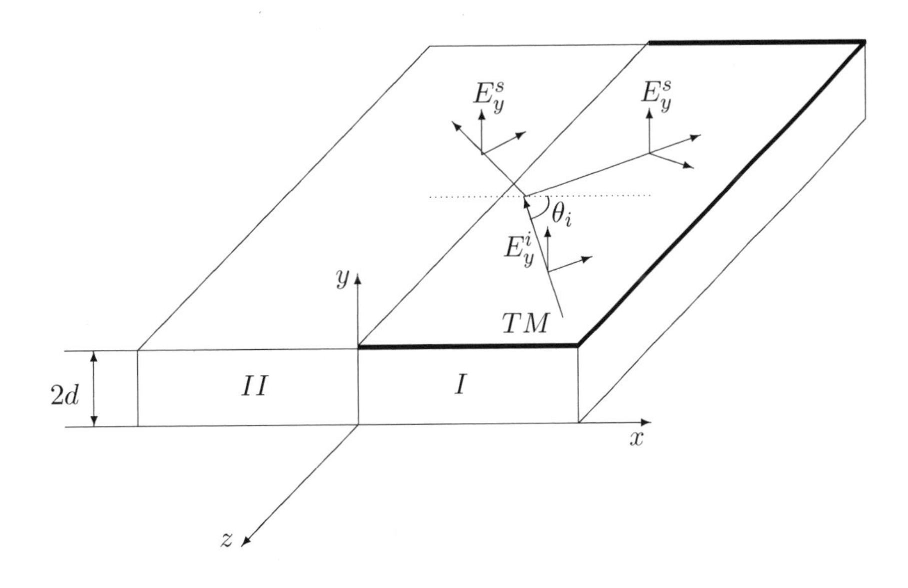

Figure 1-2 Dielectric Waveguide

Now, it is interested to point out that the normal modes in the dielectric waveguide is from the discussion of the singularities of the integrand function inside the integral (1.7). Investigate the integrand in (1.7) carefully, we found that the integrand possesses of three singularity points as follows:

1. $\lambda = \lambda_{r1}$ corresponding to a singularity point $u_0 = 0$ is a radiation mode.

2. $\lambda = \lambda_{r2}$ corresponding to a singularity point $u_n = 0$ is another radiation mode.

3. $\lambda = \lambda_m$ corresponding to a singularity point

$$\varepsilon_r u_0 + u_n \tanh(u_n k_0 2d) = 0 \tag{1.10}$$

is a guided mode inside the dielectric waveguide. In which

$$u_o = (\lambda_m^2 + \alpha^2 - 1)^{1/2}$$
$$u_n = (\lambda_m^2 + \alpha^2 - n^2)^{1/2}$$

and then the equation (1.10) becomes

$$\varepsilon_r(\lambda_m^2 + \alpha^2 - 1)^{1/2} + (\lambda_m^2 + \alpha^2 - n^2)^{1/2} \cdot \tanh[(\lambda_m^2 + \alpha^2 - n^2)^{1/2} k_0 2d] = 0 \tag{1.11}$$

Which is the eigenvalue function of the guided mode in dielectric waveguide and $n = \sqrt{\varepsilon_r}$ is the refraction index of the dielectric waveguide.

The solutions of the integral (1.7) is equal to the residue contributions of the integral at $\lambda = \lambda_{r1}$, $\lambda = \lambda_{r2}$ and $\lambda = \lambda_m$, respectively.

1.2.1 Branch point and branch cut

Come back to the integral of (1.2) and the α in (1.3), we have to discuss the branch points and branch cuts to see how we can find a integral path, which doesn't cross the *branch cut.*

Branch point:

Definition of branch point: In the mathematic field of complex analysis, a **Branch point** of a multi-value function is a point around which the function is discontinuous, namely, the value of the function at the start point is different from the end point after around which a small circle.

Example of branch point:

1. For instance, the function $f(z) = z^2$ has a branch point at $z_0 = 0$. The inverse function of f is the square root $f^{-1}(z) = z^{1/2}$. Going around the closed loop $z = e^{i\theta}$, $e^{i0/2} = 1$ at the start point $\theta = 0$. However, at the end point, $\theta = 2\pi$, $e^{i2\pi/2} = -1$.

2. 0 is also a branch point of the natural logarithm. Since e^0 and $e^{2\pi i}$ are both equal to 1, both 0 and $2\pi i$ are among the multiple value of $Log(1)$.

3. In trigonometry, since $\tan(\pi/4)$ and $\tan(5\pi/4)$ are both equal to 1, the two numbers $\pi/4$ and $5\pi/4$ are among the multiple values of $\arctan(1)$.

4. And then $\gamma = (\alpha^2 - k^2)^{1/2}$ in (1.3) has a branch point at

$$\alpha = \pm k \tag{1.12}$$

where

$$k = k_1 + ik_2, \qquad (k_1 >> k_2)$$

To make the function γ single-valued, we have to discuss the branch cut as follows.

Branch cut:

Definition of branch cut: A branch cut is a curve in the complex plane such that it is possible to define a single value surface, called the Reimann surface, for a multiple value function.

Branch cut allow one to work with a collection of single-valued functions, "glued" together along the branch cut instead of a multi-valued function.

In this case, the final result of the integral in (1.2) will be the single value as long as the integral path is on one Reimann surface. In other words, the final result of the integral in (1.2) will be the single value as long as the integral path does't around the branch point or doesn't pass across the branch cut.

Example of branch cut:

1. For example, to make the function

$$F(z) = \sqrt{z}\sqrt{1-z}$$

single valued, we can make a branch cut along the interval $[0, 1]$ on the real axis to connect the two branch points of the function.

2. The same idea can be applied to the function $\sqrt{z}$, however, we have to perceive that the point at infinity is the 'other' branch point to connect to from 0, namely along the whole negative axis.

3. A typical example of the branch cut is the complex logarithm in form of

$$log\, z = ln\, r + i\theta$$

where $z = re^{i\theta}$. In which, adding to θ any integer multiple of 2π will yields another possible angle.

And then a branch of the logarithm exists in the complement of any ray from the origin to infinity, which is a branch cut to avoid any adding to θ any integer multiple of 2π. Especially a common choice of branch cut is the negative real axis.

4. A function

$$f_a(z) = \frac{1}{z-a}$$

is a function with a single pole at $z = a$. Integrating over the location of the pole:

$$u(z) = \int_{a=-1}^{a=1} f_a z)da = \int_{a=-1}^{a=1} \frac{1}{z-a} da = log\left(\frac{z-1}{z+1}\right)$$

defines a function $u(z)$ with a branch cut from -1 to 1. [1] The branch cut can be moved around, since the integration line can be shifted without altering the value of the integral so long as the line doesn't pass across the point $a = z$.

1.2.2 Reimann surface

Now we will take an example to discuss the Reimann surface in a complex plane. To this end, we come back to see the integral (1.2) in form of

$$\phi(x,z) = \frac{1}{2\pi} \int_{-\infty}^{\infty} \Phi(\alpha) \exp(-i\alpha z) \exp(-\Upsilon x)\, d\alpha, \quad x > 0 \tag{1.13}$$

and the conditions (1.4) and (1.5) in form of

$$Re\,\Upsilon > 0 \tag{1.14}$$

$$Im\,\Upsilon \le 0 \tag{1.15}$$

Where the integral (1.13) is the solution of the equation (1.1) without any other boundary condition except the radiation condition, which results in the condition (1.14). In this case, the solution of the integral (1.13) contain the radiation modes only. And the condition (1.15) is to match the time factor of $\exp(-i\omega t)$ in this chapter.

To satisfy the condition (1.14), we have to discuss the branch cut [2] of Υ. First, look at (1.3), consider the example 1 of the branch point, the branch-points of Υ are at

$$\Upsilon = 0 \quad or \quad \alpha = \pm k \tag{1.16}$$

where

$$k = k_1 + i\,k_2 \quad (k_1 >> k_2).$$

Now we will put the branch points $\alpha = \pm k$ on the α complex plane. To this end, we have to put

$$\alpha = \sigma + i\tau \tag{1.17}$$

into (1.3), which leads to

$$\Upsilon^2 = [(\sigma^2 - \tau^2) - (k_1^2 - k_2^2)] + 2i(\sigma\tau + k_1 k_2) \tag{1.18}$$

And then the distribution of Υ^2 on α-plane is indicated in Figure 1-3.

---

[1] Since both $z = -1$ and $z = 1$ will make $log\frac{z-1}{z+1} \to \infty$.

[2] The branch cut is a curve in the complex plane across which an analytic multi-valved function is discontinuous.

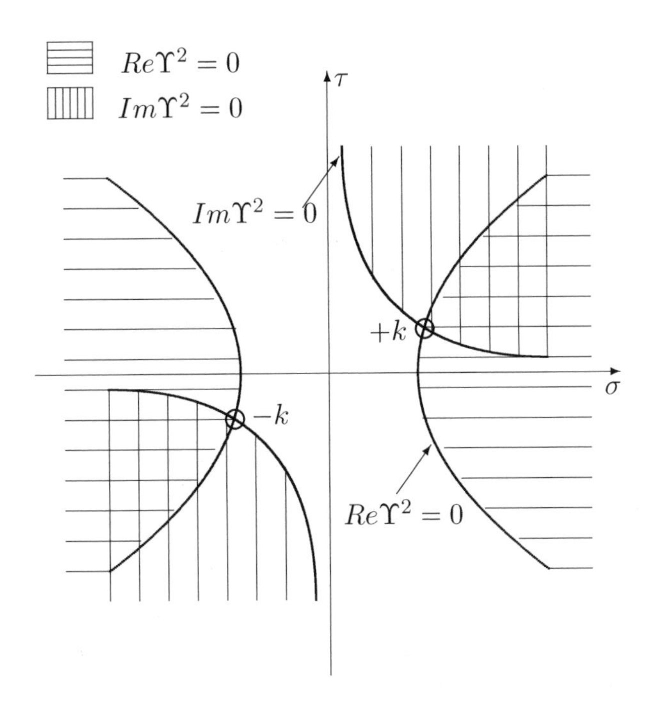

Figure 1-3 Υ^2 in the α-plane (after R. Mittra and S. W. Lee [12])

Obviously, $Re\Upsilon^2 = 0$ gives an equation in form of

$$(\sigma^2 - \tau^2) - (k_1^2 - k_2^2) = 0 \tag{1.19}$$

which is a hyperbola equation and the curves are shown in Figure 1-3. $Im\Upsilon^2 = 0$ gives another equation in form of

$$\sigma\tau + k_1 k_2 = 0 \tag{1.20}$$

which is another hyperbola equation and shown in Figure 1-3.

Now we will point out:

1. The intersection of these two kinds of hyperbola curves is at $\alpha = \pm k$, which are two branch points as indicated in figure 1-3.

2. The shadow region with horizontal lines is corresponding to $Re\Upsilon^2 > 0$ and the region with non-horizontal lines is corresponding to $Re\Upsilon^2 < 0$.

3. The shadow region with vertical lines is corresponding to $Im\Upsilon^2 > 0$ and the region with non-vertical lines is corresponding to $Im\Upsilon^2 < 0$.

Our purpose is to find the Reimann surfaces to keep Υ to be unique on each Reimann surface.

To make Υ to be unique, we can make two Reimann surfaces in the complex α-planes, in which two Reimann surfaces connected by the branch cut, and then Υ is an analytical single-valued function of α on each Reimann surface. It follows that Υ becomes discontinuous only if the Υ pass across the branch cut.

The choice of the branch cut is fairly arbitrarily in the α-plane as long as it pass through the branch points. For instance, we may define the branch cut of the Υ in the α-plane such that $Re\,\Upsilon > 0$ on the upper Reimann surface , and $Re\,\Upsilon < 0$ on the lower Reimann surface such that the branch cut of Υ located at $Re\,\Upsilon = 0$.

To assure that $Re\Upsilon > 0$ on the entire upper Reimann surface, it is required that $|\arg\Upsilon^2| < \pi$ since $\Upsilon = |\Upsilon|\exp\left(i\arg\Upsilon^2/2\right)$. If $|\arg\Upsilon^2| > \pi$, we will have $Re\Upsilon < 0$. Consequently, the branch cut is located at

$$\arg\Upsilon^2 = \pi$$

or

$$Re\Upsilon^2 < 0, \quad Im\Upsilon^2 = 0 \tag{1.21}$$

Following the condition (1.21) to investigate Figure 1-3 , we see that the branch cut are located at the dash lines in Figure 1-4. In which:

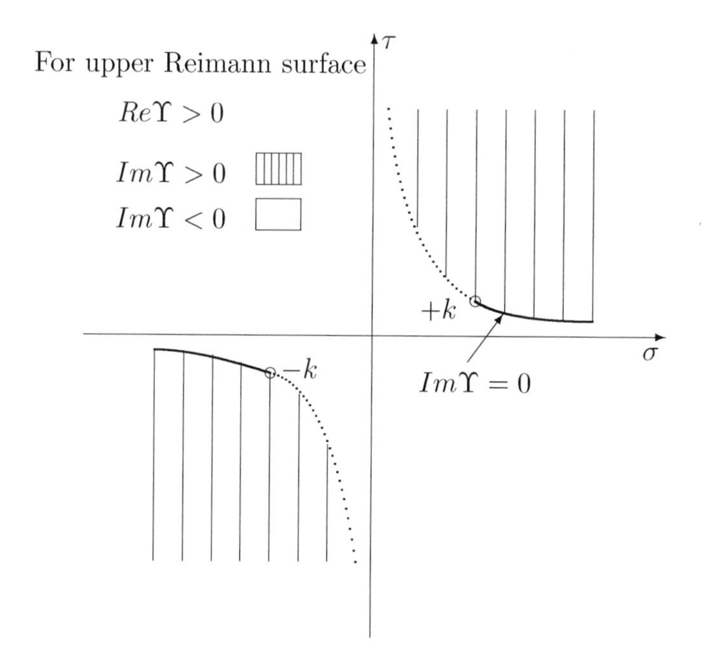

Figure 1-4 Reimann surfaces (after R. Mittra and S. W. Lee [12])

On upper Reimann surface

$Re\Upsilon > 0$
$Im\Upsilon > 0$ (in shadow region)
$Im\Upsilon < 0$ (in non-shadow region)

On lower Reimann surface

$Re\Upsilon < 0$
$Im\Upsilon > 0$ (in shadow region)
$Im\Upsilon < 0$ (in non-shadow region)

Now we will come back to see if the plane wave in (1.13) satisfies the radiation condition (1.14) and (1.15). Suppose that the integral path in (1.13) from $\alpha = -\infty$ to ∞ is realized along the real axis on the upper Reimann surface, then every α value along the integral path satisfies the radiation condition. And then the integral is fully determined.

Obviously, on upper Reimann surface, $Re\Upsilon > 0$, the wave satisfies the radiation condition. On lower Reimann surface, $Re\Upsilon < 0$, the wave doesn't satisfies the radiation condition.

Once the integral path in (1.2) is realized along the real axis on the upper Reimann surface , the integral path can be modified in terms of the *Cauchy-Reimann Integral Theorem*, and then calculated by the **Saddle point method** for the field far away from the source. The **Saddle point method** may be found, for instance, in [12].

1.3 Radiation condition and edge condition

There are two kinds of boundary value problems: One is in a closed space, another one is in an open space.

For a closed space:

The basic equations are the Maxwell equations, the material equations and the boundary conditions.

For an open space:

Except the basic equations mentioned above, we have to take the *radiation conditions* and the *edge conditions* into account. Otherwise, the solution in an open space could not be unique. Of course, like the boundary conditions, the radiation conditions and the edge conditions are also from the Maxwell equations.

Radiation condition:

In an open space, suppose that all of the sources of the fields are contained in a finite-region, then the field at infinite have to satisfy the *radiation condition*, i.e.:

(1) If the space is filled with a dissipative medium, the field components ψ at infinite has to satisfy

$$\psi|_{r\to\infty} = 0 \tag{1.22}$$

which is just the case in (1.6).

(2) If the space is filled with a non-dissipative medium, the field at infinity has to satisfy the *Sommerfeld radiation condition.*

Arnold Sommerfeld defined the radiation condition for a scalar field satisfying the Helmholtz equation as following:

Mathematically, Arnold Sommerfeld consideran inhomogeneous Helmhotz equation if form of

$$(\nabla^2 + k^2)u = -f \quad in\ \mathbf{R}^n$$

where n=2,3 is the dimension of the space, f is a given sources, which is a bounded source of energy, and $k > 0$ is a constant, called the wavenumber.

A solution u to this equation is called radiating if it satisfies the **Sommerfeld radiation condition** in form of

$$\lim_{|x|\to\infty} |x|^{\frac{n-1}{2}} \left(\frac{\partial}{\partial |x|} - ik \right) u(x) = 0 \tag{1.23}$$

For three dimensional space, $n = 3$, the the **Sommerfeld radiation condition** become

$$\lim_{r\to\infty} r\left(\frac{\partial \psi}{\partial r} - ik\psi\right) = 0 \tag{1.24}$$

where, $r = |x|$ is the distance from the source point to the investigation point in any direction, and $\psi(r) = u(x)$.

The *Sommerfeld radiation condition* says that the field far away from the source is vanishing faster than r^{-1} in magnitude with a time delay in phase.

Edge condition:

Another condition has to be considered is the geometry singularity at a sharp edge. In principle, the field at a sharp edge will approach to infinity. In this case, the results of the field could not be unique. To make the result to be unity, it is necessary to have *the edge conditions.*

The *edge conditions* says that the stored energy of the electromagnetic field in any finite space around the edge should be a finite value, namely,

$$\int_V (\epsilon |E|^2 + \mu |H|^2) dV \to 0 \tag{1.25}$$

when the volume V compacts to a neighborhood of an edge. Where, for a smooth edge, $dV = \rho\, d\rho\, d\phi\, dz$, and (ρ, ϕ, z) is the local cylinder coordinates at edge. Therefore, from equation (1.25), when $\rho \to 0$, the increasing rate of each field component $(E\,, H\,)$ can not be lager than $\rho^{-1+\tau}$ near the edge, where $\tau > 0$. The value of τ can be determined by a particular edge condition.

As shown in Figure 1-5(a), a wedge of conductor located in a dielectric medium $(\mu_0\,, \varepsilon_r\varepsilon_0\,)$, and Figure 1-5(b) is a special case of Figure 1-5(a) with $\phi_1 = \phi_0$.

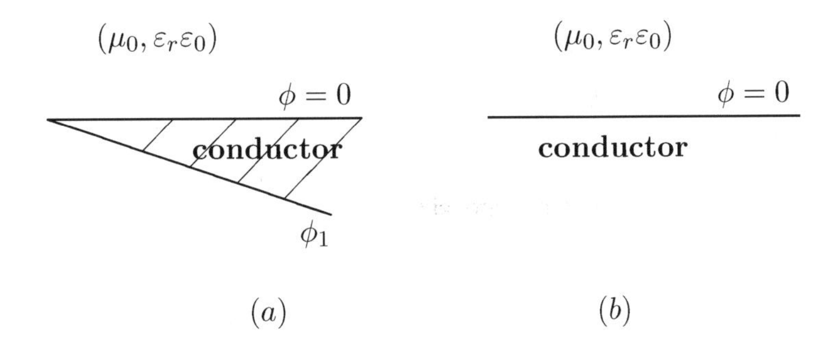

Figure 1-5 Edge condition of a wedge conductor

The edge condition has been studied by J. Meixner in the paper "The Behavior of Electromagnetic Fields at Edges" Inst. Math. Sci. Res. Rept. EM-72, New York University, New York, N.Y. Dec. 1954 [9]. And then R. Mittra and S. W. Lee have a in detail discussion in various edge conditions of conductor in the book [12]. Therefore, in the following discussion, we only concentrate to introduce that how to study the edge condition of conductor by means of the Maxwell equations, boundary conditions and the edge condition (1.25).

For a general situation, R. Mittra and S. W. Lee [12] have discuss A wedge of conductor located at the intersection point of medium 1 (μ_1, ε_1), medium 2 (μ_2, ε_2) and medium 3 (μ_3, ε_3). In this case:

1. The Maxwell equations should be written in a cylinder coordinates system (ρ, ϕ, z).

2. The components of the field near the edge can be expanded to a series of ρ, and the coefficients of the series is the functions of ϕ and z. For instance,

$$\begin{aligned} E_\rho &= \rho^{-1+\tau}[a_0^{(j)} + a_1^{(j)}\rho + a_2^{(j)} + \cdots] \\ H_\rho &= \rho^{-1+\tau}[A_0^{(j)} + A_1^{(j)}\rho + A_2^{(j)} + \cdots] \end{aligned} \tag{1.26}$$

where $j = 1, 2, 3$ denote the fields in three regions, respectively.

3. Where, from edge condition (1.25), the the increasing rate of the field component in the neighborhood of the edge could not be lager than $\rho^{-1+\tau}$. This point has been considered in 2.

4. And then the fields in the neighborhood of the edge should satisfy the Maxwell equations and the boundary conditions, which will result 30 linear equations by considering that the coefficient with the same power of ρ in both side of the equality should be equal to each other.

5. The nonzero solutions from the 30 equations require that one of the following two conditions shall be satisfied:

(1) **Condition A:**

$$a_0^{(j)} = b_0^{(j)} = c_0^{(j)} = C^{(j)} = 0, \quad j = 1, 2, 3$$

$$F_\mu(\tau) = 0 \tag{1.27}$$

where $a_0^{(j)}$, $b_0^{(j)}$ and $c_0^{(j)}$ are the first coefficients of E_ρ, E_ϕ and E_z, respectively, and $C^{(j)}$ is the first coefficient of H_z.

And then the edge conditions under the condition A are

$$\begin{aligned} H_t &= O(\rho^{-1+\tau}) \\ H_z, E &= O(\rho^{\tau}) \end{aligned} \tag{1.28}$$

when $\rho \to 0$.

(2) **Condition B:**

$$A_0^{(j)} = B_0^{(j)} = C_0^{(j)} = c_0^{(j)} = 0 \tag{1.29}$$

$$F_\varepsilon(\tau) = 0 \tag{1.30}$$

where $A_0^{(j)}$, $B_0^{(j)}$ and $C_0^{(j)}$ are the first coefficients of E_ρ, H_ϕ and H_z, respectively, and $c_0^{(j)}$ is the first coefficient of E_z.

And then the edge conditions under the condition B are

$$\begin{aligned} E_t &= O(\rho^{-1+\tau}) \\ E_z, H &= O(\rho^{\tau}) \end{aligned} \tag{1.31}$$

when $\rho \to 0$.

Then R. Mittra and S. W. Lee [12] have analyzed various particular edge conditions of conductor and gives the solution of τ for each situation. The most important one is the edge condition as indicated in Figure 1-5 (b), which is

$$\tau = \frac{1}{2} \tag{1.32}$$

Chapter 2

Conformal Transformation

2.1 Principle of Conformal Transformation

[10]

As shown in Figure 2-1, any point Z in the complex Z-plane can be expressed as

$$Z = x + i\,y = r\,\exp(i\,\theta) \tag{2.1}$$

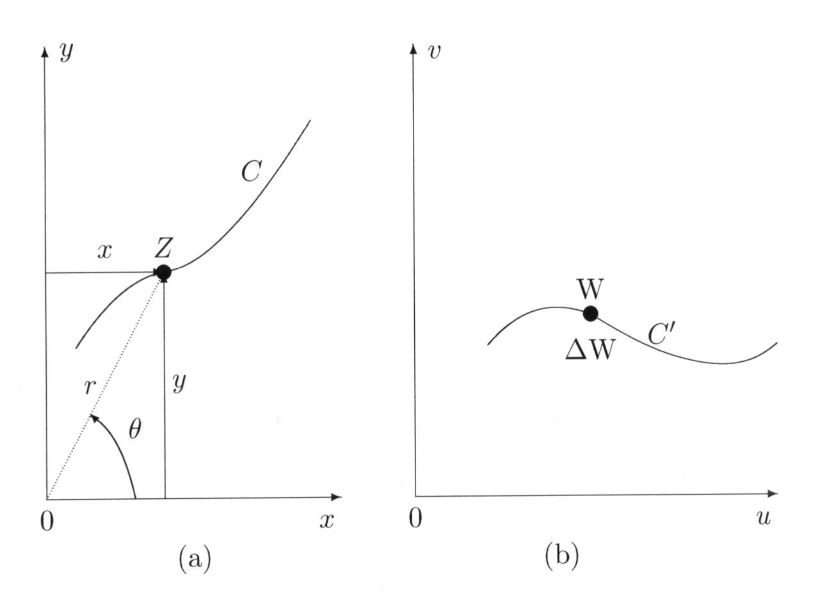

Figure 2-1 (a) Z-plane (b) W-plane

Introducing a new variable

$$W = u + iv = \rho\,\exp(i\phi) \tag{2.2}$$

where W is a function of Z,

$$W = f(Z) \tag{2.3}$$

and W will be continuous variation along the path C' in the W- plane when z is continuous variation along the path C in the Z-plane as shown in Figure 2-1(a) and Figure 2-1(b).

Following the definition of the complex functions, we have

$$\frac{dW}{dZ} = \lim_{\Delta Z \to 0} \frac{\Delta W}{\Delta Z} = \lim_{\Delta Z \to 0} \frac{f(Z+\Delta Z) - f(Z)}{\Delta Z} \tag{2.4}$$

where if $f(Z)$ is analytic in certain region, then the derivative (2.4) always tends to a value and the value is independent of the direction of ΔZ. Therefore, if $\Delta Z = \Delta x$, we have

$$\frac{dW}{dZ} = \frac{\partial W}{\partial x} = \frac{\partial}{\partial x}(u + i\,v) = \frac{\partial u}{\partial x} + i\,\frac{\partial v}{\partial x} \tag{2.5}$$

if $\Delta Z = i\,\Delta y$, we have

$$\frac{dW}{dZ} = \frac{\partial W}{\partial (iy)} = -i\frac{\partial}{\partial y}(u + i\,v) = \frac{\partial v}{\partial y} - i\,\frac{\partial u}{\partial y} \tag{2.6}$$

If $f(Z)$ is analytic in this region, then these two derivatives above should be equal to each other. And then we have

$$\frac{\partial u}{\partial x} = \frac{\partial v}{\partial y} \tag{2.7}$$

$$\frac{\partial v}{\partial x} = -\frac{\partial u}{\partial y} \tag{2.8}$$

This is the well known *the Cauchy-Reimann Equations*. Form which, it follows that we can reach to the following equations:

$$\frac{\partial^2 u}{\partial x^2} + \frac{\partial^2 u}{\partial y^2} = 0 \tag{2.9}$$

$$\frac{\partial^2 v}{\partial x^2} + \frac{\partial^2 v}{\partial y^2} = 0 \tag{2.10}$$

Where (2.9) and (2.10) are the well known *two dimensional Laplace Equations.*

As well known, the electric potential of the statistic electric field satisfies two dimensional Laplace equations. This implies that u in (2.9) (or v in (2.10)) can be used to represent the potential function of the statistic electric field. Meanwhile, the wave equations of the TEM wave in homogenous medium can be deduced to a two dimensional Laplace equations [1]. Which means that the potential function of the TEM wave also satisfies the two dimensional Laplace equations. And then the u in (2.9) (or the v in (2.10)) also can be used to represent the potential function of the TEM wave. And then the transverse components of the TEM wave will be

$$E_x = -\frac{\partial u}{\partial x}, \quad E_y = -\frac{\partial u}{\partial y} \tag{2.11}$$

if we take u as the potential function of the TEM wave.

[1] where the propagation constant of the TEM mode is $k = \omega\sqrt{\mu\varepsilon} = constant$.

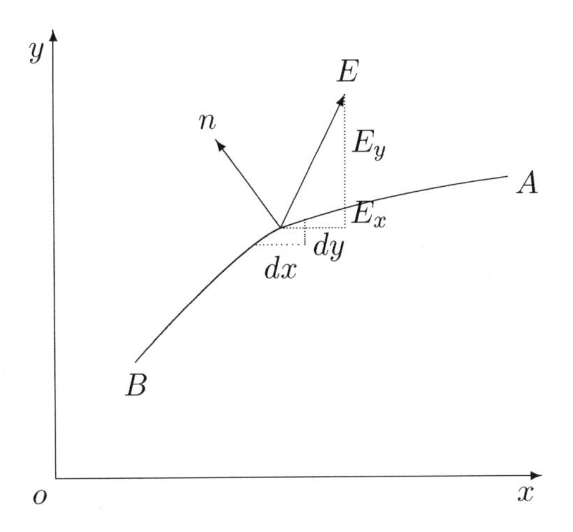

Figure 2-2 Electric Flux

As shown in Figure 2-2, suppose that there is a surface of a cylinder with a unit length along z direction in a two dimensional field. The transverse section of the cylinder is a curve between A and B in the Z-plane. Then the electric flux through the surface of the cylinder is

$$\psi = \int_A^B \mathbf{D} \cdot \mathbf{dS} = \int_A^B (D_x dy - D_y dx) \tag{2.12}$$

Where

$$D_x = \varepsilon E_x = -\varepsilon \frac{\partial u}{\partial x} = -\varepsilon \frac{\partial v}{\partial y}$$

$$D_y = \varepsilon E_y = -\varepsilon \frac{\partial u}{\partial y} = \varepsilon \frac{\partial v}{\partial x}$$

Making use of E_x and E_y in (2.12), the electric flux through the curve is

$$\psi = -\varepsilon \int_A^B (\frac{\partial v}{\partial y} dy + \frac{\partial v}{\partial x} dx) = -\varepsilon \int_A^B dv = -\varepsilon [v_B - v_A]$$

or

$$\psi = \varepsilon [v_A - v_B] \tag{2.13}$$

That means that the electric flux through the curve from A to B, ψ, is equal to the production of ε with the difference between v_A and v_B. Consequently, v is said to be the flux function. In other words, if we take $u = constant$ as the locus of a equi-potential line, then the locus of $v = constant$ is the electric flux line, vice versa. These two loci are mutually orthogonal in the complex Z-plane. In fact, in two dimensional field, we have

$$\nabla u \cdot \nabla v = \left(\mathbf{i}_x \frac{\partial u}{\partial x} + \mathbf{i}_y \frac{\partial u}{\partial y}\right) \cdot \left(\mathbf{i}_x \frac{\partial v}{\partial x} + \mathbf{i}_y \frac{\partial v}{\partial y}\right)$$

or

$$\nabla u \cdot \nabla v = \frac{\partial u}{\partial x}\frac{\partial v}{\partial x} + \frac{\partial u}{\partial y}\frac{\partial v}{\partial y} = \frac{\partial u}{\partial x}\frac{\partial v}{\partial x} - \frac{\partial v}{\partial x}\frac{\partial u}{\partial x} = 0$$

This implies that the two curve families u and v are mutually orthogonal in the complex Z-plane. Meanwhile, u and v are also mutually orthogonal in the W-plane since u and v are the abscissa and the ordinate in the complex W-plane, respectively. That implies that the transformation from the Z-plane to the W-plane (2.3) is a *conformal transformation* provided that $f(Z)$ is analytic in the Z-plane.

Mathematically, If the function $W = f(Z)$ is differentiable at point Z, and $\frac{dW}{dZ} \neq 0$, then we can write

$$\frac{dW}{dZ} = M \exp(i\,\alpha)$$

or

$$dW = M\, dZ \exp(i\,\alpha)$$

That implies that dW is M times lager than dz in magnitude and is rotating an angle α around the origin with respect to dz in phase.

It follows that two curves intersected with an angle in the Z-plane will be transformed into two curves intersected with the same angle in the W-plane. This is because that the transformation of each curve from the Z-plane to the W-plane rotates a same angle α around the origin. This transformation is said to be *the conformal transformation* and can be written as a theorem as follows:

Theorem If a function $W = f(Z)$ is differentiable at any point Z, and $f'(Z) \neq 0$, then the transformation $W = f(Z)$ at the point Z is conformal. Where arg $f'(Z)$ is the rotation angle resulting from the transformation, $|f'(Z)|$ is the magnitude factor resulting from the transformation.

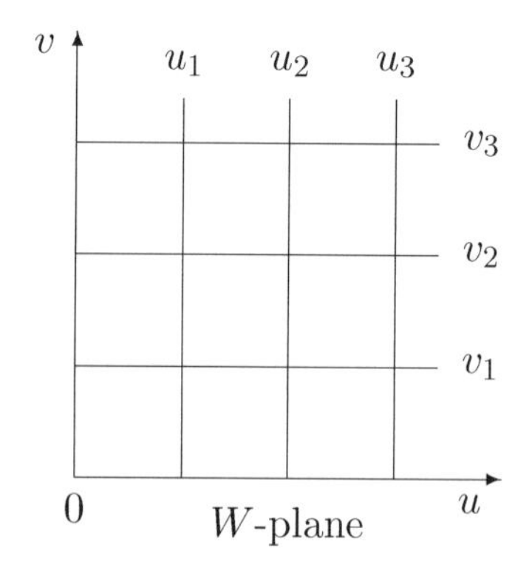

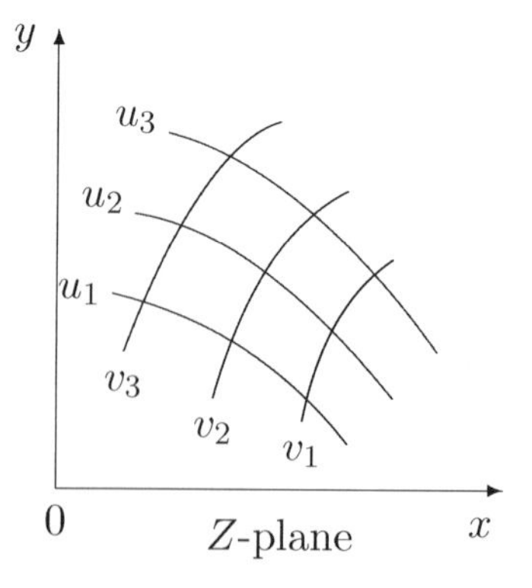

Figure 2-3 Conformal Transformation

Definition a **conformal transformation** (or conformal mapping) is a function which preserves angles. Formally, a transformation

$$W = f(Z) \tag{2.14}$$

is said to be *conformal* (or *angle-preserving*) at z_0, if it preserves oriented angle between curves through z_0, as well as their orientation, i.e. direction.

This conformal transformation may be indicated in Figure 2-3. Suppose that the equi-potential lines ($u_1 = c_1, \cdots$) in the Z-plane are in accordance with the equi-potential lines of the two dimensional field, then a complicated boundary value problem in the Z-plane can be transformed into a simple boundary value problem in the W-plane. Following examples will help us to understand step by step.

2.2 Application of the Conformal Transformation

The following examples will show us:

(1) how to obtain the solution of a two dimensional field by means of the conformal transformation method, and

(2) how to obtain the capacitance and the impedance of an unit length of the TEM transmission line,

(3) how to set up the transformation function? This is the key point.

Example 1: Find the capacitance of a coaxial cable

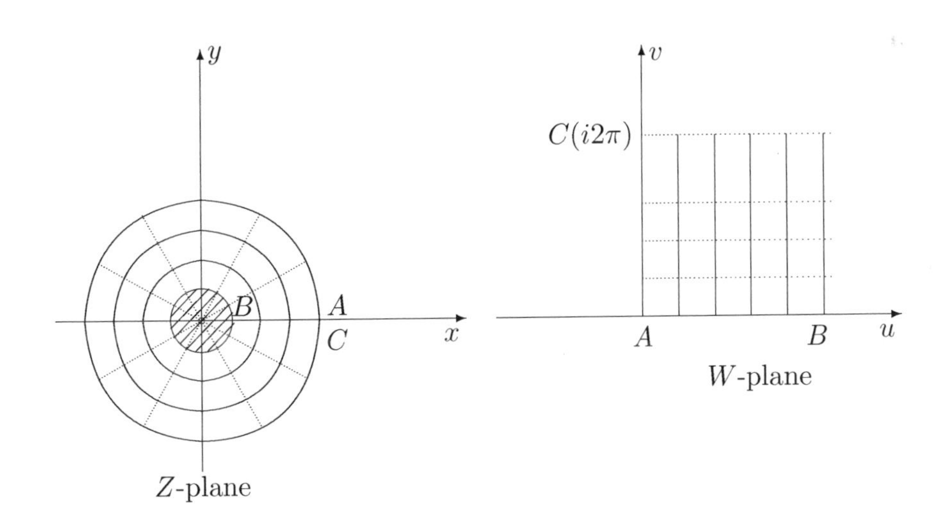

Figure 2-4 Conformal Transformation

Solve:

As shown in Figure 2-4, there is a coaxial cable in the Z-plane. Where a and b are the radius of the outer conductor and the inner conductor, respectively. The electric potential is 0 at $r = a$ and is u_0 at $r = b$. Considering that the equipotential surface of the coaxial cable is a family of circles ($r = constant$) centered at the center of the inner conductor, we may consider to take the logarithm function as the transformation function.

For instance, suppose that the transformation function is

$$W = A_1 \ln Z + A_2 \tag{2.15}$$

then, since

$$\ln Z = \ln(r\, e^{i\theta}) = \ln r + i\theta$$

we have

$$W = A_1(\ln r + i\theta) + A_2 = u + iv$$

Where

$$u = A_1 \ln r + A_2 \tag{2.16}$$

$$v = A_1 \theta \tag{2.17}$$

That means in the transformation from the Z-plane to the W-plane in Figure 2-4, the coaxial circles ($r = constant$) in the Z-plane become the parallel lines $u = contant$ in the W-plane and the radial lines in the Z-plane become another parallel lines $v = constant$ in the W-plane. Where u may be taken as the electric potential function and then v is the flux function.

Now considering that the electric potential is 0 at $r = a$ and is u_0 at $r = b$, we have, from (2.16),

$$0 = A_1 \ln a + A_2, \qquad when \quad r = a$$

$$u_0 = A_1 \ln b + A_2, \qquad when \quad r = b$$

And then we have

$$A_1 = \frac{u_0}{\ln\left(\dfrac{b}{a}\right)}, \qquad A_2 = -\frac{u_0 \ln a}{\ln\left(\dfrac{b}{a}\right)}$$

and then, from (2.16) and (2.17), the electric potential function is

$$u = u_0 \left[\ln\left(\frac{r}{a}\right) / \ln\left(\frac{b}{a}\right)\right] \tag{2.18}$$

and the flux function is

$$v = u_0 \theta / \ln\left(\frac{b}{a}\right) \tag{2.19}$$

To obtain the capacitance per unit length C_0 of the coaxial cable, we have to fine the charge per unit length q of the coaxial cable since $C_0 = q/u_0$. According to the *Causs Theorem*, the charge is equal to the electric flux to the outer conductor, namely,

$$q = \int_C^A \rho_s \, dS = \int_C^A \mathbf{D} \cdot \mathbf{dS} = \Psi = \varepsilon \left(v_C - v_A \right) \tag{2.20}$$

Where (2.13) has been used. And v_A and v_C are the values of v corresponding to $\theta = 0$ and $\theta = 2\pi$, respectively. Substituting (2.19) in (2.20), we have

$$q = \varepsilon \left[v_C - v_A \right] = \frac{2\pi\varepsilon u_0}{\ln(a/b)} \qquad (C/m)$$

And then the capacitance per unit length of the coaxial cable is

$$C_0 = \frac{q}{u_0} = \frac{2\pi\varepsilon}{\ln(a/b)} \qquad (F/m)$$

Example 2: Find the capacitance of two cylinders in parallel.

Solve:

As mentioned above, a family of coaxial circles in Z-plane can be transformed into parallel lines in the W-plane by means of the logarithm transformation function (2.15). This transformation is also suitable for the field of a single conductive cylinder.

For a linear medium, the sum of the fields from two conductive cylinders is equal to the sum of the fields from each individual conductive cylinder. As shown in Figure 2-5, suppose that there are two conductive cylinders in parallel located at $Z = d$ and $Z = -d$, respectively. Then we have a transformation formula from the Z-plane to the W-plane as follows:

$$W = A[\ln(Z - a) - \ln(Z + a)]$$

or

$$W = A \ln(\frac{Z - a}{Z + a}) \tag{2.21}$$

Where the surfaces of two cylinders in the Z-plane will be transformed into two parallel lines with $u = \mathit{constant}$ in the W-plane by means of (2.21). In which, a is a constant and $a \neq d$.

When A is a real constant, we have, from (2.21)

$$u = \frac{A}{2} \ln \left[\frac{(x - a)^2 + y^2}{(x + a)^2 + y^2} \right] \tag{2.22}$$

$$v = A \left[\tan^{-1} \left(\frac{y}{x - a} \right) - \tan^{-1} \left(\frac{y}{x + a} \right) \right] \tag{2.23}$$

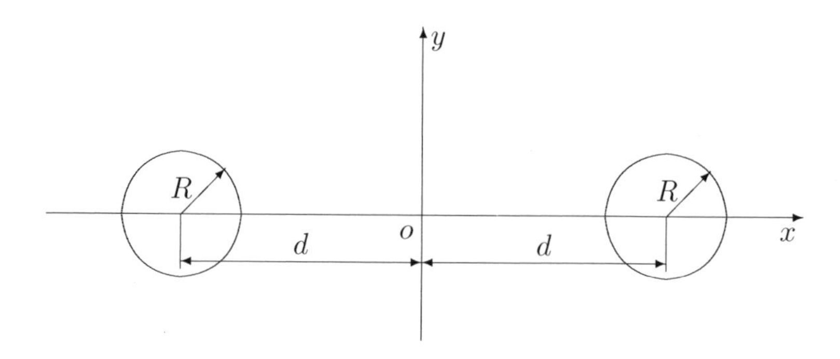

Figure 2-5 Two conductive cylinders in parallel

Where the parallel lines $u = constant$ in the W-plane correspond to a family of circles in the Z-plane as follows:

$$\frac{(x-a)^2+y^2}{(x+a)^2+y^2} = k(constant) \tag{2.24}$$

or

$$\left[x - \frac{a(1+k)}{1-k}\right]^2 + y^2 = \frac{4a^2k}{(1-k)^2} \tag{2.25}$$

In other words, the line $u = constant$ in the W-plane corresponds to a circle in the Z-plane, where the center of the circle in the Z-plane is

$$x = \frac{a(1+k)}{(1-k)}$$

and the radius of the circle is

$$R = \frac{2a\sqrt{k}}{1-k}$$

One example is shown in Figure 2-5. Where the center of the circle in the Zplane is

$$d = \frac{a(1+k_0)}{1-k_0}$$

and the radius of the circle in the Z-plane is

$$R = \frac{2a\sqrt{k_0}}{1-k_0}$$

It follows that we have

$$a = (d^2-R^2)^{1/2}, \quad \sqrt{k_0} = \frac{d}{R} - \sqrt{\left(\frac{d}{R}\right)^2 - 1} \tag{2.26}$$

Now, A in (2.22) depends on the potential of a conductive cylinder, $V_0/2$. Then, from (2.22) and (2.26), we have

$$V_0/2 = A \ln \sqrt{k_0} = A \ln \left[\frac{d}{R} + \left(\frac{d^2}{R^2} - 1 \right)^{1/2} \right]$$

or

$$A = \frac{V_0}{2 \ln \left(\frac{d}{R} + \sqrt{\frac{d^2}{R^2} - 1} \right)} = \frac{V_0}{2 \cosh^{-1}(d/R)}$$

Then from (2.22) and (2.23), we have

$$u = \frac{V_0}{4 \cosh^{-1}(d/R)} \ln \left[\frac{(x-a)^2 + y^2}{(x+a)^2 + y^2} \right], \tag{2.27}$$

$$v = \frac{V_0}{2cosh^{-1}(d/r)} \left[\tan^{-1} \frac{y}{x-a} - \tan^{-1} \frac{y}{x+a} \right] \tag{2.28}$$

Where u is the potential function and v is the electric flux function in the Z-plane. Now, according to the *Causs Theorem*, the total electric flux from the surface per unit length of the cylinder is equal to the electric charge per unit length of the cylinder. And the surface integral of (2.28) along the surface per unit length of the cylinder is 2π for the first term, and is 0 for the second term. Therefore, the total charge per unit length of the cylinder is

$$q = 2\pi \frac{\varepsilon V_0}{2 \cosh^{-1}(d/R)} \quad (C/m)$$

And the capacitance per unit length of the two cylinders is

$$C_0 = \frac{q}{V_0} = \frac{\pi \varepsilon}{\cosh^{-1}(d/R)} \quad (F/m)$$

Example 3: Find the characteristic impedance of microstrip consisting of a non-thickness metal and air medium.

Definition:

The definition of the characteristic impedance of a transmission line is

$$Z_0 = \sqrt{L_1/C_1}$$

when the transmission line is a lossless. Where, L_1 and C_1 are the inductance and the capacitance per unit length of transmission line, respectively. Consider that the phase velocity of the transmission line is $v_\phi = 1/\sqrt{L_1 C_1}$, we have

$$Z_0 = \frac{1}{v_\phi C_1}$$

If the medium inside the microstrip is air, we have $v_\phi = c = 1/\sqrt{\mu_0\varepsilon_0}$, $C_1 = C_0$. If the medium is a dielectric with $\varepsilon = \varepsilon_r\,\varepsilon_0$, then we have $v_\phi = c/\sqrt{\varepsilon_r}$, $C_1 = \varepsilon_r C_0$. And then if the medium inside the microstrip is a homogeneous medium, we have

$$Z_0 = \frac{1}{v_\phi C_1} = \frac{\sqrt{\varepsilon_r}}{c}\frac{1}{\varepsilon_r C_0} = \frac{\sqrt{\mu_0\varepsilon_0}}{\sqrt{\varepsilon_r}C_0}$$

Again,

For the air medium:

Now for a microstrip filled with air medium ($\varepsilon_r = 1$),

$$Z_0^0 = \frac{\sqrt{\mu_0\varepsilon_0}}{C_0} \tag{2.29}$$

For the homogenous medium:

For a microstrip filled with a homogeneous dielectric,

$$Z_0 = Z_0^0/\sqrt{\varepsilon_r} \tag{2.30}$$

Consequently, if we can find the distribution capacitance of the microstrip C_0, then to find the characteristic impedance of the micristrip Z_0 is straightforward.

Considering the edge condition:

Considering the edge effect of the microstrip, the characteristic impedance of the microstrip has to be modified to

$$Z_0 = Z_0^0/\sqrt{\varepsilon_e} \tag{2.31}$$

if the thickness of the metal has to be taken into account. Where ε_e is the effective dielectric constant and it is not so easy to be obtained. Therefore, we have to concentrate to find the characteristic impedance of a microstrip consisting of a non-thickness metal and air medium.

Find C_0:

As shown in Figure 2-6(a), if the microstrip is so called a "wide" microstrip, namely, the width of the microstrip b is much lager than the hight h, i.e. $b >> h$, then the electric field lines of the microstrip are almost in parallel from the meddle part to the neighborhood of the edge, where the electric lines become curves. Therefore, it is reasonable to consider that the effect of the field from another edge of the upper conductor on the field in the meddle part is negligible. And then the field in the left-hand side ($b/2$) of the "wide" microsrip can be replaced by a field in a semi-infinite plane as shown in Figure 2-6(b).

Then we try to find the distribution capacitance of the left-hand side by means of the conformal transformation and then the total capacitance of the "wide" microstrip is the twice.

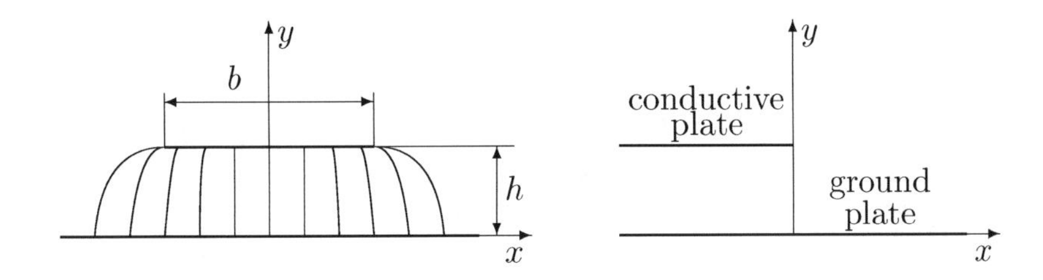

(a) Field distribution in wide microstrip (b) Approximation structure

Figure 2-6 Microstrip

From Z-plane to W-plane:

Now the transformation function

$$Z = \frac{h}{\pi}(1 + W + \ln W) \tag{2.32}$$

can be used to transform the Z-plane into the W-plane as shown in Figure 2-7. In which, the upper conductive strip EAB and $BCDE'$ in the Z-plane become the negative real axis ($u < 0,\ v = 0$) in the W-plane. And the grounded plane in the Z-plane becomes the positive real axis ($u > 0, v = 0$) in the W-plane.

In more detail:

First, the transformation of the conductive strip are as follows:

For the negative real axis in the W-plane, $u < 0, v = 0$, (2.32) becomes [2]

$$x = \frac{h}{\pi}(1 - |u| + \ln|u|), \quad y = h$$

in which:

[2] Where $W = |W|e^{i\theta}$, and then, for the negative axis in the W-plane, $u < 0$, $v = 0$, $\theta = \pi$, we have

$$z = \frac{h}{\pi}(1 - u + j0 + \ln|u| + i\pi)$$

and then

$$x = \frac{h}{\pi}(1 - |u| + \ln|u|), \quad y = h$$

$u = -\infty, \quad v = 0, \quad \rightarrow \quad x = -\infty, \quad y = h, \quad$ (point E)
$u = -1, \quad v = 0, \quad \rightarrow \quad x = 0, \quad y = h, \quad$ (point B)
$u = 0, \quad v = 0, \quad \rightarrow \quad x = -\infty, \quad y = h, \quad$ (point E')

Consequently, the negative real axis in the W-plane corresponds to the surface of the conductive strip in the Z-plane. Where $-\infty < u < -1, v = 0$ corresponds to the surface of the conductive strip EAB, and $-1 < u < 0$ corresponds to the surface of the conductive strip $BCDE'$.

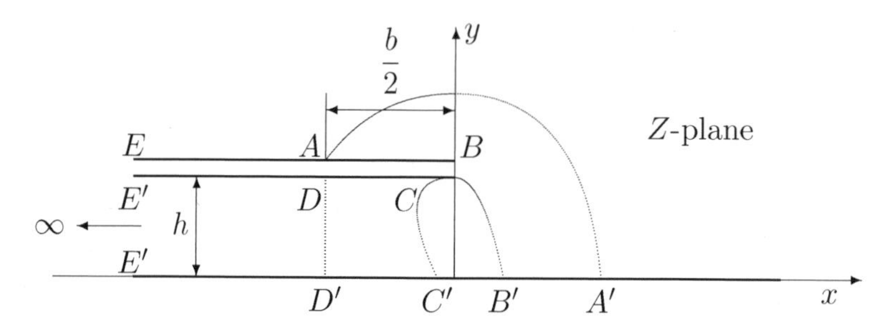

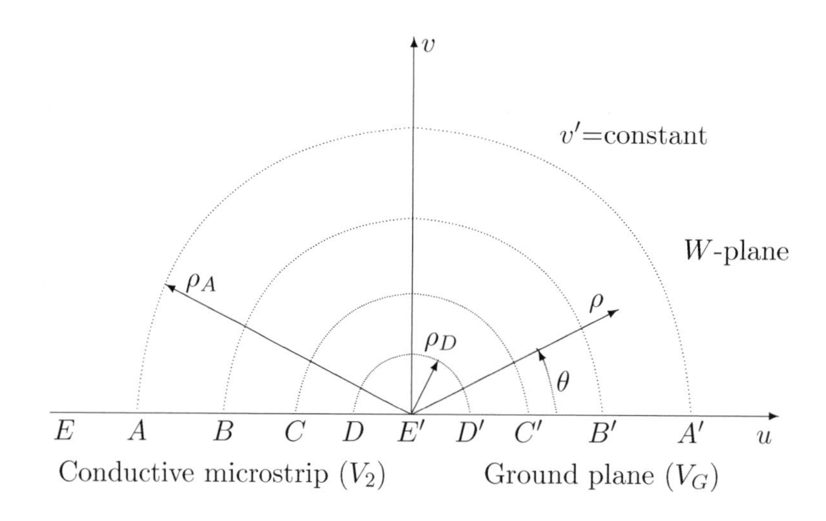

Figure 2-7 Transformation $Z = \frac{h}{\pi}(1 + W + \ln W)$

Secondary, the transformation of the grounded plate are as follows:
For the positive real axis in the W-plane, $u = |u|, v = 0$, (2.32) becomes

$$x = \frac{h}{\pi}(1 + |u| + \ln|u|), \quad y = 0$$

and then that the positive real axis in the W-plane corresponds to the grounded plate in the Z-plane. As we mentioned before that the negative real axis in the W-plane corresponds to the conductive strip in the Z-plane. These two real axes in the W-plane are separated with a small gap and form a new potential distribution.

From W-plane to W'-plane:

The following transformation function from the W-plane to the W'-plane can modify the new potential function to a more simple one, namely, if

$$W' = -iA\ln W - iB_1 + B_2 = u' + iv' \tag{2.33}$$

then, the potential function

$$u' = A\theta + B_2 \tag{2.34}$$

the flux function

$$v' = -A\ln\rho - B_1 \tag{2.35}$$

Where $W = \rho\exp(i\theta)$ has been used.

Suppose that the electric potential is V_1 on the grounded plate (where $\theta = 0$) and is V_2 on the conductive strip (where $\theta = \pi$), then, from (2.34), we may set $A = (V_2 - V_1)/\pi$, $B_2 = V_1$, and then

$$u' = \frac{V_2 - V_1}{\pi}\theta + V_1 \tag{2.36}$$

$$v' = -\frac{V_2 - V_1}{\pi}\ln\rho - B_1 \tag{2.37}$$

To make $B_1 = 0$ in (2.37), we have to consider to take $v' = 0$ at $\rho = 1$ as the reference point. In this case,

$$v' = -\frac{V_2 - V_1}{\pi}\ln\rho \tag{2.38}$$

And then the capacitance per unit length of the microstrip is

$$C_0 = \frac{2q_{AD}}{V_2 - V_1}$$

where q_{AD} is the charge per unit length of the conductive plate $ABCD$ as shown in Figure 2-7 and is

$$q_{AD} = \varepsilon[v'_D - v'_A]$$

Making use of (2.38), we have

$$q_{AD} = \varepsilon\frac{V_2 - V_1}{\pi}\ln\frac{\rho_A}{\rho_D}$$

and then the capacitance per unit length of the microstrip is

$$C_0 = \frac{2\varepsilon}{\pi}\ln(\rho_A/\rho_D)$$

Where ρ_A and ρ_D are two vector lengths at points A and D, respectively, in the W-plane. Both of them can be determined by (2.32):

For point A:

$$Z = -\frac{b}{\pi} + ih, \quad W = -\rho_A$$

and then from (2.32), we have

$$Z = -\frac{b}{2} + ih = \frac{h}{\pi}(1 - \rho_A + \ln \rho_A) + ih$$

or

$$-\frac{\pi b}{2h} = 1 - \rho_A + \ln \rho_A \tag{2.39}$$

For point D:

$$Z = -\frac{b}{2} + ih, \quad W = -\rho_D$$

and then from (2.32), we have

$$Z = -\frac{b}{2} + ih = \frac{h}{\pi}(1 - \rho_D + \ln \rho_D) + ih$$

or

$$-\frac{\pi b}{2h} = 1 - \rho_D + \ln \rho_D \tag{2.40}$$

Subtracting (2.40) from (2.39), we have

$$\ln \frac{\rho_A}{\rho_D} = \rho_A - \rho_D \tag{2.41}$$

and (2.41) can be solved by the graphic method to get $\ln(\rho_A/\rho_D)$.

2.3 The skills to find the conformal conformation

Obviously, from Example 3 above, the conformal transformation is to transform a complicated geometric in the Z-plane into the real axes in the W-plane. Therefore, the key point is to find a transformation function $W = f(Z)$. Then the complicated geometrids in the Z-plane can be transformed into the real axes in the W-plane. Here we would like to show you some skills from the book [11] as follows:

Example 1 As shown in Figure 2-8(a), suppose that there is a gap between the conductive plate BCD and the grounded plate AOA. Find the potential distribution between BCD and AOA.

Solve:

Here, the two conductive plates are in the upper Z-plane. Find a transformation function $W = f(Z)$ such that the $ABCDA$ in the Z-plane can be transformed into the real axis in the W-plane. This problem has to be separated into several steps:

First step: The transformation function

$$\omega = Z^2$$

may transform Figure 2-8(a) in the Z-plane into Figure 2-8(b) in the ω-plane. Namely, the angle $\pi/2$ in the Z-plane becomes angle π in the ω-plane. Consequently, the negative real axis in the Z-plane becomes the positive real axis in the ω-plane.

Second step: The shift transformation function

$$\omega_1 = \omega + h^2$$

may shifts the whole picture in Figure 2-8(b) to the real axis in the ω_1-plane as shown in Figure 2-8(c).

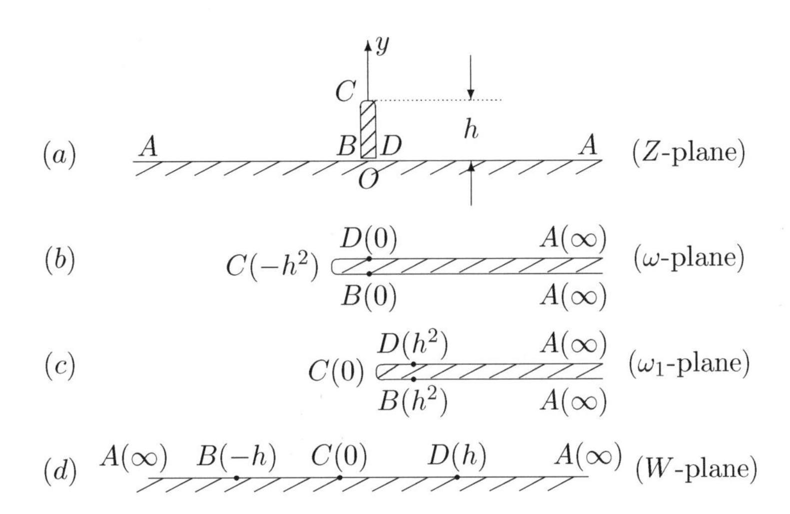

Figure 2-8 Transformation $W = \sqrt{Z^2 + h^2}$

Third step: The transformation function

$$W = \sqrt{\omega_1}$$

may transform two positive real axes in the ω_1-plane into a positive and a negative real axes in the W-plane as shown in Figure 2-8(d).

Summery, the whole transformation from the Z-plane to the W-plane is

$$W = \sqrt{\omega_1} = \sqrt{\omega + h^2} = \sqrt{Z^2 + h^2}$$

Consequently, the problem to find the potential distribution between the conductive plate BCD and AOA in the Z-plane becomes the problem to fine the potential distribution between the conductive plate BCD and two half infinity conductive plates AB and DA in the W-plane. This will be completed by reader.

Example 2 Find the potential distribution between the conductive plate located at $-\infty < x \le -1$ and the conductive plate located at $1 \le x < \infty$ (Figure 2-9(a)) [3].

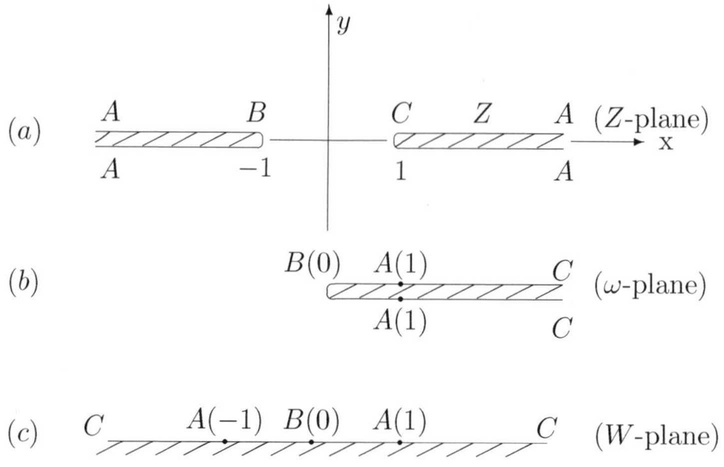

(a) A B C Z A (Z-plane) X A −1 1 A

(b) B(0) A(1) C (ω-plane) A(1) C

(c) C A(−1) B(0) A(1) C (W-plane)

Figure 2-9 Transformation $W = \sqrt{\dfrac{Z-1}{Z+1}}$

As shown in Figure 2-9(a), how to find a transformation function to transform the two plates in Figure 2-9(a) in the Z-plane into the real axes in the W-plane. To this end.

First step: The factional transformation function

$$\omega = \frac{Z+1}{Z-1}$$

will make the transformation as follows: where the point $B(Z=-1)$ in the Z-plane becomes the origin in the ω-plane, the point $C(Z=+1)$ in the Z-plane becomes point ∞ in the ω-plane, and the point A(∞) in the Z-plane becomes $\omega = 1$ in the ω-plane (Figure 2-9(b)). Therefore, the two plates in the Z-plane become two positive real axes.

Second step: The transformation function

$$W = \sqrt{\omega}$$

will transform two positive axes in the ω-plane into a positive axis and a negative real axis in the W-plane.

Consequently, the transformation function from the Z-plane to the W-plane is:

$$W = \sqrt{\omega} = \sqrt{\frac{Z+1}{Z-1}}$$

[3] Where the point $A(-\infty)$ is far away from the point $A(\infty)$ and both of them are not equipotential

which may transform the picture in the Z-plane into a positive axis and a negative real axis in the W-plane. Then the problem to find the potential distribution between two conductive plates in the Z-plane becomes to find the potential distribution between the conductive plate $A(-1)B(0)A(1)$ and two half infinity conductive plates $A(1)C$ and $A(-1)C$. This is the same as Example 1.

Example 3 As shown in Figure 2-10(a), suppose that there is a gap between two conductive plates ADC and ABC. Find the potential distribution between the two plates.

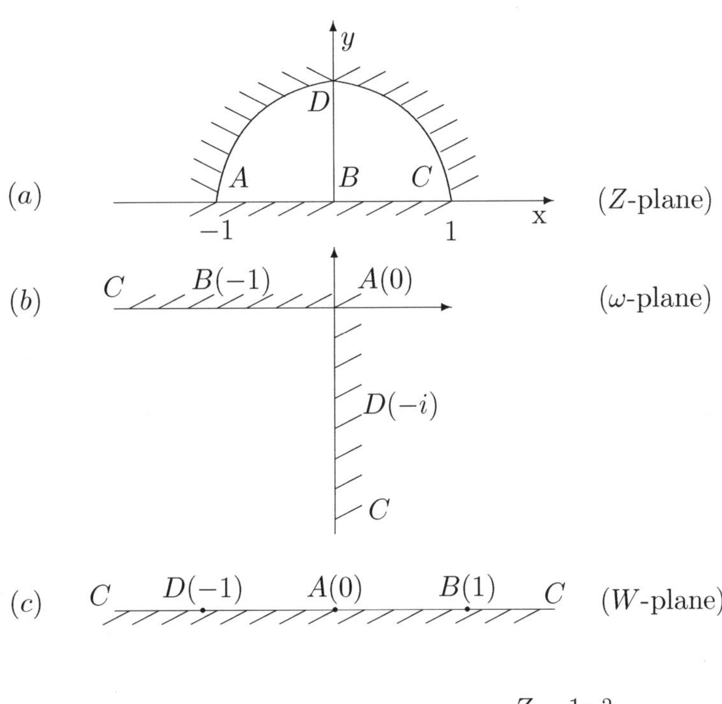

Figure 2-10 Transformation $W = \left(\frac{Z-1}{Z+1}\right)^2$

Solve:

First step: The transformation function

$$\omega = \frac{Z+1}{Z-1}$$

may transform a half circle in the Z-plane into a half line in the ω-plane and a line in the Z-plane into another half line in the ω-plane. To this end, we may move one terminal point of the half circle to infinite point in the ω-plane. For instance, the diameter of the half circle $A(Z=-1)B(Z=0)C(Z=1)$ in the Z-plane is still a line $A(\omega=0)B(\omega=-1)C(\omega=-\infty)$ in the ω-plane. And the point $D(Z=i)$ of the half circle in the Z-plane becomes $D(\omega = -i)$ in the ω-plane. The two terminal points of the half circle A and C are located at 0 and $-\infty$, respectively, in the ω-plane. Consequently, the half circle in the Z-plane becomes a negative half image axis in the ω-plane (Figure 2-10(b)).

Second step: The transformation function

$$W = \omega^2$$

may transform two lines with an angle 90^0 in the ω-plane into positive real axis and negative real axis in the ω_1-plane. Meanwhile, the third quadrant $\pi < \arg\omega < 3\pi/2$ in the ω-plane becomes the upper plane $2\pi < \arg W < 3\pi$ in the ω_1-plane (Figure 2-10(c)).

Therefore the transformation function

$$W = \omega^2 = \left(\frac{Z+1}{Z-1}\right)^2$$

may transform the half circle ADC and a diameter ABC in the Z-plane into positive real axis and negative real axis, respectively, in the W-plane. Then the potential distribution between the positive and the negative real axis in the W-plane may be obtained by means of (2.33).

Example 4 As shown in Figure 2-11(a), find the potential distribution between the conductive plate AFA and the grounded conductive plate $ABCDA$. Where $A(i\infty)$ is far away from $A(\pm\infty)$ and both of them are not equi-potential.

Solve:

First step: The transformation function

$$\omega = \frac{1}{2}\left(Z + \frac{1}{Z}\right)$$

may transform the half circle in the Z-plane (Figure 2-11(a)) to a line in the ω-plane without changing any other part as shown in Figure 2-11(b).

Second step: The transformation function

$$\omega_1 = \omega^2$$

may transform the picture in the ω-plane to a picture in the ω_1-plane as shown in Figure 2-11(c). In which the whole picture in the ω_1-plane is on the real axis only.

Third step: Following the same process as Example 2, the factional transformation function

$$\omega_2 = \frac{\omega_1 + \dfrac{9}{16}}{\omega_1} = 1 + \frac{9}{16\omega_1}$$

may transform the negative axes to the positive axes to get the picture as indicated in the ω_2-plane in Figure 2-11(d).

Fourth step: Finally, the transformation function

$$W = \sqrt{\omega_2}$$

may transform the positive lines in the ω_2-plane into a negative axis and a positive real axis in the W-plane as shown in Figure 2-11(e). Then we will find that this picture is the same as Example 1.

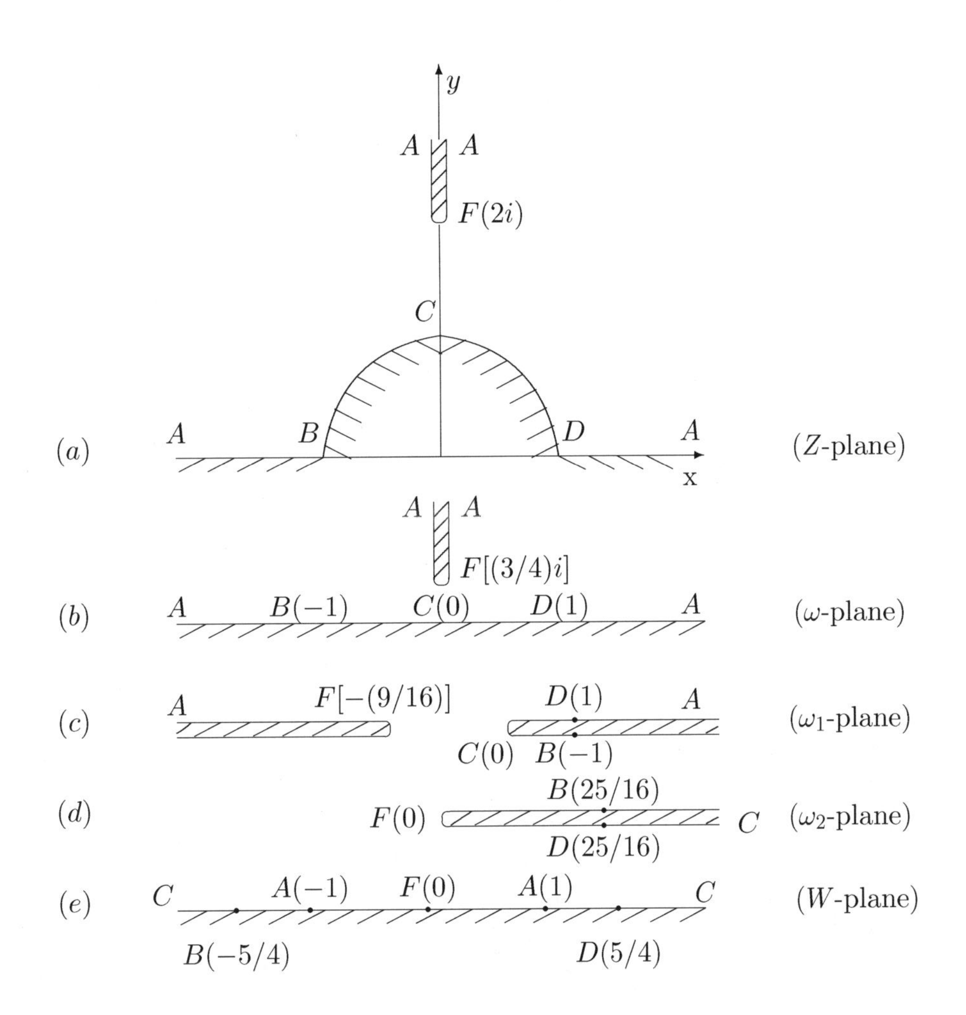

Figure 2-11 Transformation $Z - \omega - \omega_1 - \omega_2 - W$

Example 5 [*Rookolphsky transformation*] Now we will discuss a famous conformal transformation, *Rookolphsky transformation*, which is a famous conformal transformation and has been used to design the wings of an aeroplane, meanwhile this transformation skill is very interesting.

As shown in Figure 2-12:

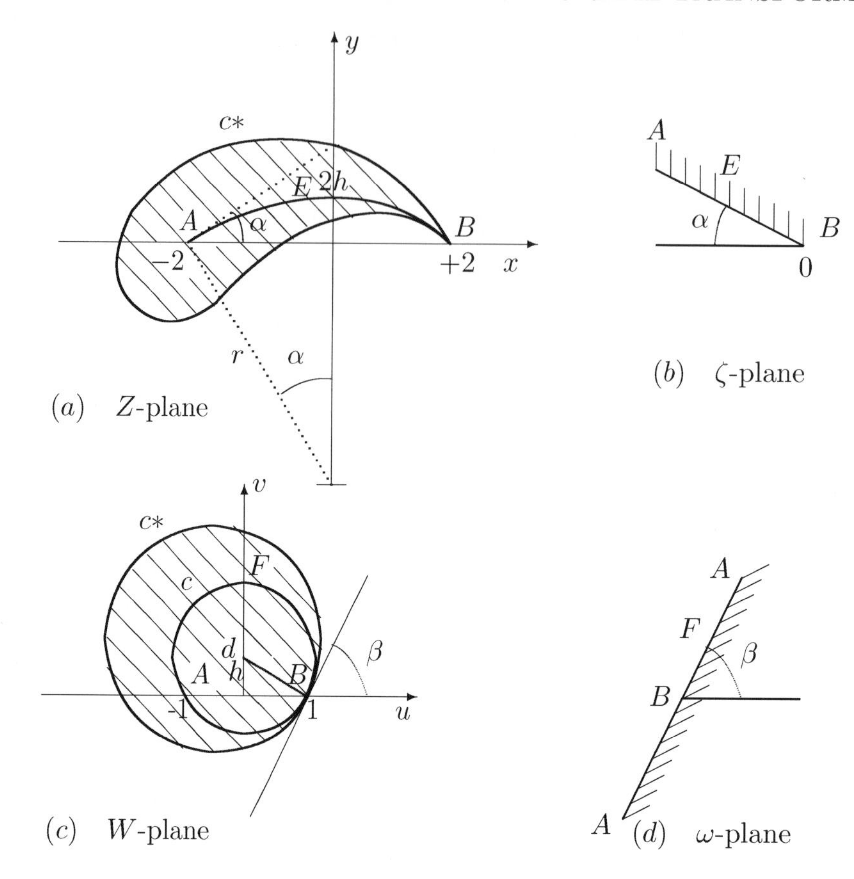

Figure 2-12 Rookolphsky Transformation

(1) First, fine a conformal transformation function to transform the arc $\widehat{AEB}$ in the Z-plane in Figure 2-12(a) to a line AEB in the ζ-plane.

(2) Secondary, find another conformal transformation function to transform the circle c in the W-plane in Figure 2-12(c) to another line $AFBA$ in the ω-plane.

(3) Then fine a conformal transformation to make the line AEB coincide with the line $AFBA$.

Now we will discuss the skill in more detail as follows:

First step: the transformation function

$$\zeta = \frac{Z-2}{Z+2} \tag{2.42}$$

may transform the arc $\widehat{AEB}$ terminated at $Z = \pm 2$ in the Z-plane into a line AEB in the ζ-plane. Obviously, from Figure 2-12(a), (b),

point A : $Z = -2$, $\rightarrow$ $\zeta \rightarrow \infty$

point B: $Z = 2$, $\rightarrow$ $\zeta = 0$

point E: $Z = i2h$, $\rightarrow$ $\zeta = e^{i(\pi - 2\arctan h)} = e^{i(\pi-\alpha)}$

which means that, in the ζ-plane, the angle between the line AEB and the negative axis is α. And the angle between AEB and the positive axis is

$$\gamma = \pi - 2\arctan h = \pi - \alpha \tag{2.43}$$

As shown in Figure 2-12(a), α is just the angle between the tangent of the arc $\widehat{AEB}$ at A and the positive axis in the Z-plane. This implies that the AEB line in the ζ-plane is in parallel with the tangent of the arc $\widehat{AEB}$ at B in the Z-plane.

Second step: The transformation function

$$\omega = \frac{W-1}{W+1} \tag{2.44}$$

may transform the circle $\widehat{AFBA}$ in the W-plane into a line $AFBA$ in the ω-plane. Where the circle $\widehat{AFBA}$ pass through $W = \pm 1$ in the W-plane. Obviously, from Figure 2-12(c),(d), that

$$\begin{array}{lll} \text{point } A: & W = -1, & \rightarrow \quad \omega \rightarrow \infty \\ \text{point } B: & W = 1, & \rightarrow \quad \omega = 0 \\ \text{point } F: & W = i[h + \sqrt{1+h^2}], & \rightarrow \quad \omega = e^{i[\pi - 2\arctan(h+\sqrt{h^2+1})]} \end{array}$$

This implies that the angle between the line $AFBA$ and positive axis in the ω-plane is

$$\beta = \pi - 2\arctan(h + \sqrt{h^2+1}) \tag{2.45}$$

Rearrange this formula we have [4]

$$\cot\beta = h \tag{2.46}$$

Which implies that β is the angle between the tangent of the circle c at point B and the positive axis u in the W-plane.

Third step: From (2.46) and (2.43) we have

$$\beta = \frac{\pi}{2} - \arctan h$$

$$2\beta = \pi - 2\arctan h = \pi - \alpha \tag{2.47}$$

Consequently, if we take

$$\zeta = \omega^2 \tag{2.48}$$

then $AFBA$ in the ω-plane will coincide with AEB in the ζ-plane.

Substituting (2.42) and (2.44) in (2.48), we have the transformation function

$$\frac{Z-2}{Z+2} = \left(\frac{W-1}{W+2}\right)^2$$

---

[4] From (2.45) we have $\tan(\frac{\pi}{2} - \frac{\beta}{2}) = h + \sqrt{h^2+1}$,

in which $\tan(\frac{\pi}{2} - \frac{\beta}{2}) = \cot\frac{\beta}{2} = \frac{1+\cos\beta}{\sin\beta} = \sqrt{1+(\cot\beta)^2} + \cot\beta.$

Making comparison between these two equations, we have $\cot\beta = h$.

or

$$Z = W + \frac{1}{W}, \qquad W = \frac{1}{2}(Z + \sqrt{Z^2 - 4}) \tag{2.49}$$

(2.49) is the well known *Rookolphsky transformation function*, which may be used to design the wing of the airplane.

The circle method:

Make a circle c^* in the W-plane, which is tangential to the circle c at point $B = 1$. After *Rookolphsky transformation* (2.49), the circle c^* in the W-plane becomes a closed curve c^* in the Z-plane. This closed curve in the Z-plane contains the arc AB and is tangential to the arc AB at point $B = 2$. The curve c^* in the Z-plane looks like a cross-section of a wing of an aeroplane. This approach technology to fine the shape of the wing is said to be *the circle method.* The shape of the wing obtained by *Roocolphsky transformation* is said to be the shape of *Roocolphsky wing* . And this approach is fairly simple.

2.4 Schwartz transformation

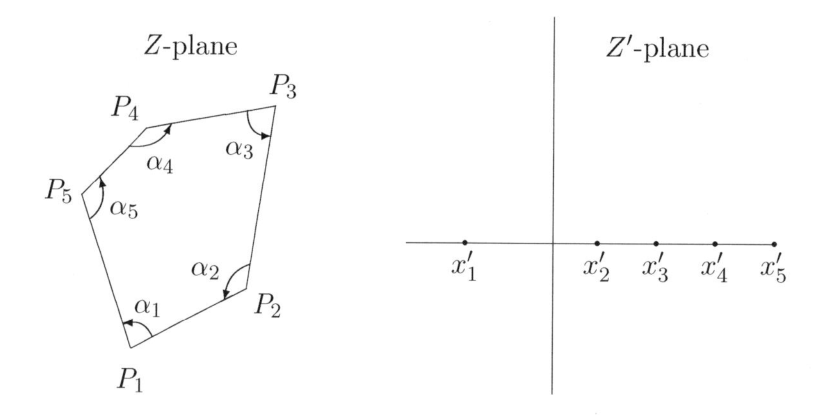

Figure 2-13 Schwarz Transformation

As shown in Figure 2-13, the Schwarz transformation is to transform a polygon region in the Z-plane into the upper half-plane in the Z'-plane, such that the vertex of the polygon P_1, P_2, P_3, P_4,P_5 in the Z-plane become x'_1, x'_2, x'_3, x'_4 and x'_5 on the real axis, respectively, in the Z'-plane.

The Schwartz Transformation Function is

$$\frac{dZ}{dZ'} = K(Z' - x_1')^{(\alpha_1/n)^{-1}}(Z' - x_2')^{(\alpha_2/n)^{-1}} \cdots (Z' - x_n')^{(\alpha_n/n)^{-1}} \tag{2.50}$$

where $\alpha_1, \alpha_2, \cdots \alpha_n$ are the inner angles of the polygon.

Now we will prove (2.50).

First, the transformation is not conformal on each vertex of the polygon, i.e. α_i doesn't become α_i, but instead π. For instance, in the proximity of x_2', we have [10]

$$\frac{\partial Z}{\partial Z'} \approx (Z' - x_2')^{\beta_2}$$

To determinate β_2, taking the argument from both sides of the equation, we have

$$\arg dZ \approx \arg dZ' + \beta_2 \arg(Z' - x_2') \tag{2.51}$$

As shown in Figure 2-13, when Z' moves from the left side of x_2' to the right side of x_2' in Z'-plane, Z moves from the left side of the point P_2 to the right side of the point P_2 in Z-plane. In this case

$$\arg dZ = (\arg Z)_{right\, side\, of\, P_2} - (\arg Z)_{left\, side\, of\, P_2} = -\alpha_2$$

$$\arg dZ' = [\arg Z']_{x_2'+0} - [\arg Z']_{x_2'-0} = -\pi$$

where the argument value is in anticlockwise. And then from (2.51), we have

$$\beta_2 \approx (\alpha_2/\pi) - 1$$

Consequently, in the proximity of x_2', we have

$$\frac{dZ}{dZ'} \approx (Z' - x_2')^{(\alpha/\pi)-1} \tag{2.52}$$

The same process acts to P_1, P_3, P_4 and P_5, respectively, will lead to (2.50).

The application of Schwarz Transformation:

One of the applications is to find the capacitance between a inner conductive line and a grounded polygon cylinder. In this case, the polygon cylinder becomes a line in the Z'-plane and the inner conductive line becomes a point in the upper Z'-plane. Then the capacitance can be obtained by the *mirror image method.*

A boundary value problem shown in Figure 2-14 is an interesting application. Where the angle between AB and CD is $\alpha_1 = 0$, and the vertex (B,C) tends to infinity. The angle is $\alpha_2 = 2\pi$ between CD and DE and is $\alpha_3 = \pi$ between DE and AB. And the vertex (E,A) tends to infinity as well.

Now, transform the vertex (B,C) in the Z-plane to the point $x_1' = 0$ with a small gap between B and C in the Z'-plane, transform D in the Z-plane to $x'_2 = 1$ in the Z'-plane, and transform the vertex (E,A) in the Z-plane to $x'_3 \to \infty$ in Z'-plane. Then (2.50) becomes the transformation function as follows:

$$\frac{dZ}{dZ'} = K(Z' - 0)^{(0/\pi)-1}(Z' - 1)^{(2\pi/\pi)-1} = K(Z' - 1)/Z' \qquad (2.53)$$

Where

$$\lim_{x'_3 \to \infty} (Z' - x'_3)^{(\alpha_3/\pi)-1} = 1$$

since $\alpha_3 = \pi$, which has been considered in (2.53). The integral of (2.53) gives

$$Z = K\int (1 - \frac{1}{Z'})dZ' = K(Z' - \ln Z') + C \qquad (2.54)$$

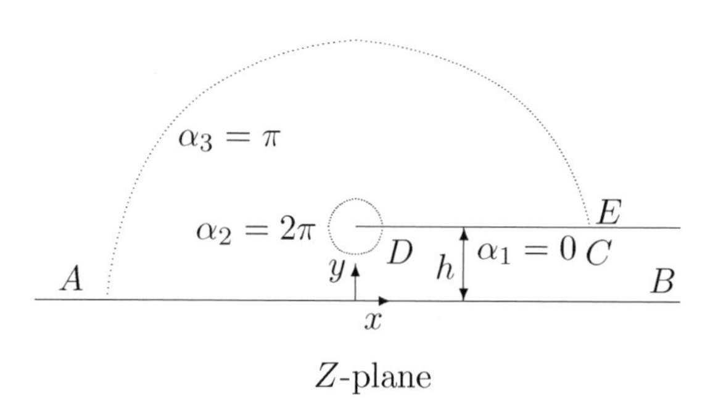

Figure 2-14 The application of the Schwarz Transformation

Where the constants K and C can be determined by the corresponding points before and after the transformation. First, when $Z' = 1$ (D point), $Z = 0 + ih$. then from (2.54), we have

$$ih = K + C$$

And then

$$Z = K(Z' - 1 + \ln Z') + ih \qquad (2.55)$$

Next, the integral

$$\int_B^C dZ = \int_B^C \frac{dZ}{dZ'} dZ' \qquad (2.56)$$

Where, the integral between B and C in the Z'-plane corespondent to $Z' \to 0$. In this case, from (2.53),

$$\frac{dZ}{dZ'} \to -\frac{K}{Z'}$$

In addition, the integral path in right-hand side of (2.56) could be a half circle with a radius r'. In this case,

$$dZ' = d(r'e^{i\theta'}) = ir'e^{i\theta'}d\theta' = iZ'd\theta'$$

and then (2.56) becomes

$$\int_{\infty+i0}^{\infty+ih} dZ = \int_{\pi}^{0} \left(-\frac{K}{Z'}\right) iZ'd\theta'$$

Or

$$ih = iK\pi$$

Consequently,

$$K = h/\pi. \tag{2.57}$$

Substituting (2.57) in (2.55), we have the transformation function from the Z-plane to the Z'-plane as follows:

$$Z = \frac{h}{\pi}(Z' - 1 - \ln Z' + i\pi) \tag{2.58}$$

In this case, the picture in the Z-plane becomes a real axis AB and another real axis CDE in the Z'-plane.

Now, as we mentioned in example 3 of section 2.2, the transformation function (2.33) from W-plane to W'-plane has been used to simply the potential function. Here, if the potential is V_0 on the real axis AB and is 0 on another real axis CDE, then, from (2.33), the following transformation function may be used to transform from the Z'-plane to the W-plane to simplify the potential function,

$$W = \frac{V_0}{\pi}\ln Z' = \frac{V_0}{\pi}(\ln\rho + i\theta) = u + iv$$

or

$$Z' = e^{\pi W/V_0} \tag{2.59}$$

Where, v is the potential function and u is the electric flux function. If point D ($\rho = 1$) is the reference point of the zero electric flux, then substituting (2.59) in (2.58), we will have the transformation function from the Z-plane to the W-plane as follows:

$$Z = \frac{h}{\pi}\left(e^{\pi W/V_0} - 1 - \frac{\pi W}{V_0} + i\pi\right)$$

Problem

2.1 As shown in Figure 2-8(a), find the potential distribution between the conductive plates BCD and AOA, and the capacitance per unit length of the two plates.

2.2 As shown in Figure 2-9(a), find the potential distribution between two plates located at $-\infty \leq x \leq -1$ and $1 \leq x < \infty$, respectively, and the capacitance per unit length of these two plates.

2.3 As shown in Figure 2-10(a), find the potential distribution between the half circle conductive plate ADC and the conductive plate ABC, and the capacitance per unit length of these two plates.

2.4 As shown in Figure 2-11(a), find the potential distribution between the conductive plate $AF(2i)A$ and the grounded plate $ABCDA$, and the capacitance unit length of these two plates.

2.5 Prove the Schwarz transformation (2.50) by means of (2.52).

2.6 As shown in Figure 2-15, following the Schwarz Transformation to prove

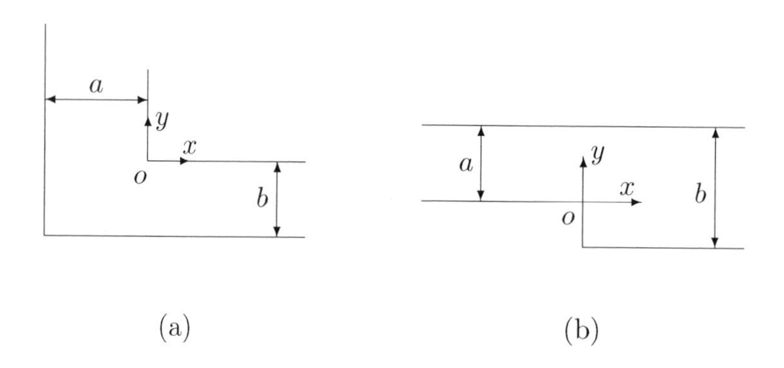

Figure 2-15 Z-plane

(1) The transformation function of Figure 2-15(a) from Z-plane to W-plane is

$$Z = \frac{b}{\pi}\left[\ln\left(\frac{1+S}{1-S}\right) - 2\alpha\tan^{-1}\left(\frac{S}{\alpha}\right)\right]$$

Where, $\alpha = \dfrac{a}{b}$, $S = \left[\dfrac{\exp(kW)+\alpha}{\exp(kW)-1}\right]^{1/2}$.

(2) The transformation function of Figure 2-15(b) from Z-plane to W-plane is

$$Z = \frac{b}{\pi}\left\{\cosh^{-1}\left[\frac{\alpha^2+1-2\alpha^2\exp(kW)}{1-\alpha^2}\right]\right.$$

$$\left. -\alpha\cosh^{-1}\left[\frac{2\exp(kW)-(\alpha^2+1)}{1-\alpha^2}\right]\right\}$$

Where, $\alpha = \dfrac{a}{b}$, $k = \dfrac{\pi}{V_0}$. In Figure 2-15(a) and Figure 2-15(b), $Z = x + iy$, $W = u + iv$, u is the electric flux function and v is the potential function.

Summery: Mathematically,

(1) The transformation function

$$Z = \frac{h}{\pi}(1 + W + \ln W)$$

may transform the microstrip in the Z-plane into two plates in the W-plane, where one of the two plates is on the positive real axis and another one is on the negative real axis as indicated in Figure 2-7.

(2) The transformation function

$$W' = -iA \ln W - iB_1 + B_2 = u' + iv'$$

may transform two plates separated by an angle π from the W-plane to the W'-plane to simply the calculation of the potential distribution as mentioned in (2.33). Where $W = \rho \exp(i\theta)$.

(3) The transformation function

$$\omega = Z^2$$

may transform the angle between two lines from $\pi/2$ in the Z-plane to π in the ω-plane as indicated in Figure 2-8(a),(b).

(4) The transformation function

$$\omega_1 = \omega + h^2$$

may make a shift h^2 from the ω-plane to the ω_1-plane as indicated in Figure 2-8(b),(c).

(5) The transformation function

$$W = \sqrt{\omega_1}$$

may transform two positive real axes in the ω_1-plane into a positive real axis and a negative real axis in the W-plane as indicated in Figure 2-8(c),(d).

(6) The transformation function

$$\omega = \frac{Z+1}{Z-1}$$

may transform the point $B(Z = -1)$ in the Z-plane into origin in the ω-plane as indicated in Figure 2-9(a),(b), and transform a half circle in the Z-plane into a half imagery axis in the ω-plane as indicated in Figure 2-10(a),(b).

(7) The transformation function

$$\omega = \frac{1}{2}\left(Z + \frac{1}{Z}\right)$$

may transform a half circle in the Z-plane into a real axis in the ω-plane as indicated in Figure 2-11(a),(b).

(8) The transformation function

$$\zeta = \frac{Z-2}{Z+2}$$

may transform an arc $\widehat{AB}$ terminated at $Z = \pm 1$ in the Z-plane into a line AEB in the ζ-plane as indicated in Figure 2-12(a),(b).

(9) The transformation function

$$\omega = \frac{W-1}{W+1}$$

may transform a circle through points $W = \pm 1$ in the W-plane into a line in the ω-plane as indicated in Figure 2-12(c),(d).

(10) The transformation function

$$\frac{dZ}{dZ'} \approx (Z' - x_2')^{(\alpha/\pi)-1}$$

may transform a vertex of polygon in the Z-plane into a positive real axis and a negative real axis in the Z'-plane as indicated in Figure 2-13.

Chapter 3

Wiener-Hopf Technology

The Wiener-Hopf Technology has been discussed by R. Mittra and S. W. Lee [12] to analyze the field in a bifurcated waveguide in Figure 3-1. In which the waveguide is a closed system. After then, the Wiener-Hopf Technology has been used to analyze the microstrip by David C. Chang and Edward F. Kuester [15], In which the microstrip is an open system. Nowadays, the Winner-Holf technology has been developed to find the solution for a semi-infinity of a cylinder dielectric waveguide [Figure 3-2-(a)] [16], [17] , and the scattering field of a limit length of gap between two dielectric slab waveguides [Figure 3-2(b)] [18], the scattering field of a lossy rectangular cylinder [Figure 3-2-(c)][19] and so on. Not only these, some authors have developed the Winner-Holf Technology to the piecewise continuous complicated boundary value problem in optical fiber and dielectric waveguide [13].

In this chapter, we will analyze the dielectric waveguide by means of the Wiener-Hopf Technology, which is another open system and is an author's study [13]. To this end, we have to introduce the principle of the Wiener-Hopf Technology based on some part of the book [12] first.

3.1 Introduction

It is well known that the Fourier Integral Method has been used for the boundary value problem in electric-magnetic field for a long time. The benefit of it is that the n-dimensional problem may be reduced to $(n-1)$-dimensional problem.

Usually the integral in the Fourier Integral Method is from $-\infty$ to ∞. However, if we use the Fourier Integral Method to a bifurcated rectangular waveguide as shown in Figure 3-1, we found that the tangential electric field along the metal thin membrane (located at $x = b, z > 0$) is zero, and then the Fourier integral becomes an integral from $-\infty$ to 0. That is different from the traditional Fourier Integral Method.

Further more , when we make use of the Fourier Integral Method to the half-infinity space problem, we may set up a single equation with two unknowns in the complex plane. In generally, there is no way to obtain the solutions of two unknowns from one single equation. However, the Winner-Hopf technology may help us to divide a single equation into two equations in the complex plane, in which, one is analytic in the upper complex plane alone and the another one is analytic in the lower complex plane alone. And then to fine the solutions of two unknowns from one single equation becomes possible. Making use of the Winner-Holf Technology we may obtain the scattering fields not only in the bifurcated rectangular waveguide shown in Figure 3-1, but also in a half-infinity space problem such as the microstrip, the dielectric waveguide and so on. Therefore, it is a powerful tool.

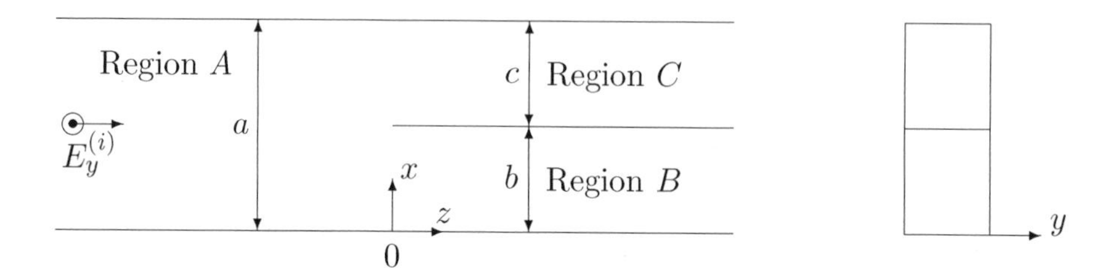

Figure 3-1 Bifurcated Rectangular Waveguide

What is the difference between the mode matching method and the Winner-Hopf technology?

The answer is as follows:

As well known, the TE_0 mode is the lowest or the dominate mode in the bifurcated rectangular waveguide. Now, suppose a known TE_0 mode is incident from the left side into the bifurcated rectangular waveguide and the non-zero components are

$$E_y^{(i)} = \phi^{(i)} = \sin(\frac{\pi}{a})\exp(-\gamma_{1a}\,z)$$
$$H_z^{(i)} = \frac{i}{\omega\mu}\frac{\partial}{\partial z}\phi^{(i)}$$
$$H_z^{(i)} = \frac{1}{i\omega\mu}\frac{\partial}{\partial x}\phi^{(i)}$$

Mode matching method:

The mode matching method considers that the incident field $\mathbf{E^i}$, $\mathbf{H^i}$ exists in region A alone and the scattering field $\mathbf{E^s}$, $\mathbf{H^s}$ exists in region A, region B and region C. And then the total fields $\mathbf{E^t} = \mathbf{E^i} + \mathbf{E^s}$ and $\mathbf{H^t} = \mathbf{H^i} + \mathbf{H^s}$ have to be matched at the transverse interface of $z = 0$.

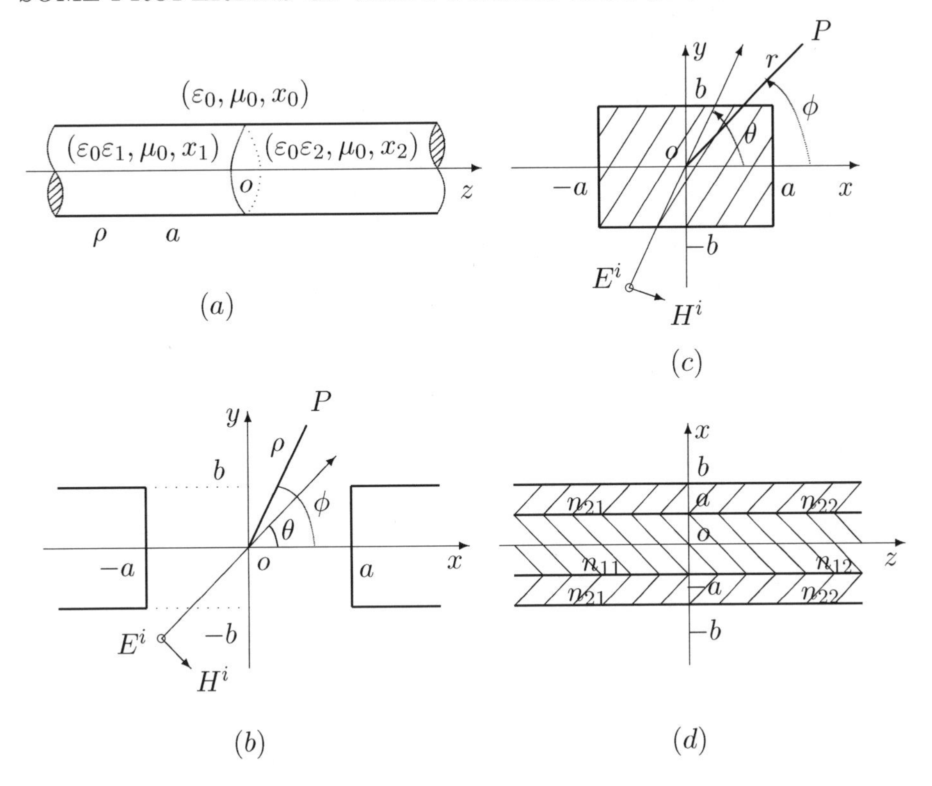

Figure 3-2 The Geometric Structure of Winner-Holf

Winner-Hopf technology:

However, the Winner-Hopf technology considers that the incident fields $\mathbf{E^i}$, $\mathbf{H^i}$ exist not only in region A, but also in region B and region C. And the scattering field $\mathbf{E^s}$, $\mathbf{H^s}$ exists in region A, region B and region C as indicated in Figure 3-1. And then the total fields $\mathbf{E^t} = \mathbf{E^i} + \mathbf{E^s}$, $\mathbf{H^t} = \mathbf{H^i} + \mathbf{H^s}$ have to satisfy the longitudinal boundary conditions (such as the metal thin membrane located at $x = b, z > 0$ and the walls at $x = 0, a$) as indicated in Figure 3-1. In this case, the total fields still are the solutions satisfied the Maxwell's equation and the original boundary conditions. Therefore, the total fields should be the same as the solutions from the mode matching method.

3.2 Some Properties of the Fourier Transform

Now, to explain the principle of the Winner-Hopf Technology, we have to discuss some important properties of the Fourier transform first.

Formula 1: The definition of the Fourier transform is

$$F(\alpha) = \frac{1}{\sqrt{2\pi}} \int_{-\infty}^{\infty} f(z) e^{i\alpha z} dz \tag{3.1}$$

Where it has been assumed that α is real, and the Fourier transform $F(\alpha)$ exists only if the integral (3.1) is Riemann integrable. [1]

Formula 2: The generalized integral

$$\int_{a}^{\infty} g(z) dz \tag{3.2}$$

is realized generally to be

$$\lim_{R \to \infty} \int_{a}^{R} g(z) dz \tag{3.3}$$

under the condition that the limit exists. Further more, if $g(z)$ approach to ∞ or tend to oscillation as $z \to b$, then

$$\int_{a}^{b} g(z) dz = \lim_{\varepsilon \to 0_+} \int_{a}^{b-\varepsilon} g(z) dz, \qquad b > a \tag{3.4}$$

under the condition that the limit exists.

After making use of the generalized integral definition, the Fourier transform is also usable even though $|f(z)\exp(iaz)|$ is non-integrable in infinity scalar .

Formula 3: The definition of the inverse Fourier transform is

$$f(z) = \frac{1}{\sqrt{2\pi}} \int_{-\infty}^{\infty} F(\alpha) e^{-i\alpha z} d\alpha \tag{3.5}$$

Formula 4: The Convolution Theorem may be expressed as

$$F\Big[\frac{1}{\sqrt{2\pi}} \int_{-\infty}^{\infty} f(z') g(z - z') dz'\Big] = F(\alpha) G(\alpha), \tag{3.6}$$

meanwhile, it may be expressed as one of following two equations as well :

$$\frac{1}{\sqrt{2\pi}} \int_{-\infty}^{\infty} f(z') g(z - z') dz' = \frac{1}{\sqrt{2\pi}} \int_{-\infty}^{\infty} F(\alpha) G(\alpha) e^{-i\alpha z} d\alpha \tag{3.7}$$

[1] Riemann integrable: A closed interval $[a, b]$ may be divided into a finite number of small pieces according to $a = a_0 < a_1 < a_2 < \cdots < a_N = b$. Each piece is denoted by η with the module $|\eta| = max \triangle x_n$, where $\triangle x_n = a_n - a_{n-1}$. The function $f(x)$ is defined as the Riemann integrable in the interval [a,b] if and only if

$$\lim_{|\eta| \to 0} \sum_{n=1}^{N} f(x_n) \triangle x_n = l, \qquad a_{n-1} \le x_n \le a_n$$

exists.

$$\frac{1}{\sqrt{2\pi}}\int_{-\infty}^{\infty} f(z)g(z)e^{i\alpha z}dz = \frac{1}{\sqrt{2\pi}}\int_{-\infty}^{\infty} F(\beta)G(\alpha-\beta)d\beta \tag{3.8}$$

Formula 5: The Fourier transform (3.1) may be expressed as the sum of two integrals, one is the integral from $-\infty$ to 0, another one is the integral from 0 to ∞. Namely,

$$F(\alpha) = F_+(\alpha) + F_-(\alpha) \tag{3.9}$$

with

$$F_+(\alpha) = \frac{1}{\sqrt{2\pi}}\int_{0}^{\infty} f(z)e^{i\alpha z}dz \tag{3.10}$$

$$F_-(\alpha) = \frac{1}{\sqrt{2\pi}}\int_{-\infty}^{0} f(z)e^{i\alpha z}dz \tag{3.11}$$

According to the definition of $F_+(\alpha)$ and $F_-(\alpha)$, sometimes, either $F_+(\alpha)$ or $F_-(\alpha)$ exists depending on the properties of f(z) at $z \to \infty$ and $z \to -\infty$. For instance, the following function is just this kind of function, namely,

$$f(z) = e^{ikz}$$

with $k = k_1 + ik_2$, and k_1, k_2 being real numbers.

Formula 6: (3.9), (3.10) and (3.11) may be realized from the other point of view. For instance , $f(z)$ may be expressed as

$$f(z) = f_-(z) + f_+(z) \tag{3.12}$$

where

$$f_+(z) = \begin{cases} f(z), & z > 0 \\ 0 \quad , & z < 0 \end{cases}$$

and

$$f_-(z) = \begin{cases} 0 \quad , & z > 0 \\ f(z), & z < 0 \end{cases}$$

And $F_+(\alpha)$ is the Fourier transform of $f_+(z)$, $F_-(\alpha)$ is the Fourier transform of $f_-(z)$. According to the definition of the inverse Fourier Transform, we have

$$F^{-1}(F_+) = \begin{cases} f(z), & z > 0 \\ 0 \quad , & z < 0 \end{cases}$$

$$F^{-1}(F_-) = \begin{cases} 0 \quad , & z > 0 \\ f(z), & z < 0 \end{cases}$$

3.3 Properties of the Fourier transform in α-plane

In section 3.2, we have discussed the Fourier transform with a real variable α. Now we will discuss the Fourier transform with a complex α in a complex α-plane so as the properties of the Fourier transform may be used to the Winner-Hopf technology.

Analytic function:

First, a function is analytic at certain point means that the function is differentiable and is unique value at this point.

If a function is analytic everywhere except some finite points in a region D, then this function is said to be an *analytic function*, and some finite points mentioned above are said to be the *singular points*.

Regular and complete function:

If a function is an analytic function without any singular point in region D, then this function is said to be regular in region D. If a function is *regular* in finite regions in entire complex plane, then the function is said to be the *complete function or integrable function.*

Theorem 3.1 A function

$$G(\alpha) = \int_a^{\infty} f(z)h(z,\alpha)dz \tag{3.13}$$

is regular in an entire region D if and only if it satisfies the following five conditions:

(1) $h(z,\alpha)$ is a continuous function with two variables z and α if α is in region D, $\alpha \le z \le R$ and R is a finite value.

(2) $h(z,\alpha)$ is regular in region D when $\alpha \le z \le R$.

(3) $f(z)$ is possessed of finite numbers of discontinuous points with non-infinite value as well as finite numbers of maximum points and finite numbers of minimum points only, when $\alpha \le z \le R$.

(4) $f(z)$ is bounded in $\alpha \le z \le R$ except at some finite points, at those points z_0, $f(z) \to \infty$ when $z \to z_0$, and then

$$G(\alpha) = \lim_{\delta \to 0}\left[\int_a^{z_0-\delta} f(z)h(z,\alpha)dz + \int_{z_0+\delta}^{\infty} f(z)h(z,\alpha)dz\right]$$

exists.

(5) For any α in region D, $G(\alpha)$ is uniformly convergent.

The proof of Theorem 3.1 is in [14] [2] if you are interested in the proof.

---

[2] E.C. Titchmarch, Theory of Functions, 2nd ed. (New York: Oxford University Press. 1939), pp. 99

3.4 Winner-Hopf Technology

Winner-Hopf Technology is a combination of the Green function method and the Fourier integral method. And it may be used to a half infinite space boundary value problems such as the bifurcated rectangular waveguide, microstrip and dielectric waveguide and so on.

The idea of the Winner-Hopf Technology is:

(1) Set up an integral equation of the field according to the Green Function Method.

(2) Making use of the Fourier transform for the integral equation to obtain an algebraic equation in a complex plane. This equation is said to be the Winner-Hopf equation. Which contains two unknowns. In general, it is unable to obtain two unknowns from one single equation.

(3) The Winner-Hopf equation not only contains a function regular in the upper complex plane alone, but also contains another function regular in the lower complex plane alone. Consequently, a single Winner-Hopf equation may be separated into two equations: one contains the functions regular in the upper complex plane alone, and another one contains the functions regular in the lower complex plane alone. This is the key point of the Winner-Hopf technology. In this case, the two unknowns may be obtained from one single equation.

Now we will explain the Winner-Hopf technology step by step.

3.4.1 Set up the field integral equation

Now we will discuss the Winner-Hopf Technology based on the paper by author [13].

Winner-Holf technology is a powerful tool to deal with a boundary value problem for a half space geometric structure such as the microstrip, and the dielectric waveguide.

As shown in Figure 3-3, the region I is a grounded dielectric planar waveguide, in which there is a metal plate at $y = 2d$ and then the electric wall is at $y = 2d$. The region II is a symmetric dielectric waveguide, in which the electric wall is at the symmetric plane $y = d$. Since the symmetric plane in region I is different from the the symmetric plane in region II, it is very difficult to obtain the solution by the mode matching method. However, it is a half space boundary value problem in the dielectric waveguide and may be solved by the Winner-Hopf technology.

Now the problem is:

Fine the reflection coefficient of TM_1 mode when a TM_1 mode is obliquely incident with an incident angle θ_i from region I to the interface between region I and region II. Where we have made an assumption that the thickness of the dielectric waveguide ($2d$) is such that the propagation mode in region I is the principle mode TM_1 alone.

To this end,

(1) Write down the incident field and the scattering fields. In which the incident field is known as TM_1 mode in the grounded dielectric waveguide and the scattering fields are unknown.

(2) Deduce the Winner-Hopf equation by means of the boundary conditions.

(3) Find the scattering field from the Winner-Hopf equation.

(4) Find the reflection coefficient of the TM_1 mode from the scattering field.

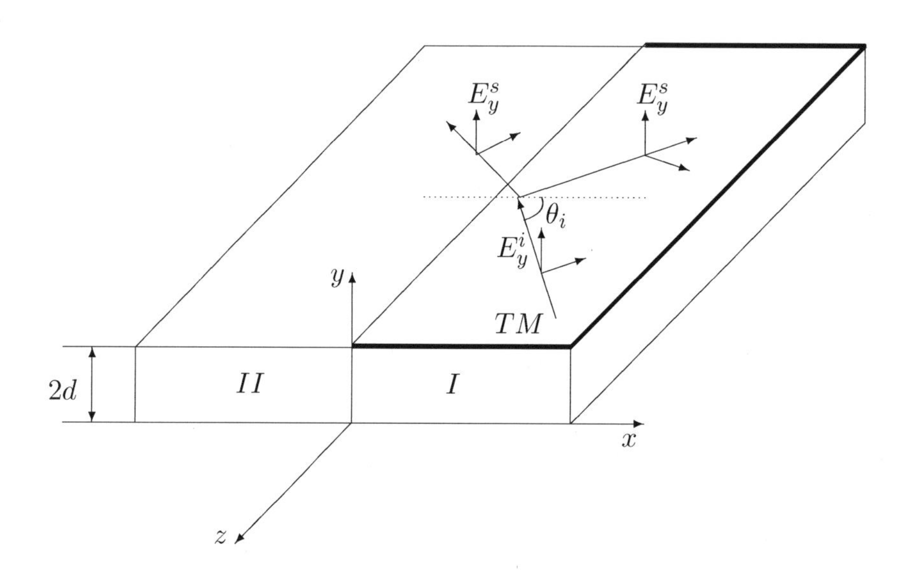

Figure 3-3 Dielectric Waveguide

The incident field and the scattering fields:

Now, as shown in Figure 3-3, the incident field is the known TM_1 mode in the grounded dielectric waveguide as follows:

In the air region ($y \leq 0, -\infty < x < \infty$)

$$
\begin{aligned}
E_y^i &= \varepsilon_r e^{ik_0\,\gamma_0\,y}\, e^{ik_0\,\alpha\,z}\, e^{ik_0\,\beta\,x} \\
\mathbf{E}_\tau^{\mathbf{i}} &= \varepsilon_r \frac{i\gamma_0}{n_{eff}^2}(\beta\mathbf{a_x} + \alpha\mathbf{a_z}) e^{ik_0\,\gamma_0\,y}\, e^{ik_0\,\alpha\,z}\, e^{ik_0\,\beta\,x} \\
\mathbf{H}_\tau^{\mathbf{i}} &= -\frac{\varepsilon_r}{\eta_0 n_{eff}^2}(-\alpha\mathbf{a_x} + \beta\mathbf{a_z}) e^{ik_0\gamma_0\,y}\, e^{ik_0\,\alpha\,z}\, e^{ik_0\,\beta\,x}
\end{aligned} \tag{3.14}
$$

In the dielectric region $(0 \leq y \leq 2d, -\infty < x < \infty)$

$$E_y^i = \frac{\cosh[k_0\gamma(y-2d)]}{\cosh[k_0\gamma 2d]} e^{ik_0\,\alpha\,z} e^{ik_0\,\beta\,x}$$

$$\mathbf{E}_\tau^\mathbf{i} = \varepsilon_r \frac{i\gamma_0}{n_{eff}^2}(\beta\mathbf{a_x} + \alpha\mathbf{a_z})\frac{\sinh[k_0\gamma(y-2d)]}{\sinh[k_0\gamma 2d]} e^{ik_0\,\alpha\,z} e^{ik_0\,\beta\,x}$$

$$\mathbf{H}_\tau^\mathbf{i} = -\frac{\varepsilon_r}{\eta_0 n_{eff}^2}(-\alpha\mathbf{a_x} + \beta\mathbf{a_z})\frac{\cosh[k_0\gamma(y-2d)]}{\cosh[k_0\gamma 2d]} e^{ik_0\,\alpha\,z} e^{ik_0\,\beta\,x}$$

(3.15)

Where $\mathbf{E}_\tau^\mathbf{i}$, $\mathbf{H}_\tau^\mathbf{i}$ are the tangential components with respect to the interface of the dielectric slab (xoz), $\alpha = n_{eff}\,sin\,\theta_i$, θ_i is the incident angle of the TM_1 mode, $\beta = (n_{eff}^2 - \alpha^2)^{1/2}$, $\gamma = (n_{eff}^2 - n^2)^{1/2}$, $n^2 = \varepsilon_r$, ε_r is the relative dielectric constant of the dielectric slab, $\eta = \sqrt{\mu_r\varepsilon_r}$, $\eta_0 = 120\pi$ is the wave impedance in the air, $\gamma_0 = (n_{eff}^2 - 1)^{1/2}$. And n_{eff} is the effective refraction index, which is from the boundary condition at the interface of xoy:

$$\begin{aligned} &\mathbf{E}_\tau^\mathbf{i}(x, 2d) = 0, \\ &\mathbf{E}_\tau^\mathbf{i}(x, 0^-) = \mathbf{E}_\tau^\mathbf{i}(x, 0^+), \\ &\varepsilon_r E_y^i(x, 0^+) = E_y^i(x, 0^-), \\ &\mathbf{H}_\tau^\mathbf{i}(x, 0^-) = \mathbf{H}_\tau^\mathbf{i}(x, 0^+). \end{aligned} \qquad (3.16)$$

This boundary condition leads to a characteristic equation as follows:

$$\varepsilon_r(n_{eff}^2 - 1)^{1/2} + (n_{eff}^2 - n^2)^{1/2}\tanh[k_0(n_{eff}^2 - n^2)^{1/2}\,2d] = 0 \qquad (3.17)$$

Obviously, the incident field (3.14) and (3.15) are the solutions satisfied the Maxwell's equations and the boundary conditions (3.16).

Mode matching method:

The mode matching method considers that the incident field $\mathbf{E^i}$, $\mathbf{H^i}$ exist in region I alone and the scattering fields $\mathbf{E^s}$, $\mathbf{H^s}$ exist both in region I and region II. And then the total field $\mathbf{E^t} = \mathbf{E^i} + \mathbf{E^s}$ matches at the transverse interface of $x = 0$.

Winner-Hopf technology:

However, the Winner-Hopf technology considers that the incident fields $\mathbf{E^i}$, $\mathbf{H^i}$ exist both in region I and region II in the same state as in region I as indicated in Figure 3-4(a). And the scattering field $\mathbf{E^s}$, $\mathbf{H^s}$ exist both in region II and region I in the same state as in region II as indicated in Figure 3-4(b). And then the total fields $\mathbf{E^t} = \mathbf{E^i} + \mathbf{E^s}$, $\mathbf{H^t} = \mathbf{H^i} + \mathbf{H^s}$ have to satisfy the original boundary conditions as indicated in Figure 3-4(c). In this case, the total fields still are the solutions satisfied the Maxwell's equation and the original boundary conditions. Therefore, the total fields should be the same as the solutions from the mode matching method.

Where the scattering field consists of the LSM modes and the LSE modes. Both of them satisfy the following Maxwell's equations.

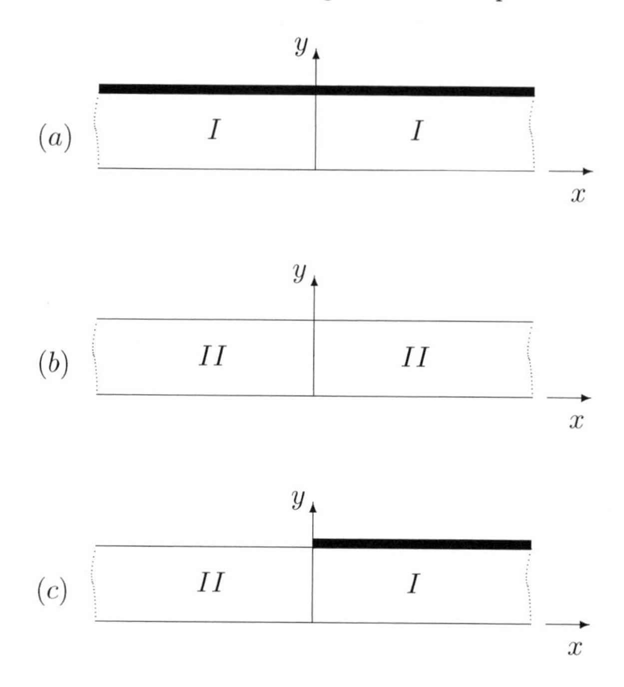

Figure 3-4 Winner-Hopf Technology

For LSM modes, we have

$$\begin{aligned} &[\nabla_{xy}^2 + k_0^2(n^2 - \alpha^2)]E_y^s = 0, \qquad (0 \leq y \leq 2d) \\ &[\nabla_{xy}^2 + k_0^2(1 - \alpha^2)]E_y^s = 0, \qquad (0 \geq y, y \geq 2d) \end{aligned} \tag{3.18}$$

For LSE modes, we have

$$\begin{aligned} &[\nabla_{xy}^2 + k_0^2(n^2 - \alpha^2)]H_y^s = 0, \qquad (0 \leq y \leq 2d) \\ &[\nabla_{xy}^2 + k_0^2(1 - \alpha^2)]H_y^s = 0, \qquad (0 \geq y, y \geq 2d) \end{aligned} \tag{3.19}$$

They satisfy the following boundary conditions:

$$\varepsilon_r E_y^s \Big|_{\substack{y=0+ \\ y=2d-}} = E_y^s \Big|_{\substack{y=0- \\ y=2d+}}$$

$$\mu_r H_y^s \Big|_{\substack{y=0+ \\ y=2d-}} = H_y^s \Big|_{\substack{y=0- \\ y=2d+}}$$

$$E_T^s \Big|_{\substack{y=0+ \\ y=2d-}} = E_T^s \Big|_{\substack{y=0- \\ y=2d+}}$$

$$H_\tau^s\Big|_{\substack{y=0+\\y=2d^-}} = H_\tau^s\Big|_{\substack{y=0^-\\y=2d^+}} \tag{3.20}$$

and the radiation conditions

$$|\mathbf{E^s}| \to 0, \qquad |\mathbf{H^s}| \to 0, \qquad when \quad |x| \to \infty \tag{3.21}$$

3.4.2 Deduce the Winner-Hopf equation

Now, making the Fourier transform of the wave equations (3.18) and (3.19) by means of the definition of

$$F(\lambda) = \frac{k_0}{2\pi}\int_{-\infty}^{\infty} f(x)\, e^{ik_0\lambda x}\, dx \tag{3.22}$$

we have the scattering fields from the wave equations as follows:

$$E_y^s(\lambda, y)_{(LSM)} = \begin{cases} E_1 \exp[-u_0 k_0 (y-2d)], & y \geq 2d \\ E_{n1}\cosh[u_n k_0 (y-2d)] & \\ \quad +E_{n2}\sinh[u_n k_0(y-2d)], & 0 \leq y \leq 2d \\ E_3 \exp[u_0 k_0 y], & y \leq 0 \end{cases} \tag{3.23}$$

and

$$H_y^s(\lambda, y)_{(LSE)} = \begin{cases} H_1 \exp[-u_0 k_0 (y-2d)], & y \geq 2d \\ H_{n1}\cosh[u_n k_0 (y-2d)] & \\ \quad +H_{n2}\sinh[u_n k_0(y-2d)], & 0 \leq y \leq 2d \\ H_3 \exp[u_0 k_0 y], & y \leq 0 \end{cases} \tag{3.24}$$

Where $u_0 = (\lambda^2 + \alpha^2 - 1)^{1/2}$, $u_n = (\lambda^2 + \alpha^2 - n^2)^{1/2}$, and the six unknowns in $E_y^s(LSM)$ and $H_y^s(LSE)$ may be obtained by the boundary conditions (3.20). It follows that:

$$\begin{aligned} E_{n1} &= [\frac{1}{\varepsilon_r}\cosh(u_n k_0 2d) + \frac{u_0}{u_n}\sinh(u_n k_0 2d)]E_3 \\ E_{n2} &= [\frac{1}{\varepsilon_r}\sinh(u_n k_0 2d) + \frac{u_0}{u_n}\cosh(u_n k_0 2d)]E_3 \\ H_{n1} &= [\frac{1}{\mu_r}\cosh(u_n k_0 2d) + \frac{u_0}{u_n}\sinh(u_n k_0 2d)]H_3 \\ H_{n2} &= [\frac{1}{\mu_r}\sinh(u_n k_0 2d) + \frac{u_0}{u_n}\cosh(u_n k_0 2d)]H_3 \end{aligned} \tag{3.25}$$

and E_3 and H_3 have to be determined by the original boundary conditions as indicated in Figure 3-4(c):

$$\mathbf{E}^{\mathbf{t}}_{\tau}(\lambda, 2d^+) = \mathbf{E}^{\mathbf{i}}_{\tau}(\lambda, 2d^-) + \mathbf{E}^{\mathbf{s}}_{\tau}(\lambda, 2d^-) = 0, \qquad (x > 0) \tag{3.26}$$

$$\mathbf{E}^{\mathbf{t}}_{\tau}(\lambda, 2d^+) = \mathbf{E}^{\mathbf{i}}_{\tau}(\lambda, 2d^-) + \mathbf{E}^{\mathbf{s}}_{\tau}(\lambda, 2d^-), \qquad (x < 0) \tag{3.27}$$

$$E^s_y(\lambda, 2d^+) - \varepsilon_r E^s_y(\lambda, 2d^-) = \varepsilon_r E^i_y(\lambda, 2d^-) = -\frac{1}{\varepsilon_0}\rho^i_-, \qquad (x > 0) \tag{3.28}$$

$$E^s_y(\lambda, 2d^+) - \varepsilon_r E^s_y(\lambda, 2d^-) = \rho^s_+/\varepsilon_0, \qquad (x < 0) \tag{3.29}$$

$$\mathbf{H}^{\mathbf{s}}_{\tau}(\lambda, 2d^-) + \mathbf{H}^{\mathbf{i}}_{\tau}(\lambda, 2d^-) = \mathbf{H}^{\mathbf{s}}_{\tau}(\lambda, 2d^+), \qquad (x < 0) \tag{3.30}$$

$$\mathbf{H}^{\mathbf{s}}_{\tau}(\lambda, 2d^+) - \mathbf{H}^{\mathbf{s}}_{\tau}(\lambda, 2d^-) = -\mathbf{a_y} \times \mathbf{J}^{\mathbf{s}}_{\tau}(\lambda, 2d^-), \quad (x > 0) \tag{3.31}$$

From which we have

$$E_3 = \frac{\varepsilon_r}{\varepsilon_r u_0 \cosh(u_n k_0 2d) + u_n \sinh(u_n k_0 2d)} F_-(\lambda) \tag{3.32}$$

$$H_3 = i\frac{u_n}{\eta_0[u_n \cosh(u_n k_0 2d) + \mu_r u_0 \sinh(u_n k_0 2d)]} G_-(\lambda) \tag{3.33}$$

Where

$$F_-(\lambda) = \frac{1}{2\pi}\int_{-\infty}^{0} \nabla_{\mathbf{t}} \cdot \mathbf{E}^{\mathbf{s}}_{\tau}(x, 2d) e^{ik_0\lambda x} dx \tag{3.34}$$

$$G_-(\lambda) = \frac{1}{2\pi}\int_{-\infty}^{0} \mathbf{a_z} \cdot \nabla_{\mathbf{t}} \times \mathbf{E}^{\mathbf{s}}_{\tau}(x, 2d) e^{ik_0\lambda x} dx \tag{3.35}$$

are the regula functions in the lower half λ-plane defined by $Im\lambda \leq \tau_0$ as shown in Figure 3-5. Here, use of the boundary condition $E^s_{\tau}(x, 2d)|_{x>0} = 0$ has been made.

Substituting E^s_y from (3.23) into (3.28) and (3.29), we have

$$\varepsilon_r E_{n1} = \varepsilon_r E^s_y(\lambda, 2d^-) \tag{3.36}$$

$$= \frac{k_0}{2\pi}\left(\int_{-\infty}^{0} + \int_{0}^{\infty}\right)\varepsilon_r E^s_y(\lambda, 2d^-) e^{ik_0\lambda x} dx$$

$$= E^s_y(\lambda, 2d^+) + \frac{1}{\varepsilon_0}[\rho^i_-(\lambda) - \rho^s_+(\lambda)]. \tag{3.37}$$

Substituting (3.25) and (3.32) into (3.37), we have

$$F_-(\lambda) = Q_e(\lambda)[\rho^s_+(\lambda) - \rho^i_-(\lambda)]/\varepsilon_0 \tag{3.38}$$

Where

$$Q_e(\lambda) = \frac{u_0 u_n[\varepsilon_r u_0 + u_n \tanh(u_n k_0 2d)]}{2\varepsilon_r u_0 u_n + (\varepsilon_r^2 u_0^2 + u_n^2)\tanh(u_n k_0 2d)} \tag{3.39}$$

$$\rho^i_-(\lambda) = \frac{k_0}{2\pi}\int_{-\infty}^{0} -\varepsilon_0\varepsilon_r E^i_y(y, 2d^-) e^{ik_0\lambda x} dx$$

$$= \frac{i\varepsilon_0\varepsilon_r}{2\pi\cosh(\gamma 2k_0 d)}\left[\frac{1}{\lambda+\beta}\right]_- \tag{3.40}$$

In which, $[\,\cdot\,]_-$ denotes that the value in the brackets is from the lower half λ-plane defined by $Im\lambda < \tau_0$.

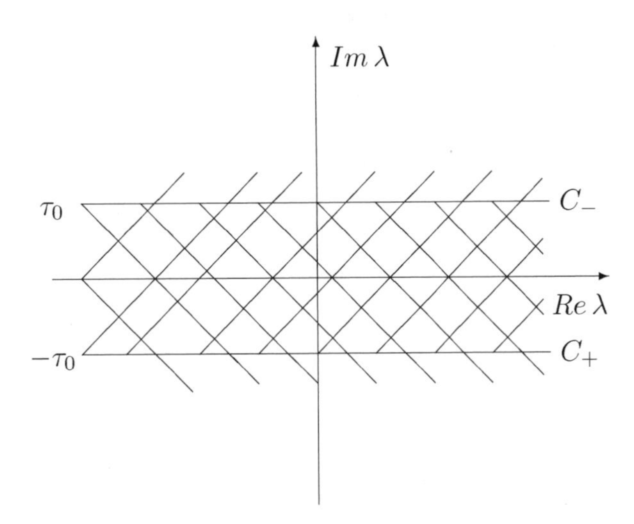

Figure 3-5 λ-plane

(3.38) is a Winner-Hopf equation with two unknowns $F_-(\lambda)$ and $\rho^s_+(\lambda)$.

Making use of the boundary conditions (3.30) and (3.31), we may have another Winner-Hopf equation as follows:

$$G_-(\lambda) = -iQ_m(\lambda)j_+(\lambda)/\omega\varepsilon_0 \tag{3.41}$$

where

$$Q_m(\lambda) = \frac{\mu_r[u_n + \mu_r u_0 \tanh(u_n k_0 2d)]}{2\mu_r u_0 u_n + (\mu_r^2 u_0^2 + u_n^2)\tanh(u_n k_0 2d)} \tag{3.42}$$

$$j_+(\lambda) = \frac{k_0}{2\pi}\int_0^\infty \mathbf{a_y}\cdot\nabla_\mathbf{t}\times\mathbf{J}^\mathbf{s}_\tau(x,2d^-)e^{iK_0\lambda x}dx \tag{3.43}$$

3.4.3 Factorization of $Q(\lambda)$ into $Q_+(\lambda)Q_-(\lambda)$

To fine the solutions of the Winner-Hopf equations (3.38) and (3.41) the key point is to make factorization of $Q(\lambda)$ into $Q_+(\lambda)Q_-(\lambda)$.

To this end, we have to come back to see the factorization formulas (3.44) and (3.45), from R. Mittra and S. W. Lee [12], [3]

[3] pp. 114 in [12]

$$G(\alpha) = G_+(\alpha)G_-(\alpha) \tag{3.44}$$

$$G_+(\alpha) = G_-(\alpha) = \sqrt{G(0)}\left(1 + \frac{\alpha}{k}\right)^{\nu/2}$$
$$\cdot \exp\left\{\frac{1}{2\pi i}\int_{-\infty-id}^{\infty-id}\left[\frac{\beta h}{(k^2-\beta^2)^{1/2}} + \frac{\beta\gamma}{(k^2-\beta^2)} + \frac{G'_\beta(\beta)}{G(\beta)}\right]\right.$$
$$\left.\cdot \ln\left(\frac{\beta-\alpha}{\beta}\right)d\beta\right\} \tag{3.45}$$

From (3.44) and (3.45) with $\nu = 0$ and $h = 0$, we have

$$Q(\lambda) = Q_+(\lambda)Q_-(\lambda) \tag{3.46}$$

$$Q_+(\lambda) = Q_-(-\lambda)$$

and

$$Q_\pm(\lambda) = \sqrt{Q(0)}\exp\left[\mp\frac{1}{2\pi i}\int_{c\pm}\frac{Q'(W)}{Q(W)}\ln\left(\frac{W-\lambda}{W}\right)dW\right]$$

From the condition (3) [12] [4]in (3.44) it is known that when $\nu = 0$ and $h = 0$, $Q(\lambda)$ is still bounded in the trip region $|Im\lambda| < \tau_0$.

Substituting (3.46) into the Winner-Hopf equation (3.38), we have

$$F_-(\lambda) = Q_{e+}(\lambda)Q_{e-}(\lambda)[\rho^s_+(\lambda) - \rho^i_-(\lambda)]/\varepsilon_0 \tag{3.47}$$

3.4.4 Decomposition of $Q_{e+}(\lambda)\rho^i_-(\lambda)$

The next step is the decomposition of $Q_{e+}(\lambda)\rho^i_-(\lambda)$.

Now, from the equation of $\rho^i_-(\lambda)$ in (3.40), $Q_{e+}(\lambda)\rho^i_-(\lambda)$ may be decomposed as follows:

$$\frac{1}{\varepsilon_0}\rho^i_-(\lambda)Q_{e+}(\lambda) = \frac{1}{\cosh(\gamma k_0 2d)}\frac{i\varepsilon_r}{2\pi}\frac{1}{[\lambda+\beta]_-}Q_{e+}(-\beta)$$
$$+\left\{\frac{1}{\cosh(\gamma k_0 2d)}\frac{i\varepsilon_r}{2\pi}\frac{1}{[\lambda+\beta]}[Q_{e+}(\lambda) - Q_{e+}(-\beta)]\right\}_+ \tag{3.48}$$

Where the subscript of $\{\ \}_+$ denotes that the value of λ in this brace is from the upper half λ-plane defined by $Im\lambda > -\tau_0$. And the $Q_{e+}(-\beta)$ in $\{\ \}_+$ is to prevent the brace $\{\ \}_+$ from the presence of a pole point at $\lambda = -\beta$ in the upper half λ-plane. Similarly, the subscript of $[\lambda+\beta]_-$ denotes that the value of λ in $[\lambda+\beta]_-$ is from the lower half λ-plane defined by $Im\lambda < \tau_0$.

[4] pp. 114 in [12]

Substituting (3.48) in (3.47), introducing a unknown c_1, the Winner-Hopf equation (3.47) may be decomposed into two equations as follows:

$$F_-(\lambda) = \frac{i\varepsilon_r}{2\pi}\left[c_1 - Q_{e+}(-\beta)\,\frac{1}{\cosh(\gamma k_0 2d)}\,\frac{1}{[\lambda+\beta]_-}\right]Q_{e-}(\lambda) \tag{3.49}$$

$$\rho^s_+(\lambda)/\varepsilon_0 = \frac{i\varepsilon_r}{2\pi}\left\{c_1 + \frac{1}{\cosh(\gamma k_0 2d)}\,\frac{1}{[\lambda+\beta]}[Q_{e+}(\lambda) - Q_{e+}(-\beta)]\right\}_+ \tag{3.50}$$

In which, (3.49) is regula in the lower half λ-plane defined by $Im\lambda < \tau_0$ alone, and (3.50) is regula in the upper half λ-plane defined by $Im\lambda > -\tau_0$ alone. As shown in Figure 3-5, these two regula regions are overlapped in the strip region defined by $|Im\lambda| < \tau_0$.

Similarly, substituting (3.46) in (3.41), we have

$$G_-(\lambda) = -\frac{i}{\omega\varepsilon_0}\,Q_{m+}(\lambda)\,Q_{m-}(\lambda) j_+(\lambda) \tag{3.51}$$

Suppose that

$$G_-(\lambda) = -\frac{\varepsilon_r}{2\pi}\,c_2\,Q_{m-}(\lambda) \tag{3.52}$$

we have, from (3.51),

$$j_+(\lambda) = \frac{i\omega\varepsilon_o\varepsilon_r}{2\pi}\,c_2\,Q^{-1}_{m+}(\lambda) \tag{3.53}$$

Where c_2 is unknown. Obviously, (3.52) is regular in the lower half λ-plane defined by $Im\lambda < \tau_0$ alone, and (3.53) is regular in the upper half λ-plane defined by $Im\lambda > -\tau_0$ alone. And then the Winner-Hopf equation (3.41) is decomposed into two equations (3.52) and (3.53).

Determination of c_1 and c_2

The next step is to determine the two unknowns c_1 and c_2. To this end, we have to come back to the boundary condition:

$$E^s_\tau(x,\,2d) = 0, \qquad (x > 0) \tag{3.54}$$

and $E^s_\tau(x, 2d)$ may be obtained from (3.23) and (3.24).

Making Fourier transform of $E^s_\tau(x, 2d)$, we have

$$\begin{aligned} E^s_{\tau-}(\lambda, 2d) &= \frac{k_0}{2\pi}\int_{-\infty}^{0} E^s_\tau(x, 2d)\,e^{ik_0x}dx \\ &= \frac{i}{\lambda^2+\alpha^2}\{[\lambda F_-(\alpha) + \alpha G_-(\alpha)]\mathbf{a_x} + [\alpha F_-(\lambda) - \lambda G_-(\lambda)]\mathbf{a_z}\} \end{aligned} \tag{3.55}$$

Obviously, $E^s_{\tau-}(\lambda, 2d)$ is regular in the lower half λ-plane defined by $Im\lambda < \tau_0$. However, the right-hand side of (3.55) contains a pole point at $\lambda = -i\alpha$ in the lower half λ-plane. To hold $E^s_{\tau-}(\lambda, 2d)$ to be regular in the lower half λ-plane, it is required that

$$[\lambda F_-(\lambda) + \alpha G_-(\lambda)]|_{\lambda=-i\alpha} = 0$$
$$[\alpha F_-(\lambda) - \lambda G_-(\lambda)]|_{\lambda=-i\alpha} = 0$$

Namely, it is required

$$F_-(-i\alpha) = -iG_-(-i\alpha) \tag{3.56}$$

Substituting (3.49) and (3.52) in (3.56), we have

$$c_1 Q_{e+}(i\alpha) + c_2 Q_{m+}(i\alpha) = \frac{1}{\cosh(\gamma k_0 2d)} \frac{1}{[\beta - i\alpha]} Q_{e+}(\beta) Q_{e+}(i\alpha) \tag{3.57}$$

In which, use of (3.46) has been made.

In addition, we have to use another two boundary conditions, i.e.

$$\rho^s + \rho^i = 0$$
$$J^s_\tau + J^i_\tau = 0$$

at $y = 0$, $x < 0$. Making the Fourier transform of these two equations, we have

$$\rho^s_-(\lambda) + \rho^i_-(\lambda) = 0$$
$$J^s_{\tau-}(\lambda) + J^i_{\tau-}(\lambda) = 0 \tag{3.58}$$

From which, we have

$$c_1 Q_{m+}(i\alpha) + c_2 Q_{e+}(i\alpha) = \frac{1}{\cosh(\gamma k_0 2d)} \frac{1}{[\beta + i\alpha]} \cdot Q_{e+}(-\beta) Q_{m+}(i\alpha) \tag{3.59}$$

And then, from (3.57) and (3.59), we have

$$c_1 = \frac{1}{\cosh(\gamma k_0 2d)} \frac{1}{n^2_{eff}} [\beta + i\alpha\Omega_1] Q_{e+}(-\beta)$$
$$c_2 = \frac{-i\alpha}{\cosh(\gamma k_0 2d)} \frac{1}{n^2_{eff}} \Omega_2 Q_{e+}(-\beta) \tag{3.60}$$

Where

$$\Omega_1 = \frac{Q^2_{e+}(i\alpha) - Q^2_{m+}(i\alpha)}{Q^2_{e+}(i\alpha) + Q^2_{m+}(i\alpha)}, \qquad \Omega_2 = \frac{2Q_{e+}(i\alpha)Q_{m+}(i\alpha)}{Q^2_{e+}(i\alpha) + Q^2_{m+}(i\alpha)} \tag{3.61}$$

3.4.5 Find the solution of the Winner-Hopf equation

Substituting (3.60) in (3.49) and (3.52), we have the solutions of the Winner-Hopf equations (3.38) and (3.41) as follows:

$$F_-(\lambda) = \frac{i\varepsilon_r}{2\pi} \frac{Q_{e+}(-\beta)}{\cosh(\gamma k_0 2d)} \Big[\frac{1}{n^2_{eff}} (\beta + i\alpha\Omega_1)$$

$$-\frac{1}{[\lambda+\beta]_-}\Big]Q_{e-}(\lambda) \quad (3.62)$$

$$G_-(\lambda) = -\frac{i\alpha\varepsilon_r}{2\pi}\frac{1}{\cosh(\gamma k_0 2d)}\frac{1}{n_{eff}^2}\Omega_2 Q_{e+}(-\beta)Q_{m-}(\lambda) \quad (3.63)$$

Where $\beta = (n_{eff}^2 - \alpha^2)^{1/2}$, $\gamma = (n_{eff}^2 - n^2)^{1/2}$.

Substituting (3.62) and (3.63) in (3.25) and (3.32), we have the scattering field $E_y^s(\lambda, y)$. And then the inverse Fourier transform of $E_y^s(\lambda, y)$ gives us

$$\begin{aligned} E_y^s(x,y) &= \int_{-\infty}^{\infty} E_y^s(\lambda,y)e^{-ik_0\lambda x}d\lambda \\ &= \int_{-\infty}^{\infty} -\Big[\frac{(u_n + \varepsilon_r u_0 \tanh u_n k_0 2d)\cosh u_n k_0(y-2d)}{u_n(\varepsilon_r u_0 + u_n \tanh u_n k_0 2d)} \\ &\quad + \frac{1}{u_0}\sinh u_n k_0(y-2d)\Big]F_-(\lambda)e^{-ik_0\lambda x}d\lambda \end{aligned} \quad (3.64)$$

Where

$$u_0 = (\lambda^2 + \alpha^2 - 1)^{1/2}, \qquad u_n = (\lambda^2 + \alpha^2 - n^2)^{1/2} \quad (3.65)$$

Mode analysis:

To analyze the modes contained in the scattering field $E_y^s(x,y)$, we have to carefully analyze the distribution of the singular points in the integrand in (3.64). Form physical point of view, the residual contribution of any singular point corresponds to a mode in the dielectric waveguide. In which, the residual contribution of the branch point $u_n = 0$ represents the radiation mode. And the residual contribution of the pole points $\lambda = \pm\lambda_m$ represent the surface modes in the dielectric waveguide. These surface waves are the reflection wave of the incident wave TM_1 (in right-hand side $x > 0$) and the transmission wave of the TM_1 (in left-hand side $x < 0$). The equation corresponding to $\lambda = \pm\lambda_m$ is as follows:

$$\varepsilon_r u_0 + u_n \tanh(u_n k_0 2d) = 0 \quad (3.66)$$

Where

$$u_0 = ({\lambda_m}^2 + \alpha^2 - 1)^{1/2}, \qquad u_n = ({\lambda_m}^2 + \alpha^2 - n^2)^{1/2}$$

Let

$$\lambda_m = (n_{neff}^2 - \alpha^2)^{1/2} = \beta, \quad (3.67)$$

then (3.66) may be rewritten as

$$\varepsilon_r(n_{eff}^2 - 1)^{1/2} + (n_{eff}^2 - n^2)^{1/2}\tanh(\sqrt{n_{eff}^2 - n^2}\, k_0 2d) = 0 \quad (3.68)$$

Obviously, (3.68) is just the eigenvalue equation (3.17) of the TM_1 mode in the grounded slab dielectric waveguide. And $k_0\lambda_m = k_0\beta$ is just the propagation constant along x direction when the TM_1 mode is oblique incident.

3.4.6 Find the reflection coefficient of the TM_1 mode

When $\lambda = +\lambda_m$, the residual contribution from (3.64) yields the reflection wave of the TM_1 wave for the case that the TM_1 wave is obliquely incident onto the interface at $x = 0$, i.e.

$$E_y^s(x,y)|_{\lambda=\lambda_m} = 2\pi \lim_{\lambda\to\lambda_m} \Big\{ -(\lambda-\lambda_m) \cdot \Big[\frac{(u_n + \varepsilon_r u_0 \tanh u_n k_0 2d)\cosh u_n k_0(y-2d)}{u_n(\varepsilon_r u_0 + u_n \tanh u_n k_0 2d)} + \frac{1}{u_n}\sinh u_n k_0(y-2d)\Big] \cdot F_-(\lambda)\Big\} e^{-ik_0\beta x}$$

$$= RE_y^i(x,y) = R\frac{\cosh[\sqrt{n_{eff}^2 - n^2}k_0(y-2d)]}{\cosh[\sqrt{n_{eff}^2-n^2}k_0 2d)]}e^{-ik_0\beta x}$$

Where R is the reflection coefficient of the TM_1 mode in oblique incidence. The residual contribution may be calculated by L'Hopital's Rule . And then the reflection coefficient is

$$R = -2\pi i\Big\{\frac{u_0(u_n^2 - \varepsilon_r^2 u_0^2)\cosh(u_n k_0 2d)}{\lambda[\varepsilon_r(u_n^2-u_0^2) + k_0 2d u_0(u_n^2 + \varepsilon_r^2 u_0^2)]} \cdot F_-(\lambda)\Big\}\Big|_{\lambda=\sqrt{n_{eff}^2-\alpha^2}} \tag{3.69}$$

Substituting $F_-(\lambda)$ from (3.62) into (3.69) and re-ranging the formula, we have the reflection coefficient of TM_1 in oblique incidence as

$$R = e^{i\psi(\alpha)} \tag{3.70}$$

with

$$\psi(\alpha) = 2\arctan\frac{\alpha\tanh\triangle}{\sqrt{n_{eff}^2-\alpha^2}} - \arctan\frac{\sqrt{n_{eff}^2-\alpha^2}}{\sqrt{\alpha^2-1}} + \frac{2\sqrt{n_{eff}^2-\alpha^2}}{\pi}\oint_0^\infty \ln q_e(\lambda)\frac{d\lambda}{\lambda^2-(n_{eff}^2-\alpha^2)} \tag{3.71}$$

Where $\oint$ represents a principle-value integral. And

$$q_e(\lambda) = -\frac{(\varepsilon_r+1)u_0^2 u_n(\varepsilon_r u_0 + u_n\tanh u_n k_0 2d)(1+\tanh^2 u_n k_0 d)}{2(\lambda^2+\alpha^2-n_{eff}^2)(\varepsilon_r u_0 + u_n \tanh u_n k_0 d)(u_n + \varepsilon_r u_0 \tanh u_n k_0 d)} \tag{3.72}$$

$$\triangle =$$
$$\frac{1}{2}\ln\frac{(\mu_r+1)n_{eff}^2}{(\varepsilon_r+1)\mu_r}$$
$$+\frac{\alpha}{\pi}\int_0^\infty\Big[\ln\frac{(\varepsilon_r+1)u_0u_n}{(\mu_r+1)(\lambda^2+\alpha^2-n_{eff}^2)}\frac{(\varepsilon_r u_0+u_n\tanh u_n k_0 2d)}{(u_n+\mu_r u_0\tanh u_n k_0 d)}$$
$$\cdot\frac{(u_n+\mu_r u_0\tanh u_n k_0 d)(\mu_r u_0+u_n\tanh u_n k_0 d)}{(u_n+\varepsilon_r u_0\tanh u_n k_0 d)(\varepsilon_r u_0+u_n\tanh u_n k_0 d)}\Big]\frac{d\lambda}{\lambda^2+\alpha^2} \quad (3.73)$$

Where $u_n = (\lambda^2+\alpha^2-n^2)^{1/2}$, $u_0 = (\lambda^2+\alpha^2-1)^{1/2}$, and $Re\, u_0 \geq 0$.

It is of interest to point out that, the formula of the refection coefficient R in (3.70) - (3.71) is similar to that of the TEM mode when the TEM mode is incident obliquely onto the boundary of the microstrip. This refection coefficient is [15]

$$R_{TEM} = e^{i\psi(\alpha)} \quad (3.74)$$

$$\psi(\alpha) = 2\arctan\frac{\alpha\tanh\triangle}{\sqrt{n^2-\alpha^2}} - \arctan\frac{\sqrt{n^2-\alpha^2}}{\sqrt{\alpha^2-1}}$$
$$+\frac{2\sqrt{n^2-\alpha^2}}{\pi}\oint_0^\infty \ln q_e(\alpha)\frac{d\alpha}{\lambda^2-(n^2-\alpha^2)} \quad (3.75)$$

Investigation carefully, we find that the difference between R_{TEM} and R_{TM1} is that n in R_{TEM} has been replaced by n_{eff} in R_{TM1}, where n is the refraction index of TEM mode in the microstrip, n_{eff} is the effective refraction index of $TM1$ mode in the dielectric waveguide. And then $q_e(\lambda)$ and $\triangle$ of the TM_1 mode are also different from those of TEM.

3.4.7 Find the propagation constant of the dielectric wave-guide

Making use of the reflection coefficient R of the TM_1 mode, it is not so difficult to obtained the propagation constant along the dielectric waveguide by means of the transverse resonance condition.

As shown in Figure 3-6, a dielectric waveguide may be viewed as the dominant mode TM_1 incident obliquely onto the boundary of the waveguide, then suffers a total refection again and again between two walls with spacing of W, such that the TM_1 mode is propagation along z direction. This is just like a LSM_1 mode propagation along x direction and suffers a total reflection again and again between two walls with spacing of W, the width of the waveguide.

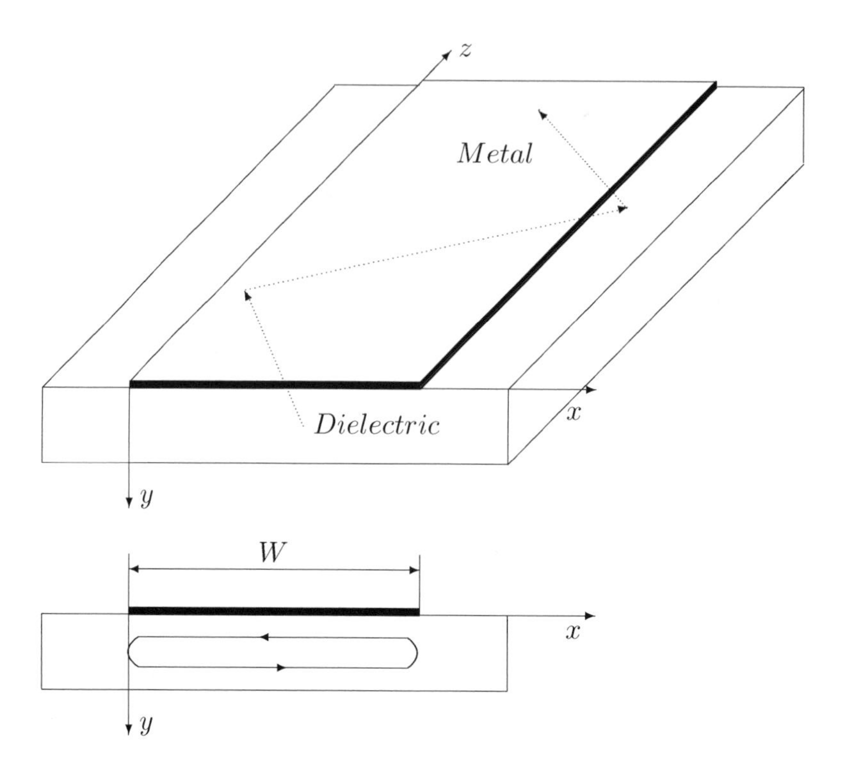

Figure 3-6 Dielectric Waveguide

For a guiding mode, it is required that the wave propagation must satisfy the *transversal resonance condition*, namely, when a LSM_1 mode is propagating along x direction from a start point and then reflection back from the right wall to $-x$ direction and then reflection back one more time along the x direction to the start point, the total phase shift from the start point to the last point must be equal to $2n\pi$. From physical point of view, it is a contracture coherent to make the guiding mode when the total phase shift is equal to $2n\pi$. Therefore the transverse resonance condition is

$$2\psi(\alpha) - 2k_{x1}W = -2\pi(q-1), \qquad q = 1, 2, \cdots \tag{3.76}$$

where $\psi(\alpha)$ is the phase of the reflection coefficient in (3.71), W is the width of the waveguide, $2k_{x1}W$ is the phase shift in the travelling of $2W$, and $k_{x1} = k_0(n_{eff}^2 - \alpha^2)^{1/2}$. In this case, we may obtain α from (3.76) by means of the successive substitution method. And then the propagation constant of the TM_1 mode along z direction is

$$k_z = k_0\alpha \tag{3.77}$$

Part II

Principle and application of the functional analysis

The functional analysis is the branch of mathematics concerned with the study of vector spaces and operators, in which a vector could be a function and then the operator could be a function of the function. Obviously, the functional analysis is a high level mathematic language. However, the applied mathematics mostly concentrates on some special physical problems , in which the scientist try to find a way to describe the special problem by means of some equations and then to find the solution from the equations under some special conditions (such as the initial condition from the experimental data). The functional analysis and the applied mathematics have been developed independently for a long time. Meanwhile, the vector spaces and the operators have displayed on some high level magazine more and more often since the applied mathematics has to face a challenge to deal with more and more complicated problem. Nowadays, the functional analysis has even been successfully used to deal with the Quantum mechanic [20] [21], those, we would like to say, are the complementary of the applied mathematics and the functional analysis.

Recently, the IT(information technology) has been developing extremely fast in these several ten years. And then the boundary value problem in the theory of electromagnetic field become much complicated than before. For instance, the optical fiber communications, the satellite communications, the mobil communications, the radio wave propagation in the forest and the plasma etc. have to deal with the boundary value problem in the nonlinear medium, the discrete medium, the random medium, the anisotropic medium, the dynamic medium in an open space. In these cases, we could not imagine how to analyze so complicated boundary value problem without the help of the functional analysis.

In deed, the functional analysis has been displaced everywhere on the high level magazine to analyze the complicated boundary value problem in the theory of electromagnetic field as we mentioned above. However, "the functional analysis" and "the theory of the electromagnetic field" are separated into two causes. And the Master degree students and the PhD students always feel it a dull lecture when they listen and study "the functional analysis".

In this book, we

1. Tries to introduce *the principle of the functional analysis and Hilbert Space* first (in Chapter 4),

2. Then to analyze the theory of the electromagnetic field *in Hilbert Space by means of the functional analysis.* Namely, making use of *the vector spaces (especially, the Hilbert Space) and the operator theory* to find the analytical solution for the boundary value problem of the electromagnetic field

(a) in *the dynamic medium* (in Chapter 5),

(b) in *the discrete medium* and *the anisotropic medium* (in Chapter 6), and

(c) in *the nonlinear medium* (in Chapter 7), respectively.

3. Then to analyze the electromagnetic field in *non-coordinates system* (in Chapter 8) and the electromagnetic field in multiple anisotropic waveguides system (in Chapter 9).

To this end, in the coming chapter, we have to introduce some concept of the functional analysis suitable to analyze the electromagnetic fields. In which, we have taken some material from Wikipedia, the free encyclopedia [22] as our reference and made some modification. We have made a note "*" as the material from Wikipedia, the free encyclopedia.

Chapter 4

Introduction to functional analysis

Functional analysis is the branch of mathematics concerned with the study of vector spaces and operators acting upon them. It has some historical roots in the study of function spaces, in particular transformation of functions, such as the Fourier Transform, as well as in the study of differential and integral equations. The functional implies a function whose argument is a function.

In modern view, functional analysis is seen as the study of *complete normed vector space* over the real or complex number. Such space are called **Banach** spaces. An important example is a **Hilbert** spaces, where the norm aries from an inner product. These space are the fundamental importance in many areas, including the mathematical formulation of quantum mechanics and now the mathematical formulation of electromagnetic field.

An important object of study in functional analysis are the continuous linear operators defined on Banach and Hilbert spaces.

4.1 Linear vector space

In linear circuits, the problem to find the unknown voltages or currents can be generalized to solve the simultaneous linear equations. In linear waveguide, the problem to find the unknowns of normal modes also can be generalized to solve the simultaneous linear equations by means the Maxwell's equations and the boundary conditions.

From functional analysis of view, those problem can be discuss in terms of the functional space. Now we will discuss the theory of the functional spaces such as the $n-dimensional$ space E_n $(n = 1, 2, 3, \cdots)$ and the infinity-dimensional space E_∞.

As well known, in linear circuits we can set up a simultaneous equation (4.1) to find the solution of the unknown voltages or currents x_1, x_2 and x_3.

$$a_{11}x_1 + a_{12}x_2 + a_{13}x_3 = b_1$$

$$a_{21}x_1 + a_{22}x_2 + a_{23}x_3 = b_2$$

$$a_{31}x_1 + a_{32}x_2 + a_{33}x_3 = b_3 \tag{4.1}$$

And then, we may use a set x in form of

$$x = (x_1,\ x_2,\ x_3)$$

to denotes the unknowns (voltages or currents).

In rectangular waveguide, there are a lot of normal modes, such as the TE_{mn} modes $(m, n = 0, 1, 2, \ldots, n)$ and the TM_{mn} modes $(m, n = 0, 1, 2, \ldots, n)$, exist independently inside the waveguide if the waveguide is filled with linear medium. In this case, we may set up a set y in form of

$$y = (y_1,\ y_2,\ \ldots, y_n)$$

to denote the normal modes inside the waveguide, in which, each y_i $i = 1, 2, \ldots, n$ represents a single normal mode.

This idea is from the linear circuit or linear normal mode point of view and can be extended to a n-dimensional space and then to an idea of sets.

Suppose that the vectors u, v, W in n-dimensional space E_n are defined as

$$\begin{aligned} u &= (u_1,\ u_2,\ \cdots,\ u_n) \\ v &= (v_1,\ v_2,\ \cdots,\ v_n) \\ W &= (\zeta_1,\ \zeta_2,\ \cdots,\ \zeta_n) \end{aligned} \tag{4.2}$$

and

$$0 = (0,\ 0,\ \cdots,\ 0)$$

Since there is no any coupling between any independent elements u_i (or v_i, or ζ_i) these vectors satisfy

$$\begin{array}{ll} (1)\ \textit{Commutative law} & v + W = W + v \\ (2)\ \textit{Associative law} & u + (v + W) = (u + v) + W \\ (3)\ \textit{Zero vector} & \textit{There exits an element } 0 \in E_n, \\ & \textit{called the zero vector, such} \\ & \textit{that } v + 0 = v \textit{ for all } v \in E_n \\ (4)\ \textit{Distributive law} & a(v + W) = av + aW \\ (5)\ \textit{Distributive law} & (a + b)v = av + bv \\ (6)\ \textit{Compatibility law} & (a(bv) = (ab)v \\ (7)\ \textit{Identity element} & Iv = v \end{array} \tag{4.3}$$

Now, if all of the vectors u, v, W are in E_n space, and the resulting vectors after the operations from (1) to (7) are also in the E_n space, then this space is said to be the *linear space*. If a, b are real numbers, then the E_n space is said to be the real linear space or **R** space. If a, b are complex numbers, then the E_n space is said to be the complex linear space or **C** space. This concepts can be extended to an infinity-dimensional space E_∞.

4.2 The orthogonal system of Hilbert space*

A **Hilbert space**, named after *David Hilbert*, is a real and complex vector space with a positive definite norm that is complete under the norm. Thus it is an inner product space, which means that it has notions of distance and angle.

At beginning, the *Hilbert space* generalizes the notion of *Euclidean space*, in which, it extends the method of vector algebra from the two-dimensional plane and three-dimensional space to infinite-dimensional spaces. In more formal terms, a Hilbert space is an inner product space — an abstract vector space in which distances and angles (especially the notion of orthogonality or perpendicularity) can be measured, which is "complete", means that if the sequence of vectors is Cauchy, then it converges to some limit in the space.

The Hilbert spaces arise frequently in mathematics, physics, and engineering typically as an infinite-dimensional function spaces. They are the power tools in the partial differential equations, quantum mechanics, and signal processing, and now they are also the power tools in the theoretical analysis of the electromagnetic field.

In Hilbert space theory, the geometric intuition plays important role. For instance, an element of a Hilbert space can be uniquely specified by its coordinates with respect to an orthonormal basis.

4.2.1 The inner product space

In the first decade of the 20th century, during David Hilbert and Echard Schmidt's study of integral equations, the first study of integral equations is that two square-integrable real-value functions f and g on an interval $[a, b]$ has an *inner product*

$$< f, g >= \int_a^b f(x)g(x)dx \tag{4.4}$$

which has many of the familiar properties of the Euclidean dot product such as following discussion:

In an abstract space $\mathbf{R}$, the inner product of two vectors x and y is a scala-valued function of these two vectors, written as $< x, y >$, which satisfies

(1) Communicative law:

$$< x, y >=< y, x > \tag{4.5}$$

(2) Associative law:

$$< \alpha_1 x_1 + \alpha_2 x_2, y >= \alpha_1 < x_1, y > +\alpha_2 < x_2, y > \tag{4.6}$$

(3) If $x \neq 0$, then

$$< x, x >> 0 \tag{4.7}$$

It follows that the length of the vector x is

$$|x| = +\sqrt{< x, x >} \tag{4.8}$$

Two vectors x and y are said to be orthogonal or perpendicular if

$$< x, y >= 0 \tag{4.9}$$

Note, the definition of the inner product in (4.4) is good for the real domain. Some problem may be caused when we make use of the definition of the inner product from the real domain to the complex domain. For instance, in the E_2 space, the vector $(1, i)$ is non-zero. However, its length is zero.

Therefore, the inner product of the vectors in the complex space is defined as

$$< f, g >= \int_a^b f(x)^* g(x) dx \tag{4.10}$$

when the real space **R** is extended to a complex space **C**. In this case, the inner product $< x, y >$ satisfies

$$\begin{aligned} &< x, y >=< y, x >^* \\ &< \alpha_1 x_1 + \alpha_2 x_2, y >= \alpha_1^* < x_1, y > + \alpha_2^* < x, y > \\ &< x, x > > 0 \qquad if\ x \neq 0 \end{aligned} \tag{4.11}$$

where " $*$ " represents the conjugate value.

Suppose that the vectors x, y in n-dimensional space E_n are defined as

$$x = (\xi_1,\ \xi_2,\ \cdots,\ \xi_n)$$

$$y = (\eta_1,\ \eta_2,\ \cdots,\ \eta_n)$$

then the definition of the inner product in E_n and E_∞ spaces are defined as

$$< x, y >= \xi_1\eta_1 + \xi_2\eta_2 + \cdots + \xi_n\eta_n \qquad (in\ E_n\ space) \tag{4.12}$$

and

$$< x, y >= \xi_1\eta_1 + \xi_2\eta_2 + \cdots \qquad (in\ E_\infty\ space) \tag{4.13}$$

respectively.

4.2.2 Hilbert Space*

Definition:

A **Hilbert space** is a real or complex inner space that is complete under the norm defined by the inner product $< \cdot, \cdot >$ in the form of

$$||x|| = \sqrt{< x, x >} \tag{4.14}$$

and the inner product of the Hilbert space is

$$< f, g >= \int_a^b f(x) g(x) dx$$

for the real functions $f(x)$ and $g(x)$ in the real space $\mathbf{R}$ and is

$$< f, g >= \int_a^b f(x)^* g(x) dx \tag{4.15}$$

for the complex functions $f(x)$ and $g(x)$ in the complex space $\mathbf{C}$.

Properties:

1. **Orthogonality:**

Based on the inner product of the Hilbert space in (4.15), the idea of an orthogonal family of function has meaning. Schmidt exploited the similarity of this inner product with the usual dot product to prove an analog of the spectral decomposition for an operator of the form

$$f(x) \rightarrow \int_a^b K(x, y) f(y) dy \tag{4.16}$$

where K is a continuous function symmetric in x and y. The resulting eigenfunction expansion express the function K as a series in the form of

$$K(x, y) = \sum_n \lambda_n \phi_n(x) \phi_n(y) \tag{4.17}$$

where the function ϕ_n are orthogonal in the sense that

$$< \phi_n, \phi_m >= 0 \quad for\ all\ n \neq m \tag{4.18}$$

2. **Pythagorean identity:**

Two vectors u and v in a Hilbert space H are orthogonal when $< u, v >= 0$. The notation for this is $u \perp v$. More generally, when S is a subset in H, the notation $u \perp S$ means that u is orthogonal to every element from S.

When u and v are orthogonal, we have

$$\begin{aligned} ||u + v||^2 &=< u + v, u + v >=<< u, u > +2Re < u, v > + < v, v > \\ &= ||u||^2 + ||v||^2 \end{aligned} \tag{4.19}$$

By introduction on n, this is extended to any family $u_1, \cdots, u_n$ of n orthogonal vectors in the form of

$$||u_1 + u_2 + \cdots + u_n||^2 = ||u_1||^2 + \cdots + ||u_n||^2 \tag{4.20}$$

3. **Completeness:**

Completeness is the key to handling infinite-dimensional examples, such as function space, and then completeness is required.

It is expressed using a form of the **Cauchy criterion** for sequences in H: a normed space H is **complete** if every Cauchy sequence converges with respect to this norm to an element in the space.

A series of vectors $\sum u_k$ converges in H when the partial sums converges to an element of H, that is called *the sum of the series*, and denoted by

$$\sum_{k=0}^{\infty} u_k = \lim_{n} \sum_{k=0}^{n} u_k \in H \tag{4.21}$$

Completeness allows to extend the Pythagorean identity to series. A series $\sum u_k$ of orthogonal vectors converges in H if the series of squares of norms converges, and

$$\left|\left|\sum_{k=0}^{\infty} u_k\right|\right|^2 = \sum_{k=0}^{\infty} ||u_k||^2 \tag{4.22}$$

4. **Direct sum:**

Two Hilbert space H_1 and H_2 can be combined into another Hilbert space, called the **(orthogonal) direct sum**, and denoted by

$$H_1 \bigoplus H_2,$$

both of them are consisting of the set of all ordered pairs (x_1, x_2) where $x_i \in H_i$, $i = 1, 2$, and the inner product defined by

$$< (x_1, x_2), (y_1, y_2) >_{H_1+H_1} = < x_1, y_1 >_{H_1} + < x_2, y_2 >_{H_2} \tag{4.23}$$

5. **Orthogonal complements and projections:**

If S is a Hilbert space H, the set of vectors orthogonal to S is defined by [1]

$$S^{\perp} = \{x \in H :< x, s >= 0, \quad \forall s \in S\} \tag{4.24}$$

Where $S^{\perp}$ is a closed subspace of H and then forms a Hilbert space.

If V is a closed subspace of H, then $V^{\perp}$ is called the *orthogonal complement* of V.

Then every x in H can be written uniquely as

$$x = v + w, \qquad with \quad v \in V, \quad w \in V^{\perp} \tag{4.25}$$

Therefore, H is the direct sum of V and $V^{\perp}$. In this case, the linear operator

$$P_V : H \to H \quad which\ maps\ x\ to\ v \tag{4.26}$$

[1] Where $\forall$ means "for all"

is called the *orthogonal projection.*

6. **Parallelogram identity and polarization:**

By definition, every Hilbert space is also a Banach space. [2] Further more, every Hilbert space is a special Banach space, in which the following Parallelogram identity holds:

$$||u+v||^2 + ||u-v||^2 = 2(||u||^2 + ||v||^2) \tag{4.27}$$

Conversely, every Banach space in which the Parallelogram identity holds is a Hilbert space.

In this case, the inner product is uniquely determined by the polarization identity.

For real Hilbert space, the polarization identity is

$$< u, v >= \frac{1}{4}\Big(||u+v||^2 - ||u-v||^2 |big) \tag{4.28}$$

For complex Hilbert space, it is

$$< u, v >= \frac{1}{4}\Big(||u+v||^2 - ||u-v||^2 + i||x+iy||^2 - i||x-iy||^2\Big) \tag{4.29}$$

4.2.3 Orthogonal basis*

Hilbert spaces allow simple geometric concepts like projection and orthogonal basis to be applied to infinite dimensional space, such as functional spaces. They provide a context with which to formalize and generalize the concepts of the Fourier series by means of arbitrary orthogonal polynomials and of the Fourier Transform, which are central concepts from functional analysis.

The notation of an orthogonal basis from the linear algebra generalizes over the case of Hilbert spaces. In a Hilbert space H, an **orthogonal basis** is a family $\{e_k\}_{k\in B}$ of elements of H satisfying the conditions:

1. **Orthogonality:** Every two different elements of B are orthogonal: $< e_k, e_j >= 0$ for all k, j in B with $k \neq j$.

2. **Normalization:** Every element of the family has norm 1: $||e_k|| = 1$ for all k in B.

3. **Completeness:** The linear span of the family e_k, $k \in B$, is dense in H.

A system of vectors satisfying the first two conditions basis is called an **orthogonal system** or an **orthonormal set** or an **orthonormal sequence** if B is countable. It can be proved that such a system is always linearly independent.

Examples of orthonormal bases include:

a. The set $\{(1,0,0), (0,1,0), (0,0,1)\}$ forms an orthonormal basis of $\mathbf{R}^3$ with the

[2] We will discuss the Banach space after then.

dot product;

b. The sequence $\{f_n : n \in \mathbf{Z}\}$ with $f_n(x) = exp(i2\pi nx)$ forms an orthonormal basis of the complex space $L^2([0,1])$;

c. The family $\{e_b : b \in B\}$ with $e_b(c) = 1$ if $b = c$ and 0 otherwise forms an orthonormal basis of $l^2(B)$.

Note that in the infinite-dimensional case, an orthonormal basis will not be a basis in the sense of linear algebra. To distinguish the two, the later basis is also called a Hamel basis,. That the span of the basis vectors is dense implies that every vector in the space can be written as the sum of an infinite series, and the orthogonality implies that this decomposition is unique.

Orthonormal basis:

1. Consider two elements x and y in S. According to the definition of a basis, both of them are able to be expressed as

$$x = \alpha_1 x_1 + \alpha_2 x_2 + \cdots + \alpha_n x_n$$

$$y = \beta_1 x_1 + \beta_2 x_2 + \cdots + \beta_n x_n$$

where the vectors $x_1, x_2, \cdots, x_n$ form a basis of S.

In general, the inner product of two vectors x and y in S is

$$< x, y >= \sum_j \sum_k \alpha_j \beta_k < x_j, x_k >, \quad j,k = 1, \cdots, n$$

If and only if

$$< x_j, x_k >= \begin{cases} 1, & j = k \\ 0, & j \neq k \end{cases} \tag{4.30}$$

namely, if and only if the vectors $x_1, x_2, \cdots, x_n$ are mutually orthogonal and the length of each vector is 1, we have

$$< x, y >= \sum_1^n \sum_1^n \alpha_j \beta_k \tag{4.31}$$

This will be the same as the usual scalar product defined in E_n, [3].

In this case, a set of vectors is said to be orthonormal (or orthogonal normal) if they satisfy (4.30).

Following the analysis above, if S is possessed of n numbers of orthogonal normal bases, then the S will be the same as E_n, namely, it will have the same multiple law , additive law and the same formula of the inner product as E_n. And then, if S is possessed of infinite numbers of orthogonal normal bases, then it will be the same as E_∞.

[3] See (4.12)

2. Based on the inner product of the Hilbert space (4.15), we may consider to express the electric field ϕ in a planar waveguide in the form of

$$\phi(y) = \sum_n C_n \phi_n(y) \tag{4.32}$$

where the function ϕ_n are orthogonal and normal in the sense that

$$< \phi_n, \phi_m >= \begin{cases} 1, & n = m \\ 0, & n \neq m \end{cases} \tag{4.33}$$

4.3 Banach space*:

Banach space, named after Polish mathematician Stefan Banach, are one of the central object of study in functional analysis.

Definition:

Banach spaces are defined as complete normed vector spaces. This means that a Banach space is a vector **V** over the real or complex numbers with a norm $||\cdot||$ such that every Cauchy sequence (with respect to the metric $d(x, y) = ||x - y||$) in **V** has a limit in **V**.

Example:

Let **K** stand for one of the fields **R** (the real space) or **C** (the complex space).

1. The Euclidean space $\mathbf{K}^n$, where the Euclidean norm of $x = (x, \cdots, x_n)$ is given by $||x|| = (\sum_{i=1\ldots n} |x_i|^2)^{1/2}$, are Banach spaces.

2. The space of all continuous function $f : [a, b] \to \mathbf{K}$ defined on a close interval $[a, b]$ becomes a Banach space if we define the norm of such a function as $||f|| = sup\{|f(x)| : x\ in\ [a, b]\}$, where $sup\{|f(x)| : x\ in\ [a, b]\}$ a supremum norm. This is indeed a norm since continuous functions defined on a close interval are bounded. The space is complete under this norm, and the resulting Banach space is defined by $\mathbf{C}[a, b]$. This example can be generalized

(a) To the space $\mathbf{C}(X)$ of all continuous functions $X \to \mathbf{K}$, where X is a compact space, or

(b) To the space of all bounded continuous functions $X \to \mathbf{K}$, where X is any topological space, or indeed to the space $B(X)$ of all bounded functions $X \to \mathbf{K}$, where X is any set.

3. If $p \geq 1$ is a real number, we can consider the space of all infinite squences $(x_x, x_2, x_3, \cdots)$ of elements in **K** such that the infinite series $\sum_i |x_i|^p$ is finite. The p-th root of this series' value is then defined to be the p-norm of the sequence. The space, together with the norm, is a Banach space denoted by l^p.

4. If $p \geq 1$ is real number, we can consider all function $f : [a, b] \to \mathbf{K}$ such that $|f|^p$ is Lebesgue integrable. The p-th root of this integral is then defined to be the

norm of f. By itself, this space is not a Banach space because there are non-zero functions whose norms are zero. We then defined an equivalence relationship as follows: f and g are equivalent if only if the norm of $f - g$ is zero. The set of equivalence classes then forms a Banach space denoted by $L^p([a, b)]$.

5. If X and Y are two Banach spaces, then we can form their direct sum $X \bigoplus Y$, which has a topological vector space structure but no canonical norm, however, it is still a Banach space for some equivalent norms such as

$$||X \bigoplus Y|| = (||X||^p + ||Y||^p)^{1/p}$$

6. If M is a closed subspace of the Banach space X, then the quotient space X/M is again a Banach space.

Relationship with the Hilbert space:

As mentioned above, any Hilbert space is a Banach space. However, not every Banach space is a Hilbert space. A necessary and sufficient condition for a Banach space $\mathbf{V}$ to be associated to an inner product, which will then necessarily make $\mathbf{V}$ into a Hilbert space, is the parallelogram identity:

$$||u + v||^2 + ||u - v||^2 = 2(||u||^2 + ||v||^2)$$

for all u and v in $\mathbf{V}$, where $|| * ||$ is the norm on $\mathbf{V}$. And then, for example, while $\mathbf{R}^n$ is a Banach space with respect to any norm defined on it, it is only a Hilbert space with respect to the Euclidean norm. Similarly, as an infinite-dimensional example, the Lebesgue space L^p is always a Banach space but is a Hilbert space only when $p = 2$.

If the norm of a Banach space satisfied the identity, the associated inner product, which makes it into a Hilbert space, is given by the polarization identity.

If $\mathbf{V}$ is a real Banach space, then the polarization identity is

$$< u, v >= \frac{1}{4}(||u + v||^2 - ||u - v||^2)$$

Whereas if $\mathbf{V}$ is a complex Banach space, then the polarization is given by

$$< u, v >= \frac{1}{4}\Big(||u + v||^2 - ||u - v||^2 + i(||u + iv)^2 - (u - iv||^2)\Big))$$

The necessity of the condition follows easily from the properties of an inner product. To see the sufficient– that the parallelogram law implies that the form defined by the polarization identity is indeed a complete inner product.

4.4 Convergence and complete space

Now we will follow the conception of convergence in function analysis to explain the convergence and the complete space in functional analysis as follows.

4.4.1 Convergence

Convergence: A sequence of vectors $x_1, x_2, \cdots$ in space S is said to be convergent to a vector x if , given $\varepsilon > 0$, there is an integer $N = N(\varepsilon)$, such that when $n > N$ we have that

$$|x - x_n| < \varepsilon \tag{4.34}$$

for all $n > N$. In this case, the vector x is said to be the limit of the sequence and is written as

$$x = \lim\ x_n$$

Cauchy Convergence : In a finite dimensional space, if a sequence vectors x_n is said to be Cauchy convergent to x if, given $\varepsilon > 0$, there is an integer $N_1 = N_1(\varepsilon)$ such that when all of $n, m > N$ we have

$$|x_n - x_m| < \varepsilon \tag{4.35}$$

Obviously, if the sequence of vectors in E_n is Cauchy convergence, then it will be convergence in the sense of (4.34).

Complete space: In this case, suppose there is a function $f(t)$ in $\mathcal{L}_2$ and whatever a sequence $f_n(t)$ in space $\mathcal{L}_2$ is Cauchy convergent to $f(t)$, then it is said that the space $\mathcal{L}_2$ is complete.

4.4.2 Domain and range

Definition of domain and range:

In the linear map (or the linear operator T [4]) in form of

$$w = Tv$$

the set of the v, which the mapping is defined, is said to be the domain of T. The set of the vectors w, which are equal to Tv for some v in the domain of T, is said to be the range of the operator.

1. An operator is linear if the mapping is such that for any vectors v_1, v_2 in the domain of T and for arbitrary scalars α_1, α_2, the vector $\alpha_1 v_1 + \alpha_2 v_2$ is in the domain of T and

$$T(\alpha_1 v_1 + \alpha_2 v_2) = \alpha_1 T v_1 + \alpha_2 T v_2 \tag{4.36}$$

2. A linear operator is bounded if its domain is the entire space V and if there is a single constant c such that

$$|T v| < c\,|v| \tag{4.37}$$

It is provable, that a linear bounded operator is continuous; that is , if a sequence of vectors v_n is convergent to v, then $T(v_n)$ is convergent to $T(v)$.

---

[4] We will discuss the linear operator T in detail after then

4.5 The subspaces and manifolds

4.5.1 Subspaces

Definition of subspace:

Theorem: Let $\mathbf{V}$ be a vector space over the field $\mathbf{K}$, and let W be a subset of $\mathbf{V}$. Then W is a subspace if only if it satisfies the following three conditions:

1. The zero vector, $\mathbf{0}$, is in W.
2. If u and v are element of W, then the sum $u + v$ is an element of W.
3. If u are elements of W and c is a scalar from $\mathbf{K}$, the the scalar product cu is an elements of W.

Proof: First, property 1 ensure W is nonempty. Property 2 and 3 above ensure closure of W under addition and scalar multiplication, so the vector space are well defined. Since elements of W are necessarily elements of $\mathbf{V}$, the additivity of addition, the commutativity of addition, the distributivity of scalar multiplication , the compatibility of scalar multiplication and the Identity element of scalar multiplication of a vector space are satisfied. By the closure of W under scalar multiplication (specifically under 0 and -1 multiplication), Identity of addition and Inverse element of addition are satisfied.

Conversely, if W is subspace of $\mathbf{V}$, then W is itself a vector space under the operations induced by $\mathbf{V}$, so property 2 and 3 are satisfied. By property 3, $-w$ is in W whenever w is, and it follows that W is closed under subtraction as well. Since W is nonempty, there is an element x in W, and $x - x = \mathbf{0}$ is in W, so property 1 is satisfied.

Properties of subspace:

A way to characterize subspace is that the subspaces are closed under linear combinations. That is, W is a subspace if and only if every linear combination of (finitely many) elements of W also belongs to W. Condition 2 and 3 for a subspace are simply the most basic kinds of linear combinations.

4.5.2 Manifolds

Definition of manifolds:

If a collection $\mathbf{X}$ of vectors in space V is such that for all scalars a and b it contain the vector $ax + by$ whenever it contains the vectors x and y, then $\mathbf{X}$ is said to be a linear manifolds.

Examples:

1. All of the $E_{n-j}(j = 0, 1, 2, \cdots, n)$ spaces in E_3 form a linear manifolds.

2. In a waveguide filled with linear medium, all of the normal modes satisfied the Maxwell's equation and the boundary conditions form a linear manifolds.

3. All of the solutions of (4.1) forms a linear manifolds if $b_1 = b_2 = b_3 = 0$.

Closed and subspace:

A linear manifold $\mathbf{X}$ is said to be closed if, whenever a vector sequence in $\mathbf{X}$, $x_1, x_2, \cdots$ is convergent to a limit and the limit of the sequence belongs to $\mathbf{X}$.

A closed linear manifold is said to be a linear subspace.

4.6 Functional:

Definition of functional:

A functional is a mapping from an arbitrary vector space into a one-dimensional space, the real-number line. In more detail, for every vector (or function) x in V, there is a scalar $f(x)$ corespondent to the x, this $f(x)$ is the functional defined in V.

From the projection theorem point of view, the functional is a projection (or a mapping) from a linear vector space x in V into an one-dimensional space or the functional is a mapping from a function (or vector) x into a scalar.

4.6.1 Properties of functional

1. If the functional $f(x)$ satisfies

$$f(\alpha x + \beta y) = \alpha f(x) + \beta f(y), \tag{4.38}$$

then the functional $f(x)$ is a linear functional.

(2) If the functional $f(x)$ satisfies

$$\lim f(x_n) = f(x), \qquad when \ \lim x_n = x, \tag{4.39}$$

then the functional $f(x)$ is a continuous functional.

(3) If the functional $f(x)$ satisfies

$$f(x) < \alpha |x| \tag{4.40}$$

for all of the x and α is a constant, then it is said that the functional $f(x)$ is a finite functional.

(4) A finite functional is definitely a continuous functional. In other words, a

continuous functional must be a finite functional.

(5) Every continuous linear functional is a inner product.

Functional with inner product:

If $f(x)$ is a continuous linear functional, then there exists a vector v in V such that

$$f(x) = < v, x > \tag{4.41}$$

Which is easy to be understand from the geometric point of view, i.e. when v in V is an arbitrary vector, the functional $f(x)$ is the projection of the vector v on x-direction. In general, a continuous linear functional $f(x)$ is a mapping from an arbitrary vector v space into an one-dimensional space (or scalar).

4.6.2 How to express a vector by means of its basis

Now we will be able to discuss the problem that how to express a vector x by means of its basis. If V is an E_n space with a finite-dimensions, let $x_1, x_2, \cdots, x_n$ be the basis of this space, then an arbitrary vector x in E_n is able to be expressed as

$$x = \alpha_1 x_1 + \cdots + \alpha_n x_n$$

where $\alpha_1, \alpha_2, \cdots, \alpha_n$ are unknown constants. First, α_1 is related to x, therefore, it is a functional of x, and is a bounded functional. In this case, from (4.41), there will be a vector v_1 such that

$$\alpha_1 = < v_1, x >$$

Similarly, there will be vectors $v_2, v_3, \cdots, v_k$, such that

$$\alpha_j = < v_j, x >, \qquad j = 1, 2, \cdots, k$$

And then, we have

$$x = x_1 < v_1, x > + x_2 < v_2, x > + \cdots + x_n < v_n, x > \tag{4.42}$$

Functional equations:

An equation $F = G$ between functionals can be read as an "function to be solved", with solutions being themselves' functions. For instance:

1. Functional additive is in the form of

$$f(x + y) = f(x) + f(y)$$

2. Functional derivative:

Functional derivative are the derivatives of functionals, i.e. they carry information on how a functional changes, when the function changes by a small amount.

which has been used in Lagrangian mechanics.

3. Functional integrals:

Functional integrals implies an integral taken over some function space. Richard Feynman used functional integrals as the central idea in his sum over the histories formulation of quantum mechanics.

4.7 Linear map or linear operator*

4.7.1 Linear map

A linear map is a function between two vectors and the operation of the vectors follow the additive law and the multiplication law. Therefore it is a linear.

The same thing can be expressed as follows. A linear operator T is a transformation from a vector V to a vector W noted by $W = TV$ and the operations of the vectors follow the additive law and the scalar multiplication law.

Meanwhile, a functional is said a transformation (or mapping) from a vector to a scalar and the functionals follow the additive law and the scalar multiplication law.

Definition of linear map:

Let V and W be vector spaces over the same field K. A function

$$f : V \to W \tag{4.43}$$

or an operator T in form of

$$W = Tv \tag{4.44}$$

is said to be a linear map (or a linear operator) if for any two vectors x and y in V and any scalar a in K, the following two conditions are satisfied:

$$f(x+y) = f(x) + f(y) \qquad \textit{additivity} \tag{4.45}$$

$$f(ax) = af(x) \qquad \textit{homogeneity of degree } 1 \tag{4.46}$$

This is equivalent to requiring that for any vectors $x_1, \cdots, x_m$ and scalars $a_1, \cdots, a_m$, the equation

$$f(a_1x_1 + \cdots + a_mx_m) = a_1f(x_1) + \cdots + a_mf(x_m) \tag{4.47}$$

holds. Then it follows that $f(0) = 0$.

Examples:

1. The *identity map* is linear.

Where an *identity map* f on M is a function that always returns the same value that was used as its argument, and the function is given by

$$f(x) = x \qquad \textit{for all elements } x \textit{ in } M.$$

2. For real numbers, the map $x \longmapsto x^2$ is not linear but a nonlinear. 3. For a real numbers, the map $x \longmapsto x + 1$ is not linear.

4. If A is a $m \times n$ matrix, then A defined a linear map (or linear operator) from $\mathbf{R}^n$ to $\mathbf{R}^m$ by sending the column vector $x \in \mathbf{R}^n$ to the column vector $Ax \in \mathbf{R}^m$. In this case, (4.44) becomes

$$\mathbf{W} = A\mathbf{v} \tag{4.48}$$

Which is a simultaneous equation with

$$A = \begin{bmatrix} a_{11} & a_{12} & \cdots & a_{1n} \\ a_{21} & a_{22} & \cdots & a_{2n} \\ \vdots & \vdots & \vdots & \vdots \\ a_{m1} & a_{m2} & \cdots & a_{mn} \end{bmatrix} \tag{4.49}$$

and

$$\mathbf{W} = \begin{pmatrix} w_1 \\ w_2 \\ \vdots \\ w_m \end{pmatrix}, \qquad \mathbf{v} = \begin{pmatrix} v_1 \\ v_2 \\ \vdots \\ v_n \end{pmatrix} \tag{4.50}$$

5. The integral yields a linear map from the space of all real-value integrable functions $v(s)$ on some interval $[a, b]$ to $\mathbf{R}$. In this case, (4.44) becomes

$$w(t) = \int_a^b K(t, s)v(s)ds \tag{4.51}$$

6. Differentiation is a linear map from the space of all of differentiable functions to the space of all functions. In this case, (4.44) becomes

$$w(t) = \left(\frac{d^2}{dt^2} + t^2 \right) v(t) \tag{4.52}$$

4.7.2 Linear operator —Matrices

When we chose the bases in the spaces V and W and V and W are finite-dimensional spaces, then every linear map T from V to W can be expressed as a matrix form A since it allows linear calculations.

If A is a real $m \times n$ matrix, then the equation

$$f(x) = Ax$$

describes a linear map $\mathbf{R}^n \to \mathbf{R}^m$.

Let $\{v_1, \cdots, v_n\}$ be a basis for V, then every vector v in V is uniquely determined by the coefficients $c_1, \cdots, c_n$ in the form of

$$v = c_1 v_1 + \cdots, c_n v_n.$$

If $f : V \to W$ is a linear map in the form of

$$f(c_1v_1 + \cdots + c_nv_n) = c_1f(v_1) + \cdots + c_nf(v_n) \tag{4.53}$$

which implies that the function f is entirely determined by the values of $f(v_1), \cdots, f(v_n)$.

Now let $\{w_1, \cdots, w_m\}$ be a basis for W. Then we can represent the values of each $f(v_j)$ as

$$f(v_j) = a_{1j}w_j + \cdots + a_{mj}w_j \tag{4.54}$$

Thus the function f is entirely determined by the values of a_{ij}.

Substitution of (4.54) into (4.53), we have

$$\begin{aligned} f(v) &= f(c_1v_1 + \cdots + c_nv_n) \\ &= c_1f(v1) + \cdots + c_nf(v_n) \\ &= c_1(a_{11}w_1 + \cdots + c_{m1}w_1) \\ &+c_2(a_{12}w_2 + \cdots + c_{m2}w_2) \\ &+\cdots \\ &+c_n(a_{1n}w_n + \cdots + a_{mn}w_n) \end{aligned}$$

Which can be expressed in the form of

$$\begin{bmatrix} a_{11}w_1 & a_{12}w_2 & \cdots & a_{1n}w_n \\ a_{21}w_1 & a_{22}w_2 & \cdots & a_{2n}w_n \\ \vdots & \vdots & \vdots & \vdots \\ a_{m1}w_1 & a_{m2}w_2 & \cdots & a_{mn}w_n \end{bmatrix} \begin{pmatrix} c_1 \\ c_2 \\ \vdots \\ c_n \end{pmatrix} = \begin{pmatrix} f(v_1) \\ f(v_2) \\ \vdots \\ f(v_n) \end{pmatrix}$$

or

$$MC = f(v) \tag{4.55}$$

Which means that if we put $a_{ij}w_j$ into a $m \times n$ matrix M, we can conveniently use it to compute the value of f for any vector v in V by means of (4.55).

Some linear transformation matrices*

1. Rotation by 90 degree clockwise:

$$A = \begin{bmatrix} 0 & 1 \\ -1 & 0 \end{bmatrix} \tag{4.56}$$

2. Rotation by θ degree clockwise:

$$A = \begin{bmatrix} cos(\theta) & sin(\theta) \\ -sin(\theta) & cos(\theta) \end{bmatrix} \tag{4.57}$$

Bounded operator and continuous operator:

A *bounded operator* is a linear transformation T between normed vectors X and Y (namely, $Y = TX$), for which the ratio of the norm of $T(v)$ to the norm of v is bounded by some number over all non-zero vectors v in X. In other words, there exists some $M > 0$ such that for all v in X, we have

$$||Tv||_Y \leq M||v||_X \tag{4.58}$$

where M is called the norm of the operator T.

It is provable: a linear operator is a *bounded operator* if and only if it is *continuous operator.*

4.8 Hermitian matrix and Self-adjoint operator

4.8.1 Hermitian matrix*

Definition of Hermitian matrix:

A Hermitian matrix A (or self-adjoint matrix) is a square matrix with complex entries, in which, the element in the i-th row and j-th column is equal to its conjugate transpose in the j-th row and i-th column for all indices i and j, i.e.

$$a_{ij} = a_{ji}^* \qquad for\ all\ i, j \tag{4.59}$$

If the conjugate transpose of a matrix A is A^*, then A is a Hermitian matrix if

$$A = A^* \tag{4.60}$$

For example

$$\begin{bmatrix} 3 & 2+i \\ 2-i & 1 \end{bmatrix}$$

is a Hermitian matrix.

It is interesting to point out, that a linear operator T is a self-adjoint operator if

$$T = T^*$$

Where T^* is the conjugate transpose of T. This will be discussed in next subsection.

Properties of Hermitian matrix:

1. The entries on the diagonal of any Hermitian matrix are necessarily real.

2. A matrix that has only real entries is Hermitian if and only if it is a symmetric matrix.

3. Every Hermitian matrix is normal.

It is said that any Hermitian matrix can be diagonalized by a unitary matrix, and the resulting diagonal matrix has only real entries. This means that all eigenvalues of a Hermitian matrix are real, and, moreover, eigenvectors corresponding to the eigenvalues are orthogonal and normal. Then it is easy to find the orthonormal basis of C^n consisting only of eigenvectors.

4. The sum of any two Hermitian matrices is Hermitian matrix as well.

5. The product of two Hermitian matrices A and B will be Hermitian if and only if A and B commute, i.e. if and only if $AB = BA$.

6. The sum of a square matrix and its conjugate transpose $(C+C^*)$ is Hermitian matrix.

7. The difference of a squire matrix and its conjugate matrix $(C - C^*)$ is skew-Hermitian (or anti-Hermitian) matrix.

8. An arbitrary squire matrix C can be written as the sum of a Hermitian matrix A and a skew Hermitian matrix B, namely,

$$C = A + B, \qquad A = \frac{1}{2}(C + C^*) \quad \textit{and} \quad B = \frac{1}{2}(C - C^*)$$

4.8.2 Self-adjoint operator

Definition of self-adjoint operator:

In a finite-dimensional inner product space, *a self-adjoint operator* is a linear operator who is equal to its own adjoint.

In other words, a linear operator T^* is the self-adjoint operator of the linear operator T if, for all x and y in V, we have

$$< y, Tx >=< T^* y, x > \tag{4.61}$$

And then, if

$$T^* = T \tag{4.62}$$

then T is said to be a self-adjoint operator. In this case,

$$< y, Tx >=< Ty, x >=< x, Ty > \tag{4.63}$$

For a self-adjoint operator, the matrix of the operator is Hermitian matrix, where a Hermitian matrix is one who is equal to its own conjugate transpose.

4.9 Spectral theorem

The *spectral theorem* provides conditions under which an operator or a matrix can be diagonalized (that is, represented as a diagonal matrix in some basis). In general, the spectral theorem identifies a class of linear operators that can be modelled by multiplication operators.

Examples of operators to which the spectral theorem applies are self-adjoint operators or more generally, the normal operators on Hilbert space.

The spectral theorem provides a canonical decomposition, called the *spectral decomposition, eigenvalue decomposition, or eigendecomposition*, of the underlying vector space on which it acts.

4.9.1 Invariant subspaces

Definition of invariant subspace:

An invariant subspace of a linear mapping

$$T : V \rightarrow V \tag{4.64}$$

from some vector space V to itself is a subspace W of V such that $T(W)$ is contained in W.

Application of invariant subspace:

1. The most important invariant subspace is the one-dimensional vector x, namely

$$Tx = \lambda x \tag{4.65}$$

which means that Tx is contained in x. Where vector x is called the eigenvector corresponding to the eigenvalue λ. That means that under the transformation T, the vector in x direction keep their direction fixed but have their lengths multiplied by λ. Therefore, Tx should be contained in x space.

2. If V is a finite-dimensional invariant subspace of T, then the effect of T to V may be represented by a matrix, namely

$$Tv = Av \tag{4.66}$$

where A is a squire matrix.

3. If $\mathcal{V}$ is a finite-dimensional subspace invariant under T, there exists an eigenvector $v(x)$ of T in $\mathcal{V}$, namely, we have

$$Tv = \lambda v \tag{4.67}$$

4.9.2 Eigenvector of self-adjoint operator

Definition: If

$$(Tu)^* = Tu^*, \tag{4.68}$$

then the operator T is said to be a real operator, where u^* is the complex conjugate of u.

Eigenvector of self-adjoint operator:

1. If T is a real self-adjoint operator, its eigenvalue are real.
Suppose

$$Tu = \lambda u$$

Then

$$(Tu)^* = Tu^* = \lambda^* u^*$$

where u^* is the complex conjugate of u, and λ^* is the eigenvalue corresponding to the eigenvector u^*. Now, since T is self-adjoint operator, we have

$$\lambda < u^*, u >=< u^*, \lambda u >=< u^*, Tu >=< Tu^*, u >= \lambda^* < u^*, u >$$

which implies that $< u^*, u >= 0$ or $\lambda = \lambda^*$. It is impossible that $< u^*, u >= 0$ since $u \neq 0$, and then $\lambda = \lambda^*$ is the only chose, which means that λ is real.

2. If T is self-adjoint operator, the eigenvectors corresponding to different eigenvalues are orthogonal.

Let λ and μ be the two different eigenvalues, and u and v are the eigenvectors corresponding to λ and μ, respectively. Consider T is a self-adjoint operator, we have

$$\lambda < u, v >=< \lambda u, v >=< Tu, v >=< u, Tv >=< u, \mu v >= \mu < u, v >$$

which implies that

$$(\lambda - \mu) < u, v >= 0$$

Now, since $\lambda \neq \mu$, we have

$$< u, v >= 0$$

namely, u is orthogonal to v.

4.9.3 Eigenvalue, Eigenvector and Eigenspace

Eigenvector:

An eigenvector of a linear transformation is a non-zero vector which, when that transformation is applied to it, may change in length, but not the direction.

Eigenvalue:

For each eigenvector of a transformation, there is a scalar value, corresponding to the eigenvector, called an eigenvalue for that vector, which determines the amount the eigenvector is scaled under the linear transformation. For example, an eigenvalue of +2 is doubled in length and points in the same direction.

Eigenvalue, Eigenvector and Eigenspace

In linear algebra, every linear transformation between finite-dimensional vector spaces can be expressed as a matrix, which is a square array of numbers arranged in rows and columns.

Many kinds of mathematical object can be treated as vectors: functions, harmonic modes, quantum states, electromagnetic fields, and frequencies, for example. Each state can be expressed as eigenfunction, eigenmode, eigenstate, and eigenfrequency. And all of the eigenmodes in one system span to an eigenspsce.

Linear transformations of a vector are vector functions. In more detail, in a vector space T, a vector function A is defined if for each vector x of T there corresponds an unique $y = A(x)$ of T.

A vector function A is linear if it follows two properties:

$$1.\, A(x+y) = A(x) + A(y), \qquad \textit{Additivity law} \tag{4.69}$$

$$2.\; A(\alpha x) = \alpha Ax, \qquad \textit{Homogeneity law} \tag{4.70}$$

Where x and y are two arbitrary vectors of the vector space T and α is an arbitrary scalar. And such a function is called a linear transformation or linear operator in the space of T.

Definition:

Given a linear transformation A, a non-zero vector x is defined to be an eigenvector of the transformation if it satisfies the eigenvalue equation

$$Ax = \lambda x \tag{4.71}$$

for a scalar λ. In this case, the scalar λ is called the eigenvalue of A corresponding to the eigenvector x.

Characteristic equation:

Now (4.71) can be expressed as

$$Ax - \lambda Ix = 0$$

and which can be rearranged to

$$(A - \lambda I)x = 0 \tag{4.72}$$

If there exists an inverse

$$(A - \lambda I)^{-1},$$

then both sides of (4.72) can be multiplied by the inverse to obtain the trivial solution: $x = 0$. Therefore, we have to assume that the determinant equation is equal to zero to avoid that the trivial solution: $x = 0$, therefore, we have

$$det(A - \lambda I) = 0 \tag{4.73}$$

This equation is called the characteristic equation of A, and the left-hand side is called the characteristic polynomial. The expansion of the characteristic polynomial gives a polynomial equation for λ.

Example:

The matrix

$$\begin{bmatrix} 2 & 1 \\ 1 & 1 \end{bmatrix}$$

defines a transformation in the real plane. The eigenvalue of this transformation are given by the characteristic equation

$$det \begin{bmatrix} 2-\lambda & 1 \\ 1 & 2-\lambda \end{bmatrix} = (2-\lambda)^2 - 1 = 0$$

The solutions of this equation are $\lambda_1 = 3$ and $\lambda_2 = 1$. Then the eigenvectors can be obtained as follows.

First, consider the eigenvalue $\lambda_1 = 3$, we have, from (4.71)

$$Ax = \begin{bmatrix} 2 & 1 \\ 1 & 2 \end{bmatrix} \begin{bmatrix} x_1 \\ x_2 \end{bmatrix} = 3 \begin{bmatrix} x_1 \\ x_2 \end{bmatrix} = \lambda_1 x$$

which will result two equations as $x_2 = x_1$ or $x_1 = x_2$. And then we may chose $x_1 = 1$ and then $x_2 = 1$ and the eigenvector is

$$\begin{bmatrix} 1 \\ 1 \end{bmatrix}$$

Now we can check this is the eigenvector by checking that

$$\begin{bmatrix} 2 & 1 \\ 1 & 2 \end{bmatrix} \begin{bmatrix} 1 \\ 1 \end{bmatrix} = 3 \begin{bmatrix} 1 \\ 1 \end{bmatrix}$$

For the eigenvalue of $\lambda_2 = 1$, the same procedure will give us the eigenvector

$$\begin{bmatrix} 1 \\ -1 \end{bmatrix}$$

4.10 Application in theory of electromagnetic field

As an example, we will discuss the electromagnetic field in Hilbert space.

As well known, the electromagnetic field in a conductive container is consist of the normal modes. In other words, it is consist of the orthogonal basis, which can be obtained by means of the Maxwell equations

$$rot\,\mathbf{B} = \mu \frac{\partial}{\partial t}\mathbf{D} + \mu \mathbf{J} + \left[\frac{\nabla \mu}{\mu} \times \mathbf{B}\right]$$

$$\begin{aligned} -rot\,\mathbf{D} &= \varepsilon\frac{\partial}{\partial t}\mathbf{B} - \left[\frac{\nabla\varepsilon}{\varepsilon}\times\mathbf{D}\right] \\ div\,\mathbf{D} &= \rho \\ div\,\mathbf{B} &= 0 \end{aligned} \tag{4.74}$$

and the boundary conditions

$$[\mathbf{n}\times\mathbf{D}]|_s = 0, \qquad [\mathbf{n}\cdot\mathbf{B}]|_s = 0 \tag{4.75}$$

Where s is the conductive cover. $\nabla\varepsilon$ and $\nabla\mu$ indicate that the container is full filled with an inhomogeneous dynamic medium.

Now we will discuss the orthogonal basis in this Hilbert space $L^2(v)$. To this end we have to introduce the **Weyl theorem**.

4.10.1 Weyl theorem

Weyl Theorem: Any Hilbert functional space $L_2(V)$ contains only the curl subspace $\dot{\mathbf{J}}$, the divergence subspace $\dot{\mathbf{G}}$ and the harmonic subspace $\mathbf{U}$, namely,

$$L_2(V) = \dot{\mathbf{J}}\oplus\mathbf{G} = \mathbf{J}\oplus\dot{\mathbf{G}} = \dot{\mathbf{J}}\oplus\dot{\mathbf{G}}\oplus\mathbf{U} \tag{4.76}$$

Where

$$\mathbf{G} = \dot{\mathbf{G}}\oplus\mathbf{U}, \qquad \mathbf{J} = \dot{\mathbf{J}}\oplus\mathbf{U}$$

The proof may be found in H. Weyl, "Methods of Orthogonal projection in Potential theory", Duke Math. Journal, 1940,7, pp.441.

From the physical point of view, the electromagnetic field contains the curl field **a** and the gradient field **b** only.

The gradient field **b** is caused by the source, consequently, the gradient field is a field with divergence (i.e. $div\,\mathbf{b}\neq 0$).

The curl of the electric field is caused by the magnetic field and the curl of the magnetic field is caused by the electric field. Consequently, the curl field is a field with non-divergence (i.e. $div\,\mathbf{a} = 0$). And then

$$\begin{array}{lcl} Curl\ field & : & \mathbf{a} = rot\,\mathbf{v}, \quad rot\,\mathbf{a}\neq 0, \quad div\,\mathbf{a} = 0 \\ Gradient\ field & : & \left\{\begin{array}{lll} \mathbf{b} = \nabla\phi & rot\,\mathbf{b} = 0, & div\,\mathbf{b} = \rho \\ \mathbf{c} = \nabla\varphi & rot\,\mathbf{c} = 0, & div\,\mathbf{c} = 0 \end{array}\right. \end{array}$$

Where **a** is the curl field, **b** is the gradient field and **c** is the harmonic field. Therefore, harmonic field is a gradient field in a non-source region. The discussion above comes to an operator explanation, that $L_2(V)$ consists of

$$\begin{array}{lcl} \dot{\mathbf{J}} & : & \mathbf{a} = rot\,\mathbf{v}, \quad div\,\mathbf{a} = 0, \quad [\mathbf{n}\times\mathbf{a}]|_s = 0 \quad or \quad [\mathbf{n}\cdot\mathbf{a}]|_s = 0 \\ \dot{\mathbf{G}} & : & \mathbf{b} = \nabla\phi, \quad \phi|_s = 0 \quad or \quad \dfrac{\partial}{\partial n}\phi\,|_s = 0 \\ \mathbf{U} & : & \mathbf{c} = \nabla\varphi, \quad \nabla^2\varphi = 0, \quad \phi|_s = C_1 \end{array}$$

namely, the sum of the three subspaces above forms a complete orthogonal system in the Hilbert functional space $L_2(V)$. Where S is a surface of a good conductor.

It follows that we may discuss the orthogonal basis in following three subspaces.

4.10.2 Orthogonal basis or the eigenvector

1. The eigenvectors in curl subspace

After some discussion in Chapter 5 we will obtain the eigenvectors in curl subspace as follows:

$$\begin{aligned} rot\,\mathbf{B}_n &= -i\,k_n\,\mathbf{D}_n \\ rot\,\mathbf{D}_n &= i\,k_n\,\mathbf{B}_n \\ [\mathbf{n}\times\mathbf{D}_n]|_s &= 0 \\ [\mathbf{n}\cdot\mathbf{B}_n]|_s &= 0 \end{aligned} \tag{4.77}$$

Where the eigenvector $\mathbf{D}_n$ and $\mathbf{B}_n$ are the orthogonal bases in Hilbert curl subspace $\dot{\mathbf{J}}$. And the orthogonal bases obey the following orthogonality

$$\begin{aligned} <\mathbf{D}_m, \mathbf{D}_n> &= \delta_m^n \\ <\mathbf{B}_m, \mathbf{B}_n> &= \delta_m^n \end{aligned}$$

and the normalization condition

$$||\mathbf{D}_n||^2_{L_2(V)} = \frac{1}{2V}\int_V \{\mathbf{D}_n\cdot\mathbf{D}_n^*\}dV = N_n = 1$$

$$||\mathbf{B}_n||^2_{L_2(V)} = \frac{1}{2V}\int_V \{\mathbf{B}_n\cdot\mathbf{B}_n^*\}dV = N_n = 1$$

2. The eigenvectors in gradient subspace

The gradient subspace is the spacial case with n=0 and $k_0 = 0$ in (4.77). Therefore, the eigenvectors in gradient subspace can be obtained by setting n=0 and $k_0 = 0$ in (4.77) as follows

$$rot\,\mathbf{D}_0 = 0 \tag{4.78}$$

$$rot\,\mathbf{B}_0 = 0 \tag{4.79}$$

$$\mathbf{n}\times\mathbf{D}_0|_s = 0 \tag{4.80}$$

$$\mathbf{n}\times\mathbf{B}_0|_s = 0 \tag{4.81}$$

From (4.78) and (4.80), we have

$$\mathbf{D}_0 = \nabla(\phi+\varphi) \tag{4.82}$$

$$[\phi+\varphi]|_s = C_1, \quad or \quad \phi|_s = 0, \ \ \varphi|_s = C_1 \tag{4.83}$$

From (4.79) and (4.81) we have

$$\mathbf{B}_0 = \nabla(\Psi+\psi) \tag{4.84}$$

$$\frac{\partial}{\partial n}[\Psi+\psi]|_s = 0, \quad or \quad \left.\frac{\partial \Psi}{\partial n}\right|_s = 0, \ \ \left.\frac{\partial \psi}{\partial n}\right|_s = 0 \tag{4.85}$$

After some discussion in chapter 5, we will have the eigenvector equation for the gradient subspace as follows

$$-\nabla^2 \phi_v = \rho_v^2 \phi_v, \quad \phi_v|_s = 0, \quad \rho_v^2 > 0 \tag{4.86}$$

Where ϕ_v is the eigenvector corresponding to $k_0 = 0$. In other words, ϕ_v is the orthogonal basis for the gradient subspace, which obeys the orthogonality

$$2 < x_v, x_{v'} >= \delta_v^{v'} \tag{4.87}$$

and the normalization condition

$$||\phi_v||^2_{L_2(V)} = N_v^2 = \frac{1}{\rho_v^2} \tag{4.88}$$

Following the same way, the characteristic equation for Ψ is

$$-\nabla^2 \Psi_\alpha = q_\alpha^2 \Psi_\alpha, \quad \frac{\partial}{\partial n}\Psi_\alpha|_s = 0, \quad q_\alpha^2 > 0 \tag{4.89}$$

and the normalization condition for Ψ is

$$||\Psi_v||^2_{L_2(V)} = N_\alpha^2 = \frac{1}{q_\alpha^2} \tag{4.90}$$

3. The eigenvectors in Harmonic subspace

The eigenvectors in harmonic subspace φ_v and ψ_α are the eigenvectors corresponding to $\rho_v^2 = 0$ in (4.86) and $q_\alpha^2 = 0$ in (4.89), respectively, namely,

$$\nabla^2 \varphi_v = 0, \qquad \varphi_v|_s = C_1 \tag{4.91}$$

$$\nabla^2 \psi_\alpha = 0, \qquad \frac{\partial}{\partial n}\psi_\alpha|_s = 0 \tag{4.92}$$

Chapter 5

Operator Theory of the Field in Dynamic medium

5.1 Introduction

In this chapter, we will discuss the problem of electromagnetic field in an inhomogeneous dynamic medium. O. A. Treyakov [23] have a good study on this subject in terms of the mathematic point of view. [1]. Today, we will study this subject in terms of the electromagnetic field theory, in which, the material parameters are:

$$\begin{aligned} \mu &= \mu(\mathbf{r}, t) \\ \varepsilon &= \varepsilon(\mathbf{r}, t) \\ \sigma &= \sigma(\mathbf{r}, t) \end{aligned} \tag{5.1}$$

Those dynamic media could be

(1) The flying air plane and the moving cloud,

(2) The sailing steamboat and the moving wave on the sea,

(3) The ionic flow in the plasma and so on.

Now we will study the electromagnetic fields in a conductive container V filled with inhomogeneous dynamic medium.

Maxwell equations:

The Maxwell equations in inhomogeneous dynamic media are

$$\begin{aligned} rot\, B &= \mu \frac{\partial}{\partial t} D + \mu J + \left[\frac{\nabla \mu}{\mu} \times B \right] \\ -rot\, D &= \varepsilon \frac{\partial}{\partial t} B - \left[\frac{\nabla \varepsilon}{\varepsilon} \times B \right] \\ div\, D &= \rho \\ div\, B &= 0 \end{aligned} \tag{5.2}$$

[1] Another paper, related to this paper, written by M. S. Antyufeeva and O. A. Tretyakov is in [24].

Where the material parameters $\mu(\mathbf{r},t)$ $\varepsilon(\mathbf{r},t)$ and $\sigma(\mathbf{r},t)$ are the slow variation functions of time t compared with the variation of the fields.

The boundary conditions on the surface of the conductive container are

$$[n \times D]\Big|_S = 0, \qquad [n \cdot B]\Big|_S = 0 \tag{5.3}$$

Where S is the surface of a conductive container V.

The energy of the field in a conductive container V is finite, namely

$$\int_V [E \cdot D + H \cdot B] dV < \infty \tag{5.4}$$

The initial conditions are

$$D(\mathbf{r}, t_0) = D(\mathbf{r}), \qquad B(\mathbf{r}, t_0) = B(\mathbf{r}) \tag{5.5}$$

Obviously, the traditional operator theory of the electromagnetic field with complex amplitude is no long suitable for those dynamic medium. However, the fields in the conductive container is consist of the normal modes even though the conductive container is filled with an inhomogeneous dynamic medium since the inhomogeneous dynamic medium still is the linear medium. From the mathematic point of view, the space in this conductive container V is a Hilbert space $L_2(V)$, and the Hilbert space is consist of the orthogonal basis. Therefore, we may find the solutions of the fields in the conductive container by means of the conception of the Hilbert space $L_2(V)$.

5.2 Hilbert space in electromagnetic field

Before discuss the orthogonal basis in Hilbert functional space $L_2(V)$, we have to discuss *Weyl Theorem.*

Weyl Theorem : Any Hilbert functional space $L_2(V)$ contains only the curl subspace $\dot{\mathbf{J}}$, the divergence subspace $\dot{\mathbf{G}}$ and the Harmonic subspace $\mathbf{U}$, namely,

$$L_2(V) = \dot{\mathbf{J}} \oplus \dot{\mathbf{G}} \oplus \mathbf{U} \tag{5.6}$$

The proof may be found in H. Weyl, "Methods of Orthogonal projection in Potential theory", Duke Math. Journal, 1940,7, pp.441.

From the physical point of view, the electromagnetic field contains the curl field $\mathbf{x}$ and the gradient field $\mathbf{y}$ only.

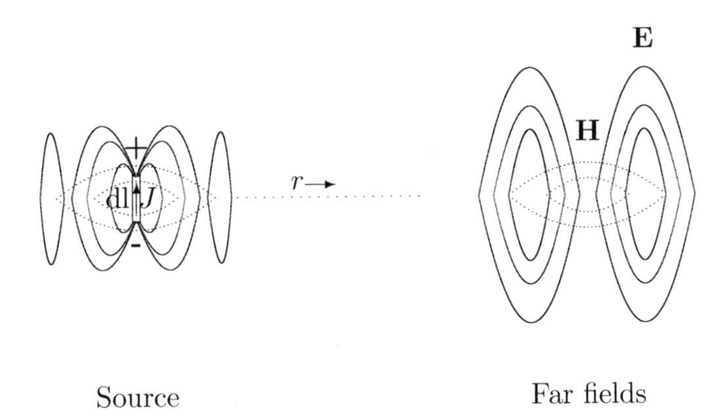

Figure 5-1 Electromagnetic field

As indicated in Figure 5-1, the electromagnetic field is consist of the curl fields $\mathbf{x}$ such as $\mathbf{x} = rot\,\mathbf{D}$ or $\mathbf{x} = rot\,\mathbf{B}$, and the divergence fields $\mathbf{y}$ such as $\mathbf{y} = div\,\mathbf{D}$.

The divergence field $\mathbf{y}$ is start from the positive charge $+Q$ and end at the negative charge $-Q$, and then $\mathbf{y} = div\,\mathbf{D} = \rho$. Therefore, For divergence field $\mathbf{y}$ we have $rot\,\mathbf{y} = 0$ and then , from vector analysis point of view, we have $\mathbf{y} = \nabla\phi$. Which means that the divergence field is the gradient field $\mathbf{y}$ in the source region and is the harmonic field $\mathbf{z}$ in the non-source region.

The curl field $\mathbf{x}$ is the field without any start point and the end point, or without any divergence, namely $div\,\mathbf{x} = 0$.

In other words:

1. The gradient field $\mathbf{y}$ is caused by the source, consequently, the gradient field is a field with divergence (i.e. $div\,\mathbf{y} = \rho \neq 0$).

2. The curl electric field, $\mathbf{x}$, is caused by the magnetic field and the curl magnetic field, $\mathbf{x}$, is caused by the electric field. Consequently, the curl field is a field with non-divergence (i.e. $div\,\mathbf{x} = 0$). And then

$$Curl\ field\ :\ \mathbf{x} = rot\,\mathbf{v},\quad rot\,\mathbf{x} \neq 0,\quad div\,\mathbf{x} = 0 \tag{5.7}$$

$$Divergence\ field\ :\ \begin{cases} \mathbf{y} = \nabla\phi & rot\,\mathbf{y} = 0, \quad div\,\mathbf{y} = \rho \\ \mathbf{z} = \nabla\varphi & rot\,\mathbf{z} = 0, \quad div\,\mathbf{z} = 0 \end{cases} \tag{5.8}$$

Where $\mathbf{x}$ is the curl field, $\mathbf{y}$ is the gradient field and $\mathbf{z}$ is the harmonic field. Therefore, the harmonic field is a gradient field in a sourceless region. The discussion above comes to an operator explanation, that $L_2(V)$ consists of the curl subspace $\dot{\mathbf{J}}$, the divergence subspace $\dot{\mathbf{G}}$ and the Harmonic subspace $\mathbf{U}$ as follows:

$$\dot{\mathbf{J}}\ :\ \mathbf{x} = rot\,\mathbf{v},\ \ div\,\mathbf{x} = 0,\ \ [\mathbf{n}\times\mathbf{x}]|_s = 0\ \ or\ \ [\mathbf{n}\cdot\mathbf{x}]|_s = 0\,(\mathbf{x} = B)$$

$$\dot{\mathbf{G}} \; : \; \mathbf{y} = \nabla\phi, \quad \nabla^2\phi = \rho, \quad \phi|_s = 0 \quad or \quad \frac{\partial}{\partial n}\phi\,|_s = 0 \; (\mathbf{y} = B)$$

$$\mathbf{U} \; : \; \mathbf{z} = \nabla\varphi, \quad \nabla^2\varphi = 0, \quad \varphi|_s = C_1$$

where S is the surface of the conductive container.

Now we will concentrate to discuss the Hilbert space in a conductive container as shown in Figure 5-2, in which, V is the space surrounded by S. V' is the space surrounded by S'.

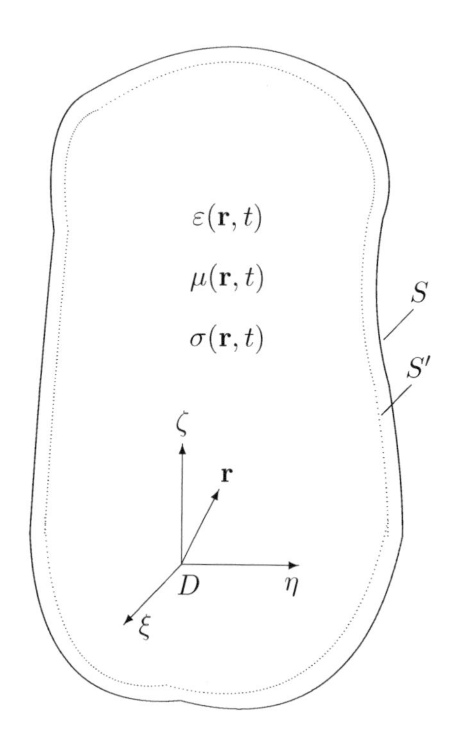

Figure 5-2 Hilbert space

5.3 The eigenfunction expansion of operator R_0

The one important feature of the Hilbert space $L_2(V)$ in a conductive container is that the Hilbert space $L_2(V)$ is consist of the orthogonal basis or the eigenvectors. From microwave point of view, the eigenvectors are just the normal modes in the conductive container.

To fine the orthogonal basis in $L_2(V)$ the best way is to discuss the function of the operator R_0 as follows.

5.3.1 The eigenvectors in $L_2(V)$ space

Define:

$$R = irot \tag{5.9}$$
$$R^* = -irot \tag{5.10}$$
$$R_0^*(D) = \{-irot\, D,\ r \in V,\ r \bar{\in} S;\ [n \times D]|_s = 0,\ r \in S\} \tag{5.11}$$
$$R_0(B) = \{irot\, B,\ r \in V,\ r \bar{\in} S;\ [n \cdot B]|_s = 0,\ r \in S\} \tag{5.12}$$

Where $r \in V,\ r \bar{\in} S$ means that the r is in V but not on the surface S, $r \in S$ means that the r is on the surface S.

Considering the vector identity

$$dis(A \times B) = B \cdot rot\, A - A \cdot rot\, B \tag{5.13}$$

we have

$$\begin{aligned}
&< D, R_0\, B > - < R_0^*(D), B > \\
&= \lim_{V' \to V} \frac{1}{V'} \int_{V'} [D \cdot (irot\, B)^* - (-irot\, D) \cdot B^*] dV' \\
&= \lim_{V' \to V} \frac{i}{V'} \int_{V'} [-D \cdot rot\, B^* + rot\, D) \cdot B^*] dV' \\
&= \lim_{V' \to V} \frac{i}{V'} \int_{V'} [div\, (D \times B^*)] dV' \\
&= \lim_{S' \to S} \frac{i}{V'} \int_{S'} (D \times B^*) \cdot n dS' \\
&= \lim_{S' \to S} \frac{i}{V'} \int_{S'} (n \times D \cdot B^*) dS' = 0
\end{aligned}$$

or

$$< D, R_0\, B > - < R_0^*\, D, B >= 0 \tag{5.14}$$

This equation is a basic equation to be used to find the eigenvectors in the Hilbert space $L_2(V)$.

5.3.2 The eigenvector of the curl subspace

Since the inhomogeneous dynamic medium is still a linear medium, we may consider to put $D = D_n$, $B = B_n$ and

$$\begin{aligned}
rot\, D_n &= ik_n B_n \\
rot\, B_n &= -ik_n D_n \\
[n \times D_n]|_s &= 0 \\
[n \cdot B_n]|_s &= 0
\end{aligned}$$

into (5.14). In this case, we have

$$R_0 B_n = k_n D_n \tag{5.15}$$
$$R_0^* (D_n) = k_n B_n \tag{5.16}$$

Substitution of (5.15)(5.16) into (5.14), we have

$$\begin{aligned} &< D_n, k_n D_n > - < k_n B_n, B_n >= \\ &\qquad k_n < D_n, D_n > -k_n < B_n, B_n >= 0 \end{aligned}$$

that means the equation (5.14) has been satisfied automatically. In which the orthonormal conditions

$$\begin{aligned} &< D_n, D_n >= 1 \\ &< B_n, B_n >= 1 \end{aligned} \tag{5.17}$$

have been considered.

Therefore, for curl subspace, the eigenvector equations are

$$rot\, D_n = ik_n B_n \tag{5.18}$$
$$rot\, B_n = -ik_n D_n \tag{5.19}$$
$$[n \times D_n]|_s = 0 \tag{5.20}$$
$$[n \cdot B_n]|_s = 0 \tag{5.21}$$

and the orthonormal conditions are

$$\begin{aligned} &< D_m, D_n >= \delta_{mn} \\ &< B_m, B_n >= \delta_{mn} \end{aligned} \tag{5.22}$$
$$\delta_m n = \begin{cases} 1 & m = n \\ 0 & m \neq n \end{cases} \tag{5.23}$$
$$\begin{aligned} ||D_n||^2 &= \frac{1}{V} \int_V D_n \cdot D_n^* dV = N_n = 1 \\ ||B_n||^2 &= \frac{1}{V} \int_V B_n \cdot B_n^* dV = N_n = 1 \end{aligned} \tag{5.24}$$

5.3.3 The eigenvector of the gradient subspace

Where $n = 0$ in (5.18) and (5.19) is corresponding to the gradient subspace. In this case, $k_0 = 0$, and then from (5.18) and (5.19), we have

$$rot\, D_0 = 0 \tag{5.25}$$
$$rot\, B_0 = 0 \tag{5.26}$$
$$div\, D_0 = \rho \tag{5.27}$$

$$div\, B_0 = 0 \tag{5.28}$$

where from (5.25) and (5.26), we have

$$D_0 = \nabla\phi, \quad \phi|_s = 0 \tag{5.29}$$
$$B_0 = \nabla\Psi, \quad \frac{\partial\Psi}{\partial n}|_s = 0 \tag{5.30}$$

and the orthonormal conditions are

$$\begin{aligned} &< D_0, B_0 >= 0 \\ &< D_n, D_0 >= 0, \quad n \neq 0 \\ &< B_n, B_0 >= 0, \quad n \neq 0 \end{aligned} \tag{5.31}$$

Define

$$\nabla_0^2\phi = \{-\nabla^2\phi, \ r \in V, \ r\bar{\in}S; \ \phi|_S = 0, \ r \in S\} \tag{5.32}$$

where

$$\begin{aligned} &< \phi_1, \phi_2 >= \frac{1}{V}\int_V \phi_1\phi_2 dV \\ &\phi_i = real\ number, \quad i = 1,2 \\ &\phi_i|_S = 0, \quad i = 1,2 \end{aligned} \tag{5.33}$$

Considering the Green formula

$$\int_V \Big\{G_2\nabla^2 G_1 - G_1\nabla^2 G_2\Big\}dV = \oint \Big\{G_2\frac{\partial}{\partial n}G_1 - G_1\frac{\partial}{\partial n}G_2\Big\}dS \tag{5.34}$$

where

$$\begin{aligned} G_1 &= \phi_1 \\ G_2 &= \Psi_2 \end{aligned}$$

we have

$$< \phi_1, \nabla_0^2\Psi_2 > - < \nabla_0^2\phi_1, \Psi_2 >= 0 \tag{5.35}$$

Obviously, from (5.35), we have

$$-\nabla^2\phi_\nu = \rho_\nu^2, \quad \rho_\nu^2 > 0 \quad \phi_\nu|_S = 0, \tag{5.36}$$

and the othogonormal condition is

$$\begin{aligned} &< D_0, D_0' >= \delta_\nu^{\nu'} \\ &\delta_\nu^{\nu'} = \begin{cases} 1 & \nu' = \nu \\ 0 & \nu' \neq \nu \end{cases} \\ &< D_0, D_0 >= \lim_{V' \to V}\frac{1}{V'}\int_{V'} \nabla\phi_\nu \cdot \nabla\phi_\nu dV' \end{aligned}$$

$$=<\nabla_0\phi_\nu, \nabla_0\phi_\nu>$$
$$=<\phi_\nu, \nabla_0^2\phi_\nu>$$
$$=\rho_\nu^2<\phi_\nu, \phi_\nu>$$
$$=\rho_\nu^2||\phi_\nu||^2_{L_2(V)}=1$$

or

$$||\phi_\nu||^2_{L_2(V)}=N_\nu^2=\frac{1}{\rho_\nu^2} \tag{5.37}$$

Similarly, from (5.35), we have

$$-\nabla^2\Psi_\alpha=q_\alpha^2, \quad q_\alpha^2>0, \quad \frac{\partial\Psi_\alpha}{\partial n}|_S=0 \tag{5.38}$$

and the othogonormal condition is

$$<B_0, B_0'>=\delta_\alpha^{\alpha'}$$
$$\delta_\alpha^{\alpha'}=\begin{cases}1 & \alpha'=\alpha\\ 0 & \alpha'\neq\alpha\end{cases}$$
$$<B_0, B_0>=\lim_{V'\to V}\frac{1}{V'}\int_{V'}\nabla\Psi_\alpha\cdot\nabla\Psi_\alpha dV'$$
$$=<\nabla_0\Psi_\alpha, \nabla_0\Psi_\alpha>$$
$$=<\Psi_\alpha, \nabla_0^2\Psi_\alpha>$$
$$=q_\alpha^2<\Psi_\alpha, \Psi_\alpha>$$
$$=q_\alpha^2||\Psi_\alpha||^2_{L_2(V)}=1$$

or

$$||\Psi_\alpha||^2_{L_2(V)}=N_\nu^2=\frac{1}{q_\alpha^2} \tag{5.39}$$

Obviously, ∇_0 in $L_2(V)$ is a self-adjoint operator and ∇_0^2 in $L_2(V)$ is a self-adjoint operator too.

5.3.4 The eigenvector of the harmonic subspace

As we knew that the harmonic subspace is a sourceless subspace and then, $\rho_\nu=0$ and $\rho_\alpha=0$. In this case, from (5.36) and (5.38), the eigenvector equations in the harmonic subspace are

$$\nabla^2\varphi_\mu=0, \quad \varphi_\mu|_S=0 \tag{5.40}$$
$$\nabla^2\psi_\beta=0, \quad \frac{\partial}{\partial n}\psi_\beta|_S=0 \tag{5.41}$$

and the orthonormal condition are

$$||\varphi_\mu||^2_{L_2(V)}=1$$
$$||\psi_\beta||^2_{L_2(V)}=1 \tag{5.42}$$

5.4 Combination of normal modes and Maxwell equations

As we mentioned before that the fields in a conductive container still is the combination of the normal modes in the curl subspace, the normal modes in the gradient subspace and the normal modes in the harmonic subspace since the inhomogeneous dynamic medium is a linear medium. And then the solutions of the fields in a linear inhomogeneous medium are the normal modes with a variation coefficient for each normal mode. In this case, the solutions of the fields can be found by mean of the substitution of all of the normal modes into the Maxwell equations in an inhomogeneous dynamic medium.

To this end, we will summery all of the normal modes and the Maxwell equations in the conductive container filled with an inhomogeneous dynamic medium as follows:

1. The normal modes of the curl subspace:

The normal modes of the curl subspace are

$$\begin{aligned} &rot\, B_n = -ik_n D_n, \\ &rot\, D_n = ik_n B_n, \\ &[n \times D_n]|_S = 0, \\ &[n \cdot B_n]|_S = 0, \qquad m, n = 1, 2, \cdots \end{aligned} \tag{5.43}$$

and the orthonormal conditions are

$$\begin{aligned} &< D_m, D_n >= \delta_{mn}, \\ &< B_m, B_n >= \delta_{mn}, \\ &< D_n, B_m >= 0, \\ &||D_n||^2_{L_2(V)} = 1, \\ &||B_n||^2_{L_2(V)} = 1, \qquad m, n = 1, 2, \cdots \end{aligned} \tag{5.44}$$

2. The normal modes of the gradient subspace:

The normal modes of the gradient subspace are corresponding to $k_0 = 0$, namely, for gradient subspace

$$\begin{aligned} &D_0 = \nabla\phi, \qquad \phi|_S = C_1 \\ &B_0 = \nabla\Psi, \qquad \frac{\partial}{\partial n}\Psi|_S = 0 \end{aligned} \tag{5.45}$$

$$\begin{aligned} &-\nabla^2\phi_\nu = \rho_\nu^2\phi_\nu, \quad \phi_\nu|_S = 0 \\ &-\nabla^2\Psi_\alpha^2 = q_\alpha^2, \qquad \frac{\partial}{\partial n}\Psi_\alpha|_S = 0 \end{aligned} \tag{5.46}$$

and the orthonormal conditions are

$$||\phi_\nu||^2 = \frac{1}{\rho_\nu^2}$$
$$||\Psi_\alpha||^2 = \frac{1}{\rho_\alpha^2} \tag{5.47}$$

3. The normal modes of the harmonic subspace:

For harmonic subspace, $\rho_0 = 0$, $q_0 = 0$ and then from (5.46), the normal modes in harmonic subspace φ_μ and ψ_β satisfy

$$\nabla^2\varphi_\mu = 0, \quad \varphi_\mu|_S = C_1$$
$$\nabla^2\psi_\beta = 0, \quad \frac{\partial}{\partial n}\psi_\beta|_S = 0 \tag{5.48}$$

and the orthonormal conditions are

$$||\varphi_\mu||^2 = 1$$
$$||\psi_\beta||^2 = 1 \tag{5.49}$$

4. The Maxwell equations in inhomogeneous dynamic medium:

The Maxwell equations in inhomogeneous dynamic medium are

$$rot\, B = \mu\frac{\partial}{\partial t}D + \mu J + \left[\frac{\nabla\mu}{\mu} \times B\right]$$
$$-rot\, D = \varepsilon\frac{\partial}{\partial t}B - \left[\frac{\nabla\varepsilon}{\varepsilon} \times B\right]$$
$$div\, D = \rho$$
$$div\, B = 0 \tag{5.50}$$

and the boundary conditions are

$$n \times D|_S = 0$$
$$n \cdot B|S = 0 \tag{5.51}$$

Note, the material parameters $\mu(\mathbf{r}, t)$ $\varepsilon(\mathbf{r}, t)$ and $\sigma(\mathbf{r}, t)$ are the slow variation functions of time t compared with the variation of the fields.

5.5 The fields in inhomogeneous dynamic medium

1. **Maxwell equations:**

$$rot\, B = \left[\mu\frac{\partial}{\partial t}D + \mu\tilde{\sigma}D + (\frac{\nabla\mu}{\mu} \times B)\right]$$

$$-rot\, D = \left[\varepsilon \frac{\partial}{\partial t} B - \frac{\nabla \varepsilon}{\varepsilon} \times D\right]$$
$$\tilde{\sigma} = \frac{\sigma}{\varepsilon} \tag{5.52}$$

2. **Unknowns:**

$$D(r,t) = \sum_{n>0} e_n(t) D_n(r) + \sum_{\nu>0} e_\nu(t) D_\nu(r) \tag{5.53}$$
$$B(r,t) = \sum_{n>0} h_n(t) B_n(r) + \sum_{\alpha>0} h_\alpha(t) B_\alpha(r) \tag{5.54}$$

Where D_n, B_n, D_ν, B_α are the normal modes in $L_2(V)$ space and all of them satisfy the Maxwell equations and the boundary conditions in V.

However, $e_n(t)$, $h_n(t)$, $e_\nu(t)$ and h_α are unknowns, which can be determinate by means of substitution of $D(r,t)$ and $B(r,t)$ into the Maxwell equations and the boundary condition.

3. **The combination of Maxwell equations and the eigenvector equations:**

The Maxwell equations (5.50) and the boundary conditions of the conductive container (5.51) have been combined into a equation (5.14) as follows:

$$< D, R_0\, B > - < R_0^*\, D, B >= 0 \tag{5.55}$$

where

$$R_0\, B = \{irot\, B,\ r \in V,\ r\bar{\in} S,\ [n \cdot B]|_S = 0,\ r \in S\} \tag{5.56}$$
$$R_0^*\, D = \{-irot\, D,\ r \in V,\ r\bar{\in} S,\ [n \times D]|_S = 0,\ r \in S\} \tag{5.57}$$

And the eigenvector equations are

$$rot\, D_n = ik_n B_n$$
$$rot\, B_n = -ik_n D_n \tag{5.58}$$

Considering that the inhomogeneous dynamic medium is a linear medium, the B in (5.55) can be a single normal mode B_n and D is still the total field D or the D can be a single normal mode D_n and the B is still the total field B. Therefore, substitution of the eigenvector equations (5.58) into (5.55) we have:

For $D = D_n$, $B = B$,

$$< D_n, R_0\, B > - < k_n B_n, B >= 0$$

or

$$< D_n, rot\, B > + ik_n < B_n, B >= 0 \tag{5.59}$$

For $B = B_n$, $D = D$,

$$< R_0(-D), B_n > - < D, k_n B_n >= 0$$

or

$$< B_n, rot\, D > -ik_n < D_n, D >= 0 \tag{5.60}$$

For the case of $k_0 = 0$ we have

$$\begin{aligned} &< D_0, rot\, B >= 0 \\ &< B_0, rot\, D >= 0 \end{aligned} \tag{5.61}$$

4. **Initial conditions:**

$$\begin{aligned} e_n(t_0) &= e_{n0} \\ h_n(t_0) &= h_{n0} \\ e_\nu(t_0) &= e_{\nu 0} \\ h_\alpha(t_0) &= h_{\alpha 0} \end{aligned} \tag{5.62}$$

5.5.1 Modes in a homogeneous isotropic stable medium

At beginning, we will discuss a general case, in which, the conductive container in Figure 5-2 is filled with a homogeneous stable medium. Obviously, V is a single connection region, $\varepsilon = \varepsilon_0$, $\mu = \mu_0$, $\sigma = 0$ and S is a conductive surface. In this case, we have

$$\begin{aligned} \nabla\varepsilon &= 0 \\ \nabla\mu &= 0 \end{aligned}$$

and then the Maxwell equations (5.52) become

$$-rot\, D = \varepsilon_0 \frac{\partial}{\partial t} B = \varepsilon_0 \dot{B} \tag{5.63}$$

$$rot\, B = \mu_0 \frac{\partial}{\partial t} \dot{D} \tag{5.64}$$

Where D and B have been given in (5.53) and (5.54). Substitution of (5.53) and (5.54) into (5.59) and (5.60), considering the orthonormal condition (5.44), we have

$$\mu_0 \dot{e}_n(t) + ik_n e_n(t) = 0 \tag{5.65}$$

$$\varepsilon_0 \dot{h}_n(t) + ik_n h_n(t) = 0 \tag{5.66}$$

$$\dot{e}_\nu = 0 \tag{5.67}$$

$$\dot{h}_\alpha = 0 \tag{5.68}$$

From (5.65) (5.66), (5.67) and (5.68), we have

$$\ddot{e}_n(t) + (\frac{k_n}{\sqrt{\mu_0 \varepsilon_0}})^2 e_n(t) = 0 \tag{5.69}$$

$$e_\nu(t) = constant \tag{5.70}$$
$$h_\alpha(t) = constant \tag{5.71}$$

and then, the solutions of the electromagnetic field are

$$D_n(r,t) = e_{n0} D_n(r) e^{-i\omega_n t}, \quad n = 1, 2, \cdots \tag{5.72}$$
$$B_n(r,t) = -\sqrt{\frac{\mu_0}{\varepsilon_0}} B_n(r) e^{-i\omega_n t}, \quad n = 1, 2, \cdots \tag{5.73}$$
$$D_\nu(r,t) = e_{n0} D_\nu(r), \quad \nu = 1, 2, \cdots \tag{5.74}$$
$$B_\alpha(r,t) = h_{n0} B_\alpha(r), \quad \alpha = 1, 2, \cdots \tag{5.75}$$
$$\omega_n = \frac{k_n}{\sqrt{\mu_0 \varepsilon_0}} \tag{5.76}$$

Obviously, the solutions above are just the normal modes in a conductive container filled with homogeneous medium. That means that all of theoretic analysis above are valid for the time being.

5.5.2 Modes in homogeneous isotropic dynamic medium

If the conductive container is filled with the homogeneous isotropic dynamic medium and the material parameters in this container are $\varepsilon = \varepsilon(t)$, $\mu = \mu(t)$, and $\sigma = \sigma(t)$, meanwhile V is a single connect area, S is a conductive surface, the Maxwell equations are, from (5.52),

$$\begin{aligned} rot\, B &= \mu \frac{\partial}{\partial t} D + \mu \tilde{\sigma} D = \mu \dot{D} + \mu \tilde{\sigma} D \\ -rot\, D &= \varepsilon \frac{\partial}{\partial t} B = \dot{B} \end{aligned} \tag{5.77}$$

Substitution of the Maxwell equations (5.77) into (5.59) and (5.60), considering the orthonormal conditions (5.44), we have

$$\begin{aligned} &\mu \dot{e}_n(t) + \frac{\mu\sigma}{\varepsilon} e_n - i k_n h_n = 0 \\ &\varepsilon \dot{h}_n - i k_n e_n = 0 \\ &\dot{e}_\nu + \frac{\sigma}{\varepsilon} e_\nu = 0 \\ &\dot{h}_\alpha = 0 \end{aligned} \tag{5.78}$$

then from (5.78) we have the time evaluation equations for homogeneous isotropic dynamic medium as follows:

$$e_n = -i \frac{\varepsilon(t)}{k_n} \dot{h}_n, \quad n = 1, 2, \cdots \tag{5.79}$$
$$\ddot{h}_n + \frac{\sigma(t)}{\varepsilon(t)} \dot{h}_n + \left(\frac{k_n}{\sqrt{\mu(t)\varepsilon(t)}} \right)^2 h_n = 0, \quad n = 1, 2, \cdots \tag{5.80}$$

$$e_\nu(t) = e_{\nu 0} \exp[-\int_{t_0}^{t} \frac{\sigma(\tau)}{\varepsilon(\tau)} d\tau \quad \nu = 1, 2, \cdots \tag{5.81}$$

$$h_\alpha(t) = h_{\alpha 0} = constant, \quad \alpha = 1, 2, \cdots \tag{5.82}$$

Where the equation (5.80) is a variable coefficient second order linear differential equation in the form of

$$y'' + P(x)y' + Q(x)y = 0 \tag{5.83}$$

and the general solution of (5.83) is not so easy to be found.

In general, if y_1 is the first particular solution of (5.83), then the second particular solution is

$$y_2 = y_1 \int \frac{1}{y_1^2} \exp\left[-\int P(x)dx\right] dx \tag{5.84}$$

and then the general solution of (5.83) is

$$Y = C_1\, y_1 + C_2\, y_2 \tag{5.85}$$

Now we have to take a special case, in which

$$\varepsilon(t) = \varepsilon_0\, \alpha(t), \quad \mu(t) = \mu_0, \alpha(t), \quad \sigma = 0 \tag{5.86}$$

Substitution of (5.86) into (5.79) $\sim$ (5.82) we have

$$\begin{aligned} &\ddot{e}_n + (\frac{\dot{\alpha}}{\alpha})\dot{e}_n + \left[\frac{k_n}{\alpha\sqrt{\mu_0\, \alpha_0}}\right]^2 e_n = 0, \quad n = 1, 2, \cdots \\ &h_n = -i\frac{\mu_0 \alpha}{k_n}\dot{e}_n \\ &e_\nu = 0 \\ &\dot{h}_\alpha = 0 \end{aligned} \tag{5.87}$$

The solutions of (5.87) are [23]

$$\begin{aligned} e_n &= e_{n0} \cos\left[\frac{k_n}{\sqrt{\mu_0 \varepsilon_0}} \int_{t_0}^{t} \frac{1}{\alpha(\tau)} d\tau\right] \\ h_n &= i e_{n0} \sin\left[\frac{k_n}{\sqrt{\mu_0 \varepsilon_0}} \int_{t_0}^{t} \frac{1}{\alpha(\tau)} d\tau\right] \\ e_{nu} &= e_{\nu 0} \\ h_\alpha &= -i h_{\alpha 0} \end{aligned} \tag{5.88}$$

Obviously, the normal modes in homogeneous isotropic dynamic medium still are independent normal modes.

5.5.3 Modes in inhomogeneous isotropic dynamic medium

For the case of inhomogeneous isotropic dynamic medium such as the case of the fields in plasma or in electronic beam, the conductivity of the medium is often a function of the space r and the time t. In this case, the normal modes are not long the independent modes but the coupling normal modes, which can be expressed by a coupling equations.

For instance, if the material parameters are

$$\sigma = \sigma(r,t), \quad \varepsilon = \varepsilon(t), \quad \mu = \mu(t)$$

then substitution of the Maxwell equations (5.52) into (5.59) and (5.60), considering the orthonormal conditions (5.44), we will have the time evolution equations as follows:

$$\begin{aligned}
&\varepsilon \dot{h}_n - ik_n e_n = 0 \\
&\mu \dot{e}_n + \mu G_n^n e_n - ik_n h_n = -\mu\Big(\sum_{n'>0} e_{n'} G_{n'}^n + \sum_{\nu'>0} e_{\nu'} G_{\nu'}^n\Big) \\
&\dot{e}_\nu + G_\nu^\nu e_\nu = -\Big(\sum_{\nu'>0} e_{\nu'} G_{\nu'}^\nu + \sum_{n'>0} e_{n'} G_{n'}^\nu\Big) \\
&\dot{h}_\alpha = 0, \\
&h_\alpha = h_{\alpha 0}
\end{aligned} \tag{5.89}$$

Where [23]

$$G_\alpha^\beta = \frac{1}{V}\int_V \Big[\frac{\sigma(r,t)}{\varepsilon(t)} D_\alpha(r)\cdot D_\beta(r)\Big] dV = G_\beta^\alpha(t) = G_\alpha^\beta(t) \tag{5.90}$$

Obviously, the inhomogeneous isotropic dynamic medium such as plasma will cause the coupling between the normal modes. The coupling coefficient G_α^β depends on the factor $\dfrac{\sigma(r,t)}{\varepsilon(t)}$. If $\sigma(r,t) = \sigma(t)$, we have, from (5.90), that $G_\alpha^\beta = 0$ and then the coupling equations (5.89) will become non-coupling equations (5.79) to (5.82). That means that the mode coupling is caused by the inhomogeneous distribution of the parameter $\sigma(r,t)$.

Chapter 6

Operator Theory of the Fields in Discrete Medium

6.1 Introduction

In this chapter, we will discuss the electromagnetic field in a discrete medium.

The writing of this chapter is involved with the radiowave propagation in the forest or in the city. In which the trees in the forest and the buildings in the city are the discrete media in the air.

The wave propagation in the discrete medium is involved in two problems:

(1) How to model the discrete medium?

(2) How to analyze the wave propagation in the discrete medium?

In the early 1960's, the research program on radiowave propagation in the environment of a forest has been developed theoretically and experimentally. The experimental date were later used by Tamir, T. [25],[26],[27] to validate a theoretical model which show that, for frequency less than 200 MHz, the principle mechanism responsible for long-distance propagation is the so-called lateral wave.

After then, the development on radiowave propagation in forest directs toward the UHF (200 - 2000 MHz) and digital spread-spectrum modulation radio signal. In UHF, the Tamir's model is no long held since the scattering of the scatter (tree trunks, leaves and branches) is ignored in Tamir's model. In 1982, R.H. Lang etc. A. Schneider, S. Seker and F. J. Altman developed a UHF model [28] in which the forest is considered to be a discrete random medium representable as a time-invariant ensemble of randomly positioned and randomly oriented discrete canonical scatters. The electromagnetic wave propagating within this medium is representable as the sum of two components: a mean component and a fluctuating field component. It is shown that with regard to the mean field, the ensemble of the discrete scatters can be represented by a equivalent dyadic permittivity which can be used to determine the wave propagation within an unbounded forest, and to define an anisotropic forest slab model for UHF in which the forests is represented as an anisotropic slab between air and the ground, and then recede the ground plane to $-\infty$ to evaluate the radio loss of the lateral wave in UHF. Which means that the ground and the scattering of the foliage are ignored in Lang's model.

The contribution of author [30] is to set up an UHF forest model with two anisotropic slabs between air and ground to represent the tree trunks and foliage, respectively, for the evaluation of the mean field in UHF. Using the parameters of scatters (tree trunks, branches, leaves) obtained by R.H. Lang et al. [28] to evaluate the radio loss of the direct wave, reflect wave and the lateral wave in the forest and then certifies that the lateral wave still is the dominant wave in forest. Meanwhile, the theoretical results of the radio loss of the UHF lateral wave is in good agreement with the experimental results.

Now we will discuss the model of the discrete medium (UHF forest model) and the wireless wave propagation in the discrete medium step by step. In which, the author's UHF model [29] [30] and the R. H. Lang's UHF model [28] [1] will be discussed.

6.2 The fields in a discrete dielectric rods

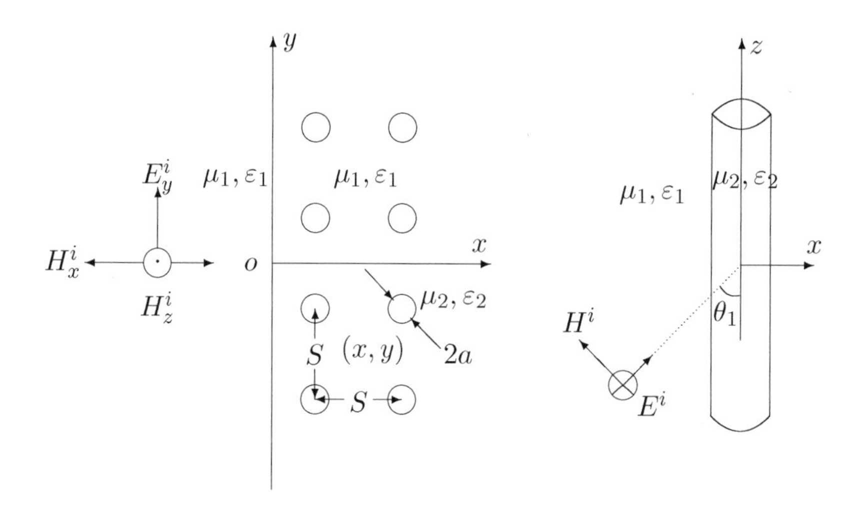

Figure 6-1 Dielectric rods in a lattice Figure 6-2 Incident wave

The start point is to set up a modelling for the forest. Following the book [31], a discrete dielectric rod may be chosen as a modelling of tree trunks as discussed in author's papers [29].

Modelling

[1] By R. H. Lang, A. Schneider, S. Seker and F. J. Altman [28]

As shown in Figure 6-1, the dielectric rods in a half space $x > 0$ is a good modelling for a forest. Which may be used to discuss the propagation of microwave in a forest. In which, the whole space is filled with the isotropic medium μ_1, ε_1 and in the half space $x \geq 0$, the cylinders with the medium parameters μ_2, ε_2 and the radius of a located at the corner of each square lattice are used to model the trunks of the tree.

Analysis

Then a plane wave with horizontal polarization is incident obliquely onto the tree as shown in Figure 6-2. The incident angle is θ_1, and then the incident wave is

$$H_z^i = \tilde{H}_0 \sin\theta_1 \exp(-i\,\lambda_1\,x)\exp(-i\,h_1\,z) \quad (x < 0)$$

$$H_x^i = \tilde{H}_0 \cos\theta_1 \exp(-i\,\lambda_1\,x)\exp(-i\,h_1\,z) \quad (x < 0)$$

$$E_y^i = \tilde{E}_0 \exp(-i\,\lambda_1\,x)\exp(-i\,h_1\,z) \quad (x < 0) \tag{6.1}$$

where

$$k_1 = \omega\sqrt{\mu_1\varepsilon_1}, \quad \tilde{E}_0 = \sqrt{\frac{\mu_1}{\varepsilon_1}}\,\tilde{H}_0 \tag{6.2}$$

and

$$\lambda_1 = k_1 \sin\theta_1, \quad h_1 = k_1 \cos\theta_1 \tag{6.3}$$

are the propagation constants along x direction and the z direction, respectively. The time factor $\exp(i\omega t)$ is omitted.

As shown in Figure 6-1, our goal is to find the effect of the filled region $(x \geq 0)$ upon the incident wave. To this end the filled region $(x \geq 0)$ is able to be equivalent to a homogeneous medium, and the parameters of the homogeneous medium are μ_H, ε_H and then the reflection coefficient of the incident electric field at the interface $x = 0$ is

$$R = \frac{(\mu_H/\varepsilon_H)^{1/2} - (\mu_1/\varepsilon_1)^{1/2}}{(\mu_H/\varepsilon_H)^{1/2} + (\mu_1/\varepsilon_1)^{1/2}} \tag{6.4}$$

and the propagation constant of the incident wave in the filled region $(x \geq 0)$ is

$$k = \omega\sqrt{\mu_H\varepsilon_H} \tag{6.5}$$

Therefore, if we can find R and k, then it is able to determine μ_H and ε_H. Meanwhile, even though the fields in the filled region is very complicated, the propagation properties such as the phase shift and the attenuation coefficient along x direction is determined by k in (6.5). Where, μ_H, ε_H are said to be the equivalent parameters of the homogeneous medium in the filled region $(x \geq 0)$ for the case that a horizontal polarization wave is incident obliquely onto the trunks.

To find R and k, it is necessary to investigate the fields at a dielectric rod located at (x, y) in the filled region. These fields consist of two part: one is the incident wave, the others are the scattering fields. As shown in Figure 6-1, the scattering fields at the rod are the sum of the scattering fields from all of the dielectric rods with this rod absent.

1. The scattering fields of a single dielectric rod .

We will find the scattering fields of a single dielectric rod at (x, y) first. As shown in Figure 6-2, the incident magnetic field H_x^i will induce a current along z direction on the trunk and the current will cause a scattering field E_z^s along z direction. Meanwhile the incident electric field E_{yi} will cause the scattering magnetic field H_z^s, where H_z^i and the scattering magnetic field H_z^s together will be continuous inside and outside the trunk.

Now we may consider that the scattering fields E^s and H^s are caused by the electric Hertz vector $\mathbf{\Pi} = \mathbf{a}_z\Pi$ and the magnetic Hertz vector $\mathbf{\Psi} = \mathbf{a}_z\Psi$, respectively, with the representations as follows: (see Appendix)

$$H^s = (\nabla\nabla \cdot +k_j^2)\mathbf{\Psi}_j + i\omega\varepsilon_j\nabla \times \mathbf{\Pi}_j \qquad (j = 1, 2)$$

$$E^s = (\nabla\nabla \cdot +k_j^2)\mathbf{\Pi}_j - i\omega\mu_j\nabla \times \mathbf{\Psi}_j \qquad (j = 1, 2) \qquad (6.6)$$

where $j = 1, 2$ denotes outside the rod $(j = 1)$ and inside the rod $(j = 2)$, respectively. Both Ψ and Π satisfy the scale wave equations and the solutions are

$$\begin{aligned}
\Psi_1 &= [A_1 H_0^{2)}(\lambda_1\rho) + B_1 H_1^{(2)}(\lambda_1\rho)\cos\phi]\exp(-ih_1 z) \quad (\rho \geq a)\\
\Psi_2 &= [A_2 J_0(\lambda_2\rho) + B_2 J_1(\lambda_2\rho)\cos\phi]\exp(-ih_1 z) \quad (\rho \leq a)\\
\Pi_1 &= D_1 H_1^{(2)}(\lambda_1\rho)\sin\phi\exp(-ih_1 z) \quad (\rho \geq a)\\
\Pi_2 &= D_2 J_1(\lambda_2\rho)\sin\phi\exp(-ih_1 z) \quad (\rho \leq a) \qquad (6.7)
\end{aligned}$$

where

$$\lambda_j^2 + h_j^2 = k_j^2, \qquad k_j = \omega\sqrt{\mu_j\varepsilon_j} \qquad (j = 1, 2)$$

and the Bessel function J_0 or J_1 satisfies that the field is finite inside the rod and the Hankel function $H_0^{(2)}$ or $H_1^{(2)}$ satisfies that the field is zero at $\rho \to \infty$. In addition, since the incident electric field is a field without z-component, the electric Hertz vector $\mathbf{\Pi} = \mathbf{a}_z\Pi$ is supported by H_x^i alone. Consequently, The distribution of Π_1 or Π_2 along ϕ is $sin\,\phi$, and ϕ is the azimuth angle started from the x axis. The magnetic Hertz vector $\mathbf{\Psi} = \mathbf{a}_z\Psi$ consists of two terms, the first one is supported by H_z^i, therefore, it is even along ϕ. The second one is supported by E_y^i, therefore the distribution along ϕ is $\cos\phi$. Where the transverse current of the rod induced by H_z^i is ignored since we assume that the rod is thin enough $(k_1\, a << 1)$.

That means the following calculation is based on the assumption that $k_1\, a << 1$ and the solution is under this condition.

Now, considering the continuity of the fields at the interface between inside and outside the rod, we may consider that the propagation constant along z direction inside and outside the rod should be equal to each other, namely $h_1 = h_2$, consequently,

$$\lambda_2 = (k_2^2 - h_1^2)^{1/2} = (k_2^2 - k_1^2 \cos\theta_1)^{1/2} \qquad (6.8)$$
$$k_2 = \omega\sqrt{\mu_2\varepsilon_2}$$

Substituting (6.7) into (6.6), we have the scattering fields

$$H_{zj}^s = \Big(\frac{\partial^2}{\partial z^2} + k_j^2\Big)\Psi_j \qquad (j=1,2)$$

$$H_{\phi j}^s = \Big(\frac{1}{\rho}\frac{\partial^2}{\partial\phi\partial z} + k_j^2\Big)\Psi_j - i\omega\varepsilon_j\frac{\partial\Pi_j}{\partial\rho} \qquad (j=1,2)$$

$$E_{zj}^s = \Big(\frac{\partial^2}{\partial z^2} + k_j^2\Big)\Pi_j \qquad (j=1,2)$$

$$E_{\phi j}^s = \Big(\frac{1}{\rho}\frac{\partial^2}{\partial\phi\partial z} + k_j^2\Big)\Pi_j + i\omega\mu_j\frac{\partial\Psi_j}{\partial\rho} \qquad (j=1,2) \qquad (6.9)$$

The boundary conditions at the surface of the rod ($\rho = a$) are:

$$H_z^i + H_{z1}^s = H_{z2}^s$$
$$E_{z1}^s = E_{z2}^s$$
$$H_\phi^i + H_{\phi1}^s = H_{\phi2}^s$$
$$E_\phi^i + E_{\phi1}^s = E_{\phi2}^s \qquad (6.10)$$

in which, from (6.1),

$$H_\phi^i = H_0\cos\theta_1\sin\phi\exp(-i\lambda_1\rho\cos\phi)\exp(-ih_1z)$$
$$E_\phi^i = E_0\cos\phi\exp(-i\lambda_1\rho\cos\phi)\exp(-ih_1z) \qquad (6.11)$$

Substituting (6.9) and (6.11) into (6.10), we have

$$A_1 = \alpha H_0\sin\theta_1$$
$$B_1 = \beta E_0$$
$$D_1 = \delta H_0\sin\theta_1 \qquad (6.12)$$

where

$$\alpha = \frac{J_1 - \dfrac{\lambda_2 a}{2}\dfrac{\mu_1}{\mu_2}\dfrac{J_0}{\sin\theta_1}}{\lambda_1^2\Big[\dfrac{\mu_1}{\mu_2}\dfrac{\lambda_2}{\lambda_1}J_0H_1 - J_1H_0\Big]}$$

$$\beta = i\frac{\lambda_1 a\sqrt{\dfrac{\varepsilon_1}{\mu_1}}J_1' - \dfrac{\lambda_2}{\omega\mu_2}J_1\Big[\dfrac{1}{\sin\theta_1} - i\dfrac{\omega\varepsilon_1}{a}\cos\theta_1\delta H_1\Big(1 - \dfrac{\lambda_1^2}{\lambda_2^2}\Big)\Big]}{\lambda_1^2\Big[H_1J_1' - \dfrac{\mu_1}{\mu_2}\dfrac{\lambda_2}{\lambda_1}J_1H_1'\Big]}\sin\theta_1$$

$$\delta = i\sqrt{\frac{\mu_1}{\varepsilon_1}}\left(1-\frac{\lambda_1^2}{\lambda_2^2}\right)\left(\frac{H_1'}{H_1}-\frac{1}{\lambda_1 a}\right)\frac{\cos\theta_1}{H_1 D} \tag{6.13}$$

in which

$$D = \frac{\cos^2\theta_1}{a^2}\left(1-\frac{\lambda_1^2}{\lambda_2^2}\right) - \lambda_1^2\left(\frac{H_1'}{H_1}-\frac{\mu_2\lambda_1}{\mu_1\lambda_2}\frac{J_1'}{J_1}\right)\left(\frac{H_1'}{H_1}-\frac{\varepsilon_2\lambda_1}{\varepsilon_1\lambda_2}\frac{J_1'}{J_1}\right) \tag{6.14}$$

and

$$\begin{aligned} H_1 &= H_1(u) \\ H_1' &= \frac{\partial}{\partial v}[H_1(u)] \\ J_1 &= J_1(v) \\ J' &= \frac{\partial}{\partial v}[J_1(v)] \end{aligned}$$

where $u = \lambda_1 a$, $v = \lambda_2 a$ and a is the radius of the dielectric rod.

Substituting (6.12) into (6.7) and then substituting the results into (6.9), we may obtain the scattering fields from a single dielectric rod, E_j^s and H_j^s $(j = 1, 2)$.

2. Find the total field at a single dielectric rod located at (x, y) in the filled region $(x \geq 0)$ as shown in Figure 6-1.

As we mentioned before, the total field is equal to the incident field plus the sum of the scattering fields from all of the scatters with the rod absent. It is known, that the location of all of the rest dielectric rods are

$$\rho_{mn} = [(ms - x)^2 + (ns - y)^2]^{1/2}$$

as shown in Figure 6-1. Where, m is the integer from o to ∞, n is the integer from $-\infty$ to $+\infty$, S is the interval between two adjoint dielectric rods. Thus, from (6.1) and (6.6), the total fields at (x, y) are

$$\begin{aligned} H_z = H(x,z)\sin\theta = H_0 \sin\theta_1 e^{-i\lambda_1 x} e^{-ih_1 z} \\ +\Big\{\sum_{\substack{0\\m}}^{\infty}{}'\sum_{\substack{-\infty\\n}}^{\infty}{}'\left(\frac{\partial^2}{\partial z^2}+k_1^2\right)\Big[\alpha H(mS,z)\sin\theta\, H_0^{(2)}(\lambda_1\rho_{mn}) \\ +\beta E(mS,z)H_1^{(2)}(\lambda_1\rho_{mn})\cos\phi\Big]\Big\} \end{aligned} \tag{6.15}$$

$$E_y = E(x,z) = E_0 e^{-i\lambda_1 x} e^{-ih_1 z}$$

$$+\sum_{\substack{0\\m}}^{\infty}{}'\sum_{\substack{-\infty\\n}}^{\infty}{}'\Big\{\Big(\frac{\partial^2}{\partial x\partial z}+k_1^2\Big)\delta H(mS,z)\sin\theta\, H_1^{(2)}(\lambda_1\rho_{mn})\sin\phi$$

$$+i\omega\mu_1\frac{\partial}{\partial x}\Big[\alpha\, H(mS,z)\sin\theta\, H_0^{(0)}(\lambda_1\rho_{mn})$$

$$+\beta E(mS,z)H_1^{(2)}(\lambda_1\rho_{mn})\cos\phi\Big]\Big\} \tag{6.16}$$

Where $\sum{}'$ denotes that the sum is the sum of the scattering fields at the rod from all of the scatters with the rod absent.

If the variation of the function in $\sum{}'$ is not so large in each interval of S, then we may consider that

$$Sf(x)\approx\int_{x-\frac{S}{2}}^{x+\frac{S}{2}}f(X)dX \tag{6.17}$$

and then we will able to replace the sum by an integral. To this end, let $mS=X$, nS=Y, then the sum

$$\sum{}'\sum{}'\equiv S^{-2}\Big[\int_0^{\infty}\int_{-\infty}^{\infty}-\int_{x-\frac{S}{2}}^{x+\frac{S}{2}}\int_{y-\frac{S}{2}}^{y+\frac{S}{2}}\Big]dXdY$$

Therefore, (6.15) and (6.16) become

$$H(x,z)\sin\theta=H_0\sin\theta_1 e^{-i\lambda_1 x}e^{-ih_1 z}$$

$$+\Big(\frac{\partial^2}{\partial z^2}+k_1^2\Big)\Big\{\frac{1}{S^2}\int_0^{\infty}\int_{-\infty}^{\infty}[\alpha H(X,z)\sin\theta H_0^{(2)}(\lambda_1\rho)$$

$$+\beta E(X,z)H_1^{(2)}(\lambda_1\rho)\cos\phi]dXdY$$

$$-\frac{1}{S^2}\int_{x-\frac{S}{2}}^{x+\frac{S}{2}}\int_{y-\frac{S}{2}}^{y+\frac{S}{2}}[\alpha H(X,z)\sin\theta H_0^{(2)}(\lambda_1\rho)$$

$$+\beta E(X,z)H_1^{(2)}(\lambda_1\rho)\cos\phi]dXdY\Big\} \tag{6.18}$$

$$E(x,z)=E_0 e^{-i\lambda_1 x}e^{-ih_1 z}+i\omega\mu_1\frac{\partial}{\partial x}\{\ \ \}$$

$$+\frac{1}{S^2}\int_0^{\infty}\int_{-\infty}^{\infty}\Big(\frac{\partial^2}{\partial y\partial z}+k_1^2\Big)\delta H(X,z)\sin\theta H_1^{(2)}(\lambda_1\rho)\sin\phi dXdY$$

$$-\frac{1}{S^2}\int_{x-\frac{S}{2}}^{x+\frac{S}{2}}\int_{y-\frac{S}{2}}^{y+\frac{S}{2}}\Big(\frac{\partial^2}{\partial y\partial z}+k_1^2\Big)\delta H(X,z)\sin\theta H_1^{(2)}(\lambda_1\rho)$$

$$\times\sin\phi dXdY \tag{6.19}$$

Where the function inside the braces { } in (6.19) is just the function inside the braces { } in (6.18).

For the sack of simplification , assume that $k_1 a << 1$, $k_1 S << 1$, then in the double integral from $(x - S/2)$ to $(x + S/2)$ and from $(y - S/2)$ to $(y + S/2)$, $H(X, z)$ and $E(X, z)$ may be considered as constants. After rearranged, (6.18) and (6.19) become

$$H(x,z)\sin\theta = H_0 \sin\theta_1 e^{-i\lambda_1 x} e^{-ih_1 z}$$

$$+\Big(\frac{\partial^2}{\partial z^2} + k_1^2\Big)\Big\{\frac{2}{(\lambda_1 S)^2}\int_0^\infty [\alpha\lambda_1 H(X,z)\sin\theta$$

$$-\beta\frac{\partial}{\partial X}E(X,z)e^{-i\lambda_1|x-X|}]dX$$

$$-\frac{2\beta}{(\lambda_1 S)^2}E(0,z)e^{-i\lambda_1 x} + i\frac{1}{\pi}[\alpha H(x,z)\sin\theta L + \frac{\beta}{\lambda_1}\frac{\partial}{\partial x}E(x,z)]\Big\} \tag{6.20}$$

$$E(x,z) = E_0 e^{-i\lambda_1 x} e^{-ih_1 z} + i\omega\mu_1 \frac{\partial}{\partial x}\{\quad\} + i\frac{2}{\lambda_1 S^2}\frac{\partial}{\partial z}\delta H(x,z)\sin\theta \tag{6.21}$$

where

$$L = i\pi + 2\ln\left(\frac{\lambda_1 S}{2}\frac{e^\gamma}{2}\right) + \ln 2 + \frac{\pi}{2} - 3$$

and $H(x, z)$ and $E(x, z)$ are the total fields that we try to find.

3. Find the propagation constant k in the filled region $(x \geq 0)$.

Obviously, (6.20) and (6.21) are two integral equations, and the unknowns are the total fields $E(x, z)$ and $H(x, z)$ at (x, y). They are the waves in the filled region $(x \geq 0)$ propagating not only along x direction but also along z direction as shown in Figure 6-1. And then these waves may be expressed as

$$\begin{aligned} H(x,z) &= Ce^{-i\lambda x}e^{-ihz} \\ E(x,z) &= De^{-i\lambda x}e^{-ihz} \end{aligned} \tag{6.22}$$

Where

$$(\lambda^2 + h^2)^{1/2} = k = \omega\sqrt{\mu_H \varepsilon_H}$$

and μ_H and ε_H are the equivalent permeability and equivalent dielectric constant of the filled region $(x \geq 0)$ for the wave of horizontal polarization.

Substituting (6.22) into (6.20) and (6.21) we have

$$a\Big(\frac{\lambda}{\lambda_1}\Big)^2 + b\Big(\frac{\lambda}{\lambda_1}\Big) - c = 0 \tag{6.23}$$

And

$$h = h_1$$

Obviously the solution of (6.23) is

$$\frac{1}{\lambda_1} = \frac{1}{2a}\left[-b \pm (b^2 + 4ac)^{1/2}\right] \tag{6.24}$$

where

$$\begin{aligned} a &= \left(1 - i\frac{\lambda_1^2 \alpha L}{\pi}\right) + \left(\frac{2}{\lambda_1 S}\right)^2 \lambda_1^2 \beta_1 \\ b &= \left(\frac{2}{\lambda_1 S}\right)^2 \lambda_1^2 \beta_1 \Phi \\ c &= \left(1 - i\frac{\lambda_1^2 \alpha L}{\pi}\right) + i\left(\frac{2}{\lambda_1 S}\right)^2 \lambda_1^2 a \\ \beta_1 &= \beta\left(\frac{\mu_1}{\varepsilon_1}\right)^{1/2} \\ \Phi &= \frac{2}{\omega \mu_1 S^2} h_1 \delta \end{aligned} \tag{6.25}$$

And then via

$$k^2 = \lambda^2 + h^2 = k_1^2 (r^2 \sin^2\theta_1 + \cos^2\theta_1)$$

we have

$$\frac{k^2}{k_1^2} = \frac{\mu_H \varepsilon_H}{\mu_1 \varepsilon_1} = r^2 \sin^2\theta_1 + \cos^2\theta_1 \tag{6.26}$$

This is the first formula to determinate μ_H and ε_H, the equivalent parameters of the homogeneous medium in the filled region ($x \geq 0$) for the case that a horizontal polarization wave is incident obliquely onto the trunks, where

$$r = \lambda/\lambda_1$$

4. Find the reflection coefficient R_H at the interface $x = 0$.

To find the second formula to determinate μ_H and ε_H, we have to find the reflection coefficient R_H at the interface $x = 0$. To this end, come back to (6.18) , in which the first term is the incident wave and the second term is the scattering wave. Suppose that x is replaced by $-x$ in second term, then the second term becomes the reflected wave . After rearranged, the second term in (6.18) becomes

$$H(x < 0, z) \sin\theta_1 = \tilde{H}_0 \sin\theta_1 R_H e^{i\lambda_1 x} e^{-ih_1 z} \tag{6.27}$$

Where

$$R_H = -\left(\frac{r-1}{r+1}\right) \frac{i\left(\alpha/r\right) + \beta_1\left(1 + \Phi/r\right)}{i\left(\alpha/r\right) - \beta_1\left(1 + \Phi/r\right)} \tag{6.28}$$

is the reflection coefficient of the magnetic field at $x = 0$, $z = 0$, and each parameter in (6.28) has been given by (6.25) already.

Now we may determinate μ_H and ε_H by means of (6.28) and (6.26). As shown in Figure 6-1, suppose that the filled region ($x \geq 0$) may be equivalent to a homogeneous medium with μ_H and ε_H, then the reflection coefficient of the magnetic field at $x = 0, z = 0$ is

$$R_H = \frac{Y_1 - Y_H}{Y_1 + Y_H} = \frac{\lambda_1/\omega\mu_1 - \lambda/\omega\mu_H}{\lambda_1/\omega\mu_1 + \lambda/\omega\mu_H} \tag{6.29}$$

Where Y_1 is the wave admittance in the unfilled region ($x < 0$), and Y_H is the wave admittance in the filled region ($x \geq 0$).

Making comparison between (6.28) and (6.29) we have

$$\frac{\mu_H}{\mu_1} = \frac{(i\alpha - \beta_1)^2 r^2 - \beta\Phi r}{i\alpha - \beta_1(r + \Phi)r} \tag{6.30}$$

Then substituting (6.30) into (6.26), we have

$$\frac{\varepsilon_H}{\varepsilon_1} = (r^2 \sin^2\theta_1 + \cos^2\theta_1)\frac{i\alpha - \beta_1(r + \Phi)r}{(i\alpha - \beta_1)^2 r^2 - \beta_1\Phi r} \tag{6.31}$$

All of the parameters in the right-hand side of (6.30) and (6.31) are known. Therefore, the analytical solution of the equivalent parameters μ_H and ε_H in the filled region ($x \geq 0$) have been obtained.

Note , that μ_H and ε_H are the equivalent parameters of the filled region ($x \geq 0$) for the case that a horizontal polarization wave is incident obliquely onto the trunks.

5. Find the equivalent parameters for the perpendicular polarization wave, μ_ν, ε_ν.

If the incident wave is a perpendicular polarization wave incident obliquely onto the trunks, then the same process above is able to be used to find the equivalent parameters in the filled region ($x \geq 0$), μ_ν and ε_ν, as

$$\frac{\varepsilon_\nu}{\varepsilon_1} = 1 + i\frac{k_1^2\gamma}{\left(k_1 S/2\right)^2\left(1 - ik_1^2\gamma G/\pi\right)^2} \tag{6.32}$$

$$\frac{\mu_\nu}{\mu_1} = 1 \tag{6.33}$$

where

$$G = 2\Big[\ln\frac{k_1 S}{\sqrt{2}} - \frac{3}{2} + \frac{\pi}{4}\Big]$$

$$\gamma = \frac{-J_{ON}}{k_1^2 H_0^{(2)}(u)}\,\frac{(\varepsilon_2 J_1/\varepsilon_1 k_2 a J_0) + (J_{1N}/k_1 a J_{ON})}{(\varepsilon_2 J_1/\varepsilon_1 k_2 a J_0) + (H_1/k_1 a H_0^{(2)}(u))}$$

and

$$J_{0N} = J_0(u), \qquad J_{1N} = J_1(u)$$
$$H_1 = H_1^{(2)}(u), \qquad u = \lambda_1 a$$

Obviously, μ_ν, ε_ν in (6.33) , (6.32) are different from μ_H, ε_H in (6.30) , (6.31), respectively. Which means that the filled region ($x \geq 0$) with dielectric rods represents an anisotropic properties, where μ_ν, ε_ν are for the incident wave with perpendicular polarization and μ_H, ε_H are for the incident wave with horizontal polarization. And then the propagating properties for the perpendicular polarization incident wave is different from that for the horizontal polarization incident wave. This is due to the fact that the dielectric rods are placed along the perpendicular direction. Consequently, the scattering wave caused by the perpendicular polarization incident wave is different from that caused by the horizontal polarization incident wave.

Problem As shown in Figure 6-1, the whole space is a homogeneous medium with μ_1, ε_1. In the half space $x \geq 0$, there are a lot of small balls placed at each edge of the square cell as shown in Figure 6-1, the medium parameters of the ball is μ_2, ε_2 and the radius of the ball is a. Find the equivalent medium parameters μ, ε for the filled region ($x \geq 0$).

6.3 The fields in a random anisotropic medium

To discuss the UHF ($f = 200 \sim 2000 MHz$) propagation loss in forest, we have to take a few words to introduce an UHF forest model developed by R. H. Lang, A. Schneider, S. Seker and F. J. Altman in 1982 [28], in which, the forest is considered to be a discrete random medium, which is representable as a time-invariant ensemble of randomly position and randomly oriented discrete canonical scatters. And then the field in the forest is a random field.

The random field:

First, the random field ψ in the forest is able to be expressed as the sum of the average field $<\psi>$ and the fluctuation field $\tilde{\psi}$, i.e.

$$\psi = <\psi> + \tilde{\psi} \tag{6.34}$$

and the correlation function of the field is

$$<\psi\psi^*> = <\psi><\psi^*> + <\tilde{\psi}\tilde{\psi}^*> \tag{6.35}$$

Where ψ^* is the complex conjugate of ψ. Obviously, the correlation function consists of two components: one is the average component $<\psi><\psi^*>$, another one is the fluctuation component $<\tilde{\psi}\tilde{\psi}^*>$. The average component represents the absorptive effect of the field in the forest and the fluctuation component represents the scattering effect of the field in the forest. In low frequency, the absorptive effect of the fields is the domain effect, and the value of the fluctuation component is very small. However, as the frequency is increasing, the scattering effect of the forest is getting larger and larger, and then the value of the fluctuation component is getting larger and larger.

In the following discussion, we will concentrate to discuss the average component since the attenuation of the total fields is dominated by the average component.

6.3.1 Maxwell equations in a discrete medium

First, we will discuss the scattering fields of the paralleling rods, in which the scattering fields are symmetry in plane, and then this problem may be discussed by means of the scale equation.

To model the trees, suppose that there are N number of dielectric rods with the same size in parallel with z direction. The relative dielectric constant of each rod is ε_p, and the transverse area of each rod is S_p. All of the dielectric rods are contained in a area S.

Maxwell equations in a discrete medium

In this case, the time harmonic Maxwell equations is

$$\begin{aligned} \nabla \times \mathbf{E} &= -i\omega\mu_0\mathbf{H} \\ \nabla \times \mathbf{H} &= i\omega\varepsilon_0\varepsilon(\chi_\mathbf{t})\mathbf{E} + \mathbf{J}_s \end{aligned} \tag{6.36}$$

Where μ_0, ε_0 are the permeability and the dielectric constant in the free space, respectively. The time factor $\exp(i\omega t)$ is omitted. Suppose that the location of each rod is at $\chi_\mathbf{t}$, which is a transverse vector, and then the relative dielectric constant of the scatters (or rods)

$$\varepsilon(\chi_\mathbf{t}) = 1 + \sum_{j=1}^{N} \chi_j(\chi_\mathbf{t}) \tag{6.37}$$

is a function of the location $\chi_\mathbf{t}$ and $\chi_\mathbf{t}$ is a transverse vector of the location. Obviously, $\varepsilon(\chi_\mathbf{t})$ is the relative dielectric constant of a discrete medium. And the electric polarizability of the jth scatter is

$$\chi_j(\chi_\mathbf{t}) = \begin{cases} \chi_l & \chi_\mathbf{t} \in \mathbf{S_p} \\ 0 & \chi_\mathbf{t} \overline{\in} \mathbf{S_p} \end{cases} \tag{6.38}$$

Where

$$\chi_l = \varepsilon_p - 1$$

and $\chi_\mathbf{t} \in \mathbf{S_p}$ denotes that $\chi_\mathbf{t}$ is in the transverse cross section $\mathbf{S_p}$, $\chi_\mathbf{t} \overline{\in} \mathbf{S_p}$ denotes that $\chi_\mathbf{t}$ is not in $\mathbf{S_p}$.

Analysis

When the electric field is incident onto a trunk, i.e. when

$$\mathbf{e}_i = \mathbf{a_z} E_z$$

is incident on a trunk, the current density on the trunk is

$$\mathbf{J_s} = \mathbf{a_z} J_z(\chi_\mathbf{t})$$

since the scatters (trunks) are all in parallel with z direction. In this case, the total field may be expressed as a general form with the sum of the transverse component and a z component as follows:

$$\begin{aligned} \mathbf{E} &= \mathbf{E_t} + \mathbf{a_z} E_z \\ \mathbf{H} &= \mathbf{H_t} + \mathbf{a_z} H_z \end{aligned} \tag{6.39}$$

Substituting (6.39) into (6.36) reveals that E_z and $\mathbf{H_t}$ components are excited by J_z, and E_z satisfies

$$[\nabla_t^2 + k_0^2 \,\varepsilon\,(\chi_\mathbf{t})]\, E_z(\chi_\mathbf{t}) = i\omega\,\mu_0\, J_z\,(\chi_\mathbf{t}) \tag{6.40}$$

Where ∇_t is a transverse operator and $k_0^2 = \omega^2\,\mu_0\,\varepsilon_0$.

The other components may be obtained by substituting E_z into the Maxwell equations (6.36). Consequently, all of the equations are scale equations.

The operator method

Introduce an operator

$$L = -(\nabla_t^2 + k_0^2) \tag{6.41}$$

and

$$V_j = k_0^2\,\chi_j\,(\chi_\mathbf{t}) \tag{6.42}$$

$$\psi = E_z, \quad g = -i\,\omega\,\mu_0\,J_z \tag{6.43}$$

Then the wave equation (6.40) may be expressed as

$$\left(L - \sum_{j=1}^{N} V_j \right) \psi = g \tag{6.44}$$

This is the scale wave equation in a discrete medium. Where the total field ψ is the sum of the incident field ψ_i and the scattering field ψ_s, i.e.

$$\psi = \psi_i + \psi_s \tag{6.45}$$

For the case of non-scatter, $V_j = 0$, and $\psi_s = 0$, then from (6.44) and (6.45), we have

$$L\,\psi_i = g \tag{6.46}$$

This is the wave equation in a homogeneous medium.

6.3.2 The scattering amplitude and the transform operator

To find the solution of (6.44), R. H. Lang, A. Schneider, S. Seker and F. J. Altman [28] have proposed a conception about the scattering amplitude and the transform operator of a single scatter.

The transform operator

Now we will start to discuss this idea from a single scatter, in this case, (6.44) become

$$(L - V)\psi = g \tag{6.47}$$

Now, in the case of non-scatter, the incident wave satisfies

$$L\psi_i = g \tag{6.48}$$

Substituting (6.48) into (6.47), we have

$$L\psi_s = g_{eq} = V\psi \tag{6.49}$$

Where, both ψ_s and g_{eq} are unknown. However, both ψ_s and g_{eq} are induced by ψ_i. Consequently, we may define a *transform operator of a single scatter* T as

$$g_{eq} = T\psi_i \tag{6.50}$$

Then if T is known, it is possible to find the induced equivalent source g_{eq} from the known ψ_i. Since the incident wave ψ_i is bounded, the induced g_{eq} is bounded, and then T should be a bounded operator.

In fact, g_{eq} is induced by ψ_i via the scattering body, therefore, the operator T should be expressed as the integral form as

$$g_{eq}(\chi_{\mathrm{t}}) = \int t(\chi_{\mathrm{t}}, \chi'_{\mathrm{t}})\, \psi_i(\chi'_{\mathrm{t}})\, d\chi'_{\mathrm{t}} \tag{6.51}$$

Where χ'_t is the location vector of the scattering body and χ_t is the coordinate vector of the observation point. Obviously, $t(\chi_{\mathrm{t}}, \chi'_{\mathrm{t}})$ is not zero only if $\chi_{\mathrm{t}} \in S_p$ and $\chi'_{\mathrm{t}} \in S_p$ and the equation (6.51) may be used to define *the transform operator kernel of a single scatter* $t(\chi_{\mathrm{t}}, \chi'_{\mathrm{t}})$. That means that the equivalent source g_{eq} is in the scattering body.

In low frequency, the wavelength is much larger than the size of the scattering body, and then the equivalent source is able to be viewed as a line source. In that case, *the transform operator kernel of a single scatter* is

$$t(\chi_{\mathrm{t}}, \chi'_{\mathrm{t}}) = k_0^2\, \alpha\, \delta(\chi_{\mathrm{t}})\, \delta(\chi'_{\mathrm{t}}) \tag{6.52}$$

where the constant α is the polarization coefficient. Obviously, g_{eq} in (6.51) is the combination of a large amount of line sources.

Till now, the problem to find the scattering field ψ_s has been transformed to find the transform operator kernel $t(\chi_{\mathbf{t}}, \chi'_{\mathbf{t}})$.

The scattering amplitude

Now, the scattering field may be expressed as the far field of the scattering body either. Assume that the incident wave with unit amplitude is

$$\psi_i(\chi_{\mathbf{t}}) = \exp(-i\, k_0\, \mathbf{i} \cdot \chi_{\mathbf{t}}) \tag{6.53}$$

Then the scattering field in the radiation region will take the form of

$$\psi_s(\chi_{\mathbf{t}}) = f(\mathbf{o}, \mathbf{i})\, \frac{\exp(-i\, k_0\, \chi_t)}{\sqrt{\chi_t}} \qquad (\chi_t = |\chi_{\mathbf{t}}|) \tag{6.54}$$

Where $\mathbf{i}$ denotes the unit vector along the incident direction of the incident wave, $\mathbf{o}$ denotes the unit vector along the direction to the observer point. Thus (6.54) becomes an equation to define *the scattering amplitude of a single scatter* $f(\mathbf{o}, \mathbf{i})$.

6.3.3 The relative dielectric tensor and the effective polarizability tensor

After a long discussion by R. H. Lang, A. Schneider, S. Seker and F. J. Altman [28], they obtain

$$\overline{\overline{\varepsilon}} = \overline{\overline{I}} + \frac{(2\pi)^2}{k_0^2}\, \rho\, \overline{\overline{\tilde{t}}}\,(\mathbf{k}, \mathbf{k}) \tag{6.55}$$

This is the representative of the relative dielectric tensor for the forest in the case that the forest is considered to be an equivalent continuous medium. Since $\overline{\overline{\varepsilon}}$ is a tensor, the equivalent medium is considered to be an anisotropic medium. Where $\overline{\overline{\tilde{t}}}\,(\mathbf{k}, \mathbf{k})$ is the Fourier transform of $\overline{\overline{t}}\,(\chi_{\mathbf{t}}, \chi'_{\mathbf{t}})$.

Considering that the relationship between the relative dielectric tensor $\overline{\overline{\varepsilon}}$ and the effective polarizability tensor $\overline{\overline{\chi}}$ is

$$\overline{\overline{\varepsilon}} = \overline{\overline{I}} + \overline{\overline{\chi}} \tag{6.56}$$

and then the effective polarizability tensor of the forest is

$$\overline{\overline{\chi}} = \frac{(2\pi)^2}{k_0^2}\, \rho\, \overline{\overline{\tilde{t}}}\,(\mathbf{k}, \mathbf{k}) \tag{6.57}$$

Obviously, the effective polarizability of the forest is proportional to ρ, the density of the discrete scattering bodies (trunks, branches, leave).

Since the time factor is $exp(i\omega t)$, the definition of the real part $\overline{\overline{\chi'}}$ and the imaginary part $\overline{\overline{\chi''}}$ have to satisfy

$$\overline{\overline{\chi}} = \overline{\overline{\chi'}} - i\, \overline{\overline{\chi''}} \tag{6.58}$$

And $\overline{\overline{\chi}}$ may be expressed as the component form as

$$\overline{\overline{\chi}} = \sum_{\alpha^0\beta^0} \chi_{\alpha\beta}\alpha^0\beta^0 \qquad \alpha^0,\, \beta^0 \in \{\mathbf{h}^0, \mathbf{v}^0, \mathbf{i}^0\} \tag{6.59}$$

Obviously, there are nine components, in which, only four components χ_{hh}, χ_{vh}, χ_{hv} and χ_{vv} are important, other components are either zero or almost no contribution to the approximation value. These four components may be determined by the corresponding component of the tensor of the scattering amplitude one by one. It has been proved by R. H. Lang, A. Schneider, S. Seker and F. J. Altman [28] that

$$\chi_{pq} = \frac{(2\pi)^2\rho}{\gamma\, k_0^2} f_{pq}(\mathbf{i},\, \mathbf{i}), \qquad p,\, q \in \{h,\, v\} \tag{6.60}$$

This means that χ_{pq} is proportional to the corresponding component of the scattering amplitude f_{pq}.

When the wave is incident perpendicular onto the trunks, i.e. $\theta_i = \pi/2$,

$$f_{hv}(\mathbf{i},\, \mathbf{i}) = -f_{vh}(\mathbf{i},\, \mathbf{i}) = 0$$

and then

$$\chi_{hv} = \chi_{vh} = 0 \tag{6.61}$$

In this case, the equivalent continuous medium is an uniaxial anisotropic medium. the ensemble of each forest component (trunks, branches or leaves) can be represented as an uniaxial anisotropic slab, with an effective tensor permittivity in each slab $\overline{\overline{\varepsilon}}_i$ as follows:

$$\overline{\overline{\varepsilon}}_i = \begin{bmatrix} \varepsilon_{hci} & & \\ & \varepsilon_{hci} & \\ & & \varepsilon_{vci} \end{bmatrix} \qquad (i = 1, 2) \tag{6.62}$$

6.4 UHF propagation loss in forest modeled by four layered media

This part is based on author's research in University of Colorado, Boulder, USA. [30]

When radiowave propagation through the forest, the trees present two behaviors: absorbtion and scattering, when the frequency $f > 200MHz$. In this case the forest components are randomly located and randomly oriented, and the the field ψ (such as E or H) is a random variable. As such, it can be broken up into a mean component $< \psi >$ and a fluctuating component $\tilde{\psi}$, namely,

$$\psi = < \psi > + \tilde{\psi} \tag{6.63}$$

where the ensemble average of $\tilde{\psi}$, $<\tilde{\psi}>$, is equal to zero. However, the correlation function of ψ can be written as

$$<\psi,\psi^*>=<\psi><\psi^*>+<\tilde{\psi},\tilde{\psi}^*> \tag{6.64}$$

Where $*$ stands for complex comjugate. Here, $<\psi><\psi^*>$ is the mean part and $<\tilde{\psi},\tilde{\psi}^*>$ is the fluctuating part. Comparison of $<\psi><\psi^*>$ to $<\tilde{\psi},\tilde{\psi}^*>$ will show us whether the absorbtion or the scattering is more important.

Now we only concentrate on the mean fields in the stochastic model as as indicated in Figure 6-3. In which the forest is modelled as as two anisotropic slabs between the air and the ground to to represent the tree trunks and foliage, respectively, for the evaluation of the main field in UHF.

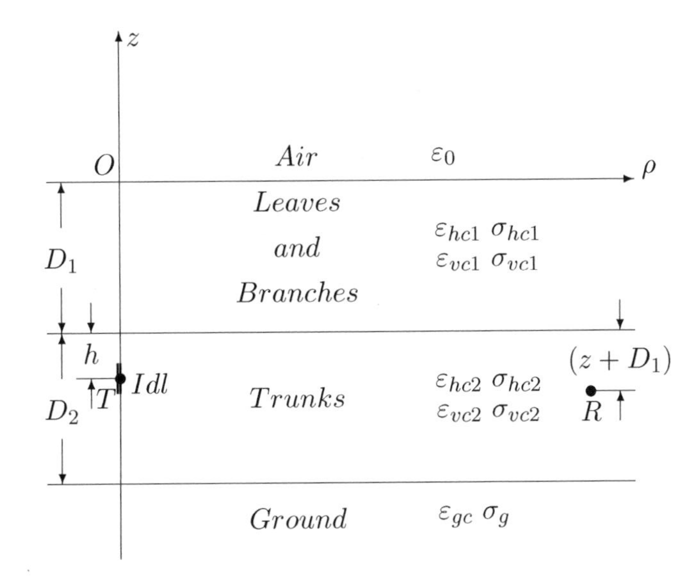

Figure 6-3 Forest model in UHF

In other words, the ensemble of discrete scatters (tree trunks, branches or leaves) can be represented by an equivalent continuous medium characterized by an effective tensor permitivity $\tilde{\varepsilon}_i$. Each slab possesses a tensor effective permittivity in form of

$$\overline{\overline{\varepsilon_i}} = \begin{bmatrix} \varepsilon_{hci} & & \\ & \varepsilon_{hci} & \\ & & \varepsilon_{vci} \end{bmatrix} \qquad (i=1,2) \tag{6.65}$$

where

$$\varepsilon_{hci} = \varepsilon_{hi} + \sigma_{hi}/i\omega\varepsilon_0$$

$$\varepsilon_{vci} = \varepsilon_{vi} + \sigma_{vi}/i\omega\varepsilon_0$$

and the free space permitivity is μ_0. Where 1 and 2 stand for the foliage and tree trunks, respectively. Both of the source dipole (T) and the receiver (R) are located in the trunks slab. The region $z > 0$ is the free space with permitivity ε_0 and zero conductivity. The region $z < -(D_1 + D_2)$ is the ground with permitivity ε_g and conductivity σ_g and

$$\varepsilon_{gc} = \varepsilon_g + \sigma_g / i\omega\varepsilon_0$$

6.4.1 The field of a vertical electric dipole in an uniaxial anisotropic medium

Before further discussion of the model in Figure 6-3, we need to find out the field of a vertical dipole in an uniaxial anisotropic medium.

The Maxwell equations in this medium are

$$\nabla \times \overline{E} = -j\omega\mu_0\overline{H} \tag{6.66}$$
$$\nabla \times \overline{H} = j\omega\varepsilon_0\overline{\overline{\varepsilon}} \cdot \overline{E} + \overline{J} \tag{6.67}$$
$$\nabla \cdot (\overline{\overline{\varepsilon}} \cdot \overline{E}) = 0 \tag{6.68}$$
$$\nabla \cdot \overline{H} = 0 \tag{6.69}$$

with

$$\overline{\overline{\varepsilon}} = \overline{\overline{a}}_x\varepsilon_{hc} + \overline{\overline{a}}_y\varepsilon_{hc} + \overline{\overline{a}}_z\varepsilon_{vc} \tag{6.70}$$

According to the vector identity $\nabla \cdot (\nabla \times \overline{\Pi}_e) = 0$ and $\nabla \cdot \overline{H} = 0$, we can introduce an electric Hertz vector $\overline{\Pi}_e$ so that

$$\overline{H} = i\omega\varepsilon_0\varepsilon_{hc}\nabla \cdot \overline{\Pi}_e \tag{6.71}$$

Substitution of (6.71) into (6.66) yields

$$\overline{E} = k_0^2\varepsilon_{ha}\overline{\Pi}_e + \nabla\phi \tag{6.72}$$

Where ϕ is an unknown. For the sake of simplification, set

$$\phi = \nabla \cdot \overline{\Pi}_e \tag{6.73}$$

If the current source is a vertical electric dipole Idl located at $(0, 0, h)$, then $\overline{\Pi}_e$ has only a z component, namely,

$$\overline{\Pi}_e = \overline{a}_z\Pi_z \tag{6.74}$$

Substitution of (6.71)-(6.74) into (6.67) yields an inhomogeneous wave equation in form of

$$\nabla^2\Pi_z + k_0^2\varepsilon_{Vc}\Pi_z + \frac{\varepsilon_{vc} - \varepsilon_{hc}}{\varepsilon_{hc}}\frac{\partial^2\Pi_z}{\partial z^2} = \frac{Idl}{i\omega\varepsilon_0\varepsilon_{hc}}\delta(x)\delta(y)\delta(z-h) \tag{6.75}$$

If the unbounded medium is air, then $\varepsilon_{vc} = \varepsilon_{hc} = 1$, $\Pi_z = \Pi_0 z$, and then, from (6.75),

$$\nabla^2 \Pi_z + k_0^2 \Pi_z = \frac{Idl}{i\omega\varepsilon_0}\delta(x)\delta(y)\delta(z-h) \tag{6.76}$$

We can find the solution of (6.76), from [34], as follows:

$$\Pi_{0z} = \frac{Idl}{4\pi i\omega\varepsilon_0}\int_0^\infty e^{-u_0|z-h|}\frac{\lambda}{u_0}J_0(\lambda\rho)d\lambda \tag{6.77}$$

with

$$u_o = (\lambda^2 - k_0^2)^{1/2}, \quad R_e(u_0) \geq 0$$

Now the introduction of

$$z' = (\kappa)^{1/2} z, \qquad \kappa = \frac{\varepsilon_{hc}}{\varepsilon_{vc}} \tag{6.78}$$

will change (6.75) into the form of

$$\nabla'^2 \Pi_z + k_0^2 \varepsilon_{vc} \Pi_z = \frac{Idl}{i\omega\varepsilon_0\varepsilon_{hc}}\delta(x)\delta(y)\delta(z-h) \tag{6.79}$$

with

$$\nabla'^2 = \nabla_t^2 + \frac{\partial^2}{\partial z'^2}$$

The solution of (6.79) is

$$\Pi_z = \frac{Idl}{4\pi i\omega\varepsilon_0\varepsilon_{hc}}\int_0^\infty e^{-V|z-h|}\frac{\lambda}{V}J_0(\lambda\rho)d\lambda \tag{6.80}$$

with

$$V = (\lambda^2\kappa - k_0^2\varepsilon_{hc})^{1/2}, \qquad Re(V) \geq 0$$

Substitution of (6.80) into (6.71) yields the nonzero field components as follows:

$$\begin{aligned} H_\phi &= \frac{Idl}{4\pi}\int_0^\infty e^{-V(z-h)}\frac{\lambda^2}{V}J_1(\lambda\rho)d\lambda \\ E_\rho &= \frac{Idl}{4\pi i\omega\varepsilon_0\varepsilon_{hc}}\int_0^\infty e^{-V(z-h)}\frac{\lambda^2}{V}J_1(\lambda\rho)d\lambda \\ \frac{E_\rho}{H_\phi} &= \frac{V}{i\omega\varepsilon_0\varepsilon_{hc}} = K \end{aligned} \tag{6.81}$$

which means that the wave in an uniaxial anisotropic medium can be viewed as a plane wave propagating along the uniaxial axis, z, and the wave number is V, the characteristic impedance is K.

The field of a vertical electric dipole located in trunk region

For the vertical electric dipole located in the trunk region, $-(D_1 + D2) < z < -D_1$, as shown in Figure 6-3, the field everywhere can be derived from an electric Hertz Vector with z component Π_{iz} only.

In the air region, $z > 0$, from [34],

$$\Pi_{0z} = \frac{Idl}{4\pi i\omega\varepsilon_0} \int_0^\infty T_0(\lambda)e^{-u_0 z}\frac{\lambda}{u_0} J_0(\lambda\rho)d\lambda \tag{6.82}$$

and the nonzero fields in the air region are given by, from (6.72),(6.73),(6.71),

$$E_{0\rho} = \frac{\partial^2\Pi_{0z}}{\partial\rho\partial z}, \quad E_{0z} = (k_0^2 + \frac{\partial^2}{\partial z^2})\Pi_{0z}, \quad H_{0\phi} = -i\omega\varepsilon_0\frac{\partial\Pi_{0z}}{\partial\rho} \tag{6.83}$$

In the ground region, $z < -(D_1 + D_2)$,

$$\Pi_{gz} = \frac{Idl}{4\pi i\omega\varepsilon_0\varepsilon_{gc}} \int_0^\infty T_g(\lambda)e^{u(z+D_1+D_2)}\frac{\lambda}{u} J_0(\lambda\rho)d\lambda \tag{6.84}$$

and, from (6.71), (6.71),

$$E_{g\rho} = \frac{\partial^2\Pi_{gz}}{\partial\rho\partial z}, \quad E_{gz} = (k_0^2 + \frac{\partial^2}{\partial z^2})\Pi_{gz}, \quad H_{g\phi} = -i\omega\varepsilon_0\varepsilon_{gc}\frac{\partial\Pi_{gz}}{\partial\rho} \tag{6.85}$$

In trunks region, $-(D_1 + D_2) < z < -D_1$, from (6.80),

$$\Pi_{2z} = \frac{Idl}{4\pi i\omega\varepsilon_0\varepsilon_{hc_2}} \int_0^\infty \left[e^{-(V_2|z+D_1-h|)} + A_2(\lambda)e^{V_2(z+D_1)} + B_2(\lambda)e^{-V_2(z+D_1)}\right]\frac{\lambda}{V_2} J_0(\lambda\rho)d\lambda \tag{6.86}$$

Where the first term is the direct wave, the second and the third terms are the waves propagating along $+z$ and ρ direction, respectively, since the trunk slab has two boundaries.

The fields in trunks, from (6.71), (6.73), (6.66), take the forms of

$$H_{2\phi} = \frac{Idl}{4\pi} \int_0^\infty J_1(\lambda\rho)\left(e^{-(V_2|z+D_1-h|)} + A_2(\lambda)e^{V_2(z+D_1)} + B_2(\lambda)e^{-V_2(z+D_1)}\right)\frac{\lambda^2}{V_2}d\lambda$$
$$E_{2\rho} = \frac{\partial^2\Pi_{2z}}{\partial\rho\partial z}, \quad E_{2z} = (k_0^2\varepsilon_{hc_2} + \frac{\partial^2}{\partial z^2})\Pi_{2z} \tag{6.87}$$

In the foliage slab, $-D_1 < z < 0$,

$$\Pi_{1z} = \frac{Idl}{4\pi i\omega\varepsilon_0\varepsilon_{hc_1}} \int_0^\infty \left(A_1(\lambda)e^{-V_1(z+D_1)} + B_1(\lambda)e^{V_1(z+D_1)}\right)\frac{\lambda^2}{V_1} J_0(\lambda\rho)d\lambda \tag{6.88}$$

and, from (6.71), (6.73), (6.67),

$$H_{1\phi} = \frac{Idl}{4\pi} \int_0^\infty J_1(\lambda\rho)\Big(A_1(\lambda)e^{-V_1(z+D_1)} + B_1(\lambda)e^{V_1(z+D_1)}\Big)\frac{\lambda^2}{V_1} d\lambda \tag{6.89}$$

$$\tag{6.90}$$

where h is negative and

$$\begin{aligned} V_1 &= (\lambda^2\kappa_1 - k_0^2\varepsilon_{hc_1})^{1/2}, \quad \kappa_1 = \frac{\varepsilon_{hc_1}}{\varepsilon_{vc_1}}, \quad Re(V_1) \geq 0 \\ V_2 &= (\lambda^2\kappa_2 - k_0^2\varepsilon_{hc_2})^{1/2}, \quad \kappa_2 = \frac{\varepsilon_{hc2}}{\varepsilon_{vc2}}, \quad Re(V_2) \geq 0 \\ u_0 &= (\lambda^2 - k_0^2)^{1/2}, \quad k_0 = \omega\sqrt{\mu_0\varepsilon_0}, \quad Re(u_0) \geq 0 \\ u &= (\lambda^2 - k_0^2\varepsilon_{gc})^{1/2}, \quad Re(u) \geq 0 \end{aligned} \tag{6.91}$$

The six unknowns $T_0(\lambda)$, $T_g(\lambda)$, $A_1(\lambda)$, $B_1(\lambda)$, $A_2(\lambda)$ and $B_2(\lambda)$, can be determined by the continuity of the tangential electric and the magnetic fields at the tree interfaces, namely,

$$\begin{aligned} (E_{0\rho} - E_{1\rho})|_{z=0} &= 0, & (H_{0\phi} - H_{1\phi})|_{z=0} &= 0 \\ (E_{1\rho} - E_{2\rho})|_{z=-D_1} &= 0, & (H_{1\phi} - H_{2\phi})|_{z=-D1} &= 0 \\ (E_{2\rho} - E_{g\rho})|_{z=-(D_1+D2)} &= 0, & (H_{2\phi} - H_{g\phi})|_{z=-(D_1+D_2)} &= 0 \end{aligned} \tag{6.92}$$

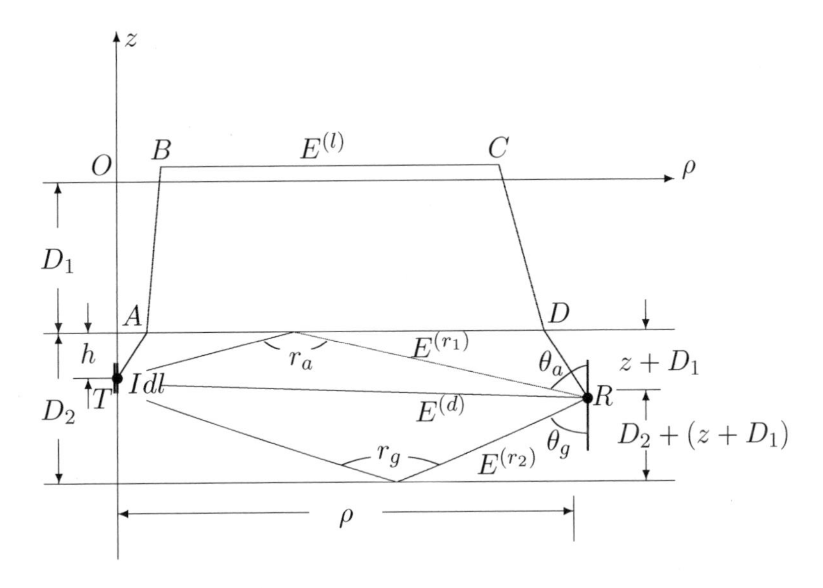

Figure 6-4 Propagation wave in forest model in UHF

Substitution of (6.83), (6.85), (6.87),(6.90) into (6.92) yields

$$A_1(\lambda) = \left(\frac{V_1}{V_2 S}\right)\frac{1 + R_{10}}{1 + R_0 e^{-2V_1 D_1}} R_0 e^{V_2 h}\left[1 + R_g e^{-V_2(2D_2+2h)}\right]$$

$$B_1(\lambda) = \left(\frac{V_1}{V_2 S}\right)\frac{1+R_{10}}{1+R_0 e^{-2V_1 D_1}} R_0 e^{-V_2(2D_2-h)}\left[1+R_g e^{-V_2(2D_2+2h)}\right]$$
$$A_2(\lambda) = R_{10} e^{V_2 h}\left[1+R_g e^{-V_2(2D_2+2h)}\right]\Big/S$$
$$B_2(\lambda) = R_g e^{-V_2(2D_2+h)}\left[1+R_{10} e^{2V_2 h}\right]\Big/S$$
$$T_0(\lambda) = \left(\frac{u_0}{V_2 S}\right)\frac{(1+R_0)(1+R_{10})}{1+R_0 e^{-V_1 D_1}} e^{-V_1 D_1} e^{V_2 h}\left[1+R_g e^{-V_2(2D_2+2h)}\right]$$
$$T_g(\lambda) = \frac{u(1+R_g)}{V_2 S} e^{-V_2(D_2+h)}\left[1+R_{10} e^{2V_2 h}\right]$$
$$S = 1 - R_{10} R_g e^{-2V_2 D_2}$$

where R_0, R_{10}, R_g are the reflection coefficients at the interfaces $z = 0$, $z = -D_1$, $z + -(D_1 + D_2$, respectively, as shown in Figure 6-4, namely,

$$R_0 = \frac{K_1 - K_0}{K_1 + K_0}, \qquad K_1 = \frac{V_1}{i\omega\varepsilon_0\varepsilon_{hc1}}, \qquad K_0 = \frac{u_0}{i\omega\varepsilon_0}$$
$$R_g = \frac{K_2 - K_g}{K_2 + K_g}, \qquad K_2 = \frac{V_2}{i\omega\varepsilon_0\varepsilon_{hc2}}, \qquad K_g = \frac{u}{i\omega\varepsilon_0\varepsilon_{gc}}$$
$$R_{10} = \frac{K_2 - Z_{10}}{K_2 + Z_{10}}, \qquad Z_{10} = K_1\frac{K_0 + K_1\tanh(V_1 D_1)}{K_1 + K_0\tanh(V_1 D_1)} \tag{6.93}$$

Substitution of (6.93) into (6.86) yields

$$\Pi_{2z} = \frac{Idl}{4\pi i\omega\varepsilon_0\varepsilon_{vc2}} \int_0^\infty \left(e^{-V_2|z+D_1-h|} + A_2(\lambda)e^{V_2(z+D_1)} + B_2(\lambda)e^{-V_2(z+D_1)}\right)\frac{\lambda}{V_2}J_0(\lambda\rho)d\lambda \tag{6.94}$$

For the sake of the computation, the Bessel function in (6.94) can be replaced by the Hankel function as follows:

$$\Pi_{2z} = \frac{Idl}{4\pi i\omega\varepsilon_0\varepsilon_{hc2}} \int_0^\infty \frac{1}{2}\left(e^{-V_2|z+D_1-h|} + A_2(\lambda)e^{V_2(z+D_1)} + B_2(\lambda)e^{-V_2(z+D_1)}\right)\frac{\lambda}{V_2}H_0^{(2)}(\lambda\rho)d\lambda \tag{6.95}$$

Where, from (6.93),

$$A_2(\lambda) = \frac{\frac{K_2 - Z_{10}}{K_2 + Z_{10}} e^{V_2 h}\left[1 + \frac{K_2 - K_g}{K_2 + K_g} e^{-V_2(2D_2+2h})\right]}{1 - \frac{K_2 - Z_{10}}{K_2 + Z_{10}}\frac{K_2 - K_g}{K_2 + K_g} e^{-2V_2 D_2}}$$

$$B_2(\lambda) = \frac{\frac{K_2 - K_g}{K_2 + K_g} e^{-V_2(2D_2+h)} \left[1 + \frac{K_2 - Z_{10}}{K_2 + Z_{10}} e^{2V_2 h}\right]}{1 - \frac{K_2 - Z_{10}}{K_2 + Z_{10}} \frac{K_2 - K_g}{K_2 + K_g} e^{-2V_2 D_2}}$$

$$Z_{10} = K_1 \frac{K_0 + K_1 \tanh (V_1 D_1)}{K_1 + K_0 \tanh (V_1 D_1)} \tag{6.96}$$

It is easy to show, from [34], that in (6.94),

$$\frac{e^{(-ik_0\sqrt{\varepsilon_{vc_2}}R_d)}}{\sqrt{\kappa_2}R_d} = \int_0^\infty e^{-V_2|z+D_1-h|} \frac{\lambda}{V_2} J_0(\lambda\rho) d\lambda$$

where

$$R_d = [\rho^2 + \kappa_2(z + D_1 - h)^2]^{1/2}$$

therefore,

$$\Pi_{2z} = \Pi_{2z}^{(d)} + \Pi_{2z}^{(r_1)} + \Pi_{2z}^{(r_2)}$$

$$\Pi_{2z}^{(d)} = \frac{Idl}{4\pi i\omega\varepsilon_0\varepsilon_{hc2}\sqrt{\kappa_2}R_d} e^{-ik_0\sqrt{\varepsilon_{vc_2}}R_d}$$

$$\Pi_{2z}^{(r_1)} = \frac{Idl}{4\pi i\omega\varepsilon_0\varepsilon_{hc2}} \int_0^\infty \frac{1}{2} A_2(\lambda) e^{V_2(z+D_1)} \frac{\lambda}{V_2} H_0^{(2)}(\lambda\rho) d\lambda$$

$$\Pi_{2z}^{(r_2)} = \frac{Idl}{4\pi i\omega\varepsilon_0\varepsilon_{hc2}} \int_0^\infty \frac{1}{2} B_2(\lambda) e^{-V_2(z+D_1)} \frac{\lambda}{V_2} H_0^{(2)}(\lambda\rho) d\lambda \tag{6.97}$$

Where
(1) $\Pi_{2z}^{(d)}$ is the direct wave from the transmitter (T) to the receiver (R),
(2) $\Pi_{2z}^{(r_1)}$ is the reflected wave from the lower foliage-trunk interface,
(3) $\Pi_{2z}^{(r_2)}$ is the reflected wave from the upper trunk-ground interface.
Substitution of (6.86) into (6.87) gives us the direct electric field from T to R as follows:

$$E_{2z}^{(d)} \approx \frac{Idl}{4\pi\sqrt{\kappa_2}R_d} e{-}ik_0\sqrt{\varepsilon_{vc_2}}R_d \tag{6.98}$$

where the terms of R_d^{-3}, R_d^{-4} and R_d^{-5} have been neglected in (6.98) since the assumption of $\rho >> (D_1 + D_2)$.

Obviously, from (6.95), (6.96), there are two pairs of branch points located at $\lambda = \pm k_0(u_0 = 0)$ and $\lambda = \pm k_g(u = 0)$. The contribution of the integral (6.95) at the branch point $\lambda = k_0$ gives the lateral wave in air region since the relative permittivity in foliage is larger than that in air. However, the contribution of the integral (6.95) at the branch point $\lambda = k_g$ doesn't give the later wave in the ground region since the relative permittivity in trunks is less than that in ground, in gives the radiation wave in the ground and the radiation wave suffer a considerable attenuation since the ground is lossy.

It seems like, from (6.95), (6.96), that there is a pair of poles located at $\lambda = \pm\lambda_p$, which satisfy the equation

$$1 - R_{10}(\lambda_p)R_g(\lambda_p)e^{-2V_2D_2} = 0 \tag{6.99}$$

Numerical results reveal that there is no any pole in integral (6.95) since the amplitude of the second term in (6.99) is less than one. Indeed, equation (6.99) will give a pair of poles at low frequencies since the forest can be viewed as a single layer slab, at low frequencies the ground can be viewed as of good conductivity and the total reflection will occur at both interfaces in the forest slab. At UHF, however, the forest must be viewed as two layers of slab and the relative permittivity in trunks is somewhat less than that in foliage, and less than that in ground. Therefore, no total reflection occurs at both the trunk-ground interface and the trunk-foliage interface, and then the amplitude of the second term in (6.99) is less than one for any λ.

Therefore, we only consider the direct wave, the reflected wave and the lateral wave in our model.

6.4.2 Reflected wave of a vertical dipole located in trunks

Substitution of (6.97) into (6.87) yields that reflection wave from the lower foliage-trunk interface and from the upper trunk-ground interface, respectively, in form of

$$E_{2z}^{(r_1)} = E_0^{(r)} \int_0^\infty A_2(\lambda)E^{V_2(z+D_1)}\frac{\lambda^3}{V_2}H_0^{(2)}(\lambda\rho)d\lambda$$
$$E_{2z}^{(r_2)} = E_0^{(r)} \int_0^\infty B_2(\lambda)E^{-V_2(z+D_1)}\frac{\lambda^3}{V_2}H_0^{(2)}(\lambda\rho)d\lambda \tag{6.100}$$

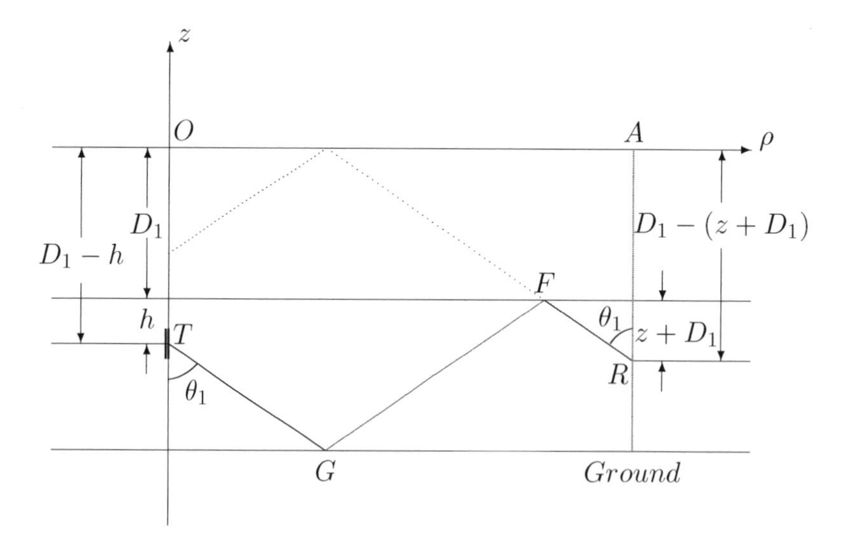

Figure 6-5 $r_1 = TGFR$, $\rho = OA$

where, from (6.97),

$$A_2(\lambda) = \frac{R_{10}}{S(\lambda)} e^{V_2 h}\left[1 + R_g e^{-V_2(2D_2+2h)}\right]$$
$$B_2(\lambda) = \frac{R_g}{S(\lambda)} e^{-V_2(2D_2+h)}\left[1 + R_{10} e^{2V_2 h}\right]$$
$$S(\lambda) = 1 - R_{10} R_g e^{-2V_2 D_2}$$
$$E_0^{(r)} = \frac{Idl}{8\pi i\omega\varepsilon_0\varepsilon_{hc2}} \qquad (6.101)$$

and R_{10}, R_g are given in (6.93).

The integral (6.100) is involved in the conformal transformation of the integrand from the λ-plane to the W-plane by using the transformation

$$\lambda = k_0\sqrt{\varepsilon_{hc2}} Sin\, W \qquad (6.102)$$

and then to obtain the final solution by the SPD (the steepest-descent method) as the form of

$$E_{2z}^{(r_1)} \approx \frac{-i\omega\mu_0 Idl}{4\pi\kappa_2^{3/2}}\left[\Gamma(\theta_a)\frac{e^{-ik_0\sqrt{\varepsilon_{vc_2}}r_a}}{r_a} + \Gamma(\theta_1)\frac{e^{-ik_0\sqrt{\varepsilon_{vc_2}}r_1}}{r_1}\right]$$
$$E_{2z}^{(r_2)} \approx \frac{-i\omega\mu_0 Idl}{4\pi\kappa_2^{3/2}}\left[\Gamma(\theta_g)\frac{e^{-ik_0\sqrt{\varepsilon_{vc_2}}r_g}}{r_g} + \Gamma(\theta_2)\frac{e^{-ik_0\sqrt{\varepsilon_{vc_2}}r_2}}{r_2}\right] \qquad (6.103)$$

where

$$\Gamma(\theta_a) = \frac{R_{10}(\theta_a)Sin^2\theta_a}{S(\theta_a)}, \qquad \Gamma(\theta_1) = \frac{R_{10}(\theta_1)R_g(\theta_1)Sin^2\theta_1}{S(\theta_1)}$$
$$\Gamma(\theta_g) = \frac{R_{10}(\theta_g)Sin^2\theta_g}{S(\theta_g)}, \qquad \Gamma(\theta_2) = \frac{R_{10}(\theta_2)R_g(\theta_2)Sin^2\theta_2}{S(\theta_2)} \qquad (6.104)$$

and

$$R_{10} = \frac{n\ Cos\ \theta - \sqrt{\varepsilon_{vc_2}}Z_{10}(\theta)}{n\ Cos\ \theta + \sqrt{\varepsilon_{vc_2}}Z_{10}(\theta)}$$
$$Z_{10} = \frac{n}{\sqrt{\varepsilon_{hc_1}}}\left(1 - \frac{\varepsilon_{vc2}}{\varepsilon_{vc_1}}\ Sin^2\theta\right)^{1/2} \cdot$$
$$\frac{\sqrt{\varepsilon_{hc_1}}\left(1 - \varepsilon_{vc_2}\ Sin^2\theta\right)^{1/2} + i\left(1 - \frac{\varepsilon_{vc_2}}{\varepsilon_{vc_1}}\ Sin^2\theta\right)^{1/2} tan\ \theta_0}{\left(1 - \frac{\varepsilon_{vc_2}}{\varepsilon_{vc_1}}\ Sin^2\theta\right)^{1/2} + i\sqrt{\varepsilon_{hc_1}}\left(1 - \varepsilon_{vc_2}\ Sin^2\theta\right)^{1/2} tan\ \theta_0}$$
$$\theta_0 = k_0\sqrt{\varepsilon_{hc_1}}\left(1 - \frac{\varepsilon_{vc_2}}{\varepsilon_{vc_1}}\ Sin^2\theta\right)^{1/2} D_1$$
$$R_g(\theta) = \frac{\sqrt{\varepsilon_{gc}}\ Cos\ \theta - \sqrt{\varepsilon_{hc_2}}\left(1 - \frac{\varepsilon_{vc_2}}{\varepsilon_{gc_2}}\ Sin^2\theta\right)^{1/2}}{\sqrt{\varepsilon_{gc}}\ Cos\ \theta + \sqrt{\varepsilon_{hc_2}}\left(1 - \frac{\varepsilon_{vc_2}}{\varepsilon_{gc_2}}\ Sin^2\theta\right)^{1/2}} \qquad (6.105)$$

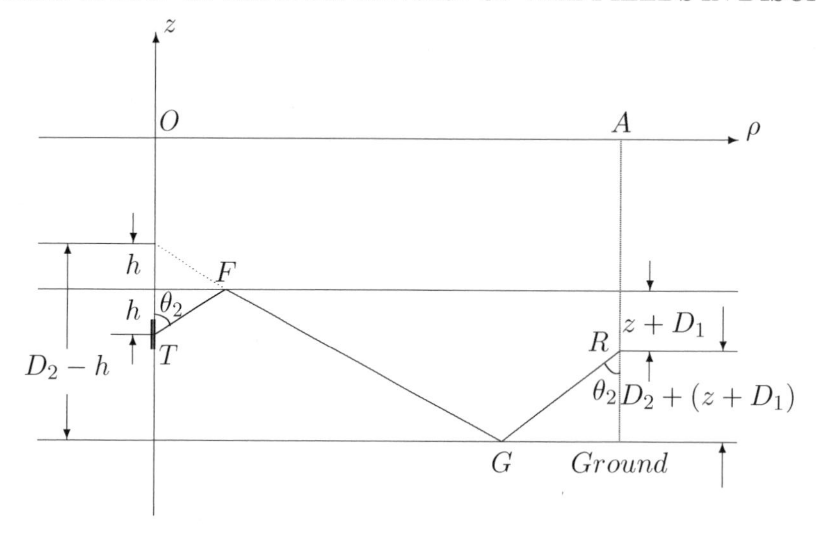

Figure 6-6 $r_2 = TGFR, \quad \rho = OA$

where θ will be θ_a, θ_1, θ_g, θ_2 for $\Gamma(\theta_a)$, $\Gamma(\theta_1)$, $\Gamma(\theta_g)$ $\Gamma(\theta_2)$, respectively. r_a, θ_a, r_1, θ_1, r_g, θ_g and r_2, θ_2 can be obtained by the geometric parameters from Figure 6-4 and Figure 6-5 and Figure 6-6 as follows:

$$\begin{aligned}
-\sqrt{\kappa_2}(z + D_1 + h) &= r_a \; Cos \; \theta_a \\
-\sqrt{\kappa_2}(z + D_1 - 2D_2 - h) &= r_1 \; Cos \; \theta_1 \\
-\sqrt{\kappa_2}(z + D_1 + 2D_2 + h) &= r_g \; Cos \; \theta_g \\
-\sqrt{\kappa_2}(z + D_1 + 2D_2 - h) &= r_2 \; Cos \; \theta_2.
\end{aligned} \tag{6.106}$$

Note, h and $(z + D_1)$ are negative and $S(\lambda)$ is given in (6.101).

Here, the solution of the reflected wave as given in (6.103) is obtained under the condition of

$$k_0|\sqrt{\varepsilon_{vc2}}|r >> 1, \qquad r = r_a, \; r_1, \; r_g \; or \; r_2 \tag{6.107}$$

Which is reasonable that we can use the steepest-descent method to obtain the solution.

6.4.3 Lateral wave of a vertical dipole located in the trunk region

As discussed last subsection, the contribution of the integral (6.95) at the branch point $\lambda = k_0$ gives the lateral wave in the air region ($z > 0$). To find the contribution, the integral (6.95) may be deformed around the branch point $\lambda = k_0$ as shown in Figure 6-7, namely,

$$\Pi_{2z}^{(l)} = \frac{Idl}{4\pi i\omega\varepsilon_0\varepsilon_{hc2}} \int_{C_0} \frac{1}{2}\left[A_2(\lambda)e^{V_2(z+D_1)} + B_2(\lambda)e^{-V_2(z+D_1)}\right]\frac{\lambda}{V_2}H_0^{(2)}(\lambda\rho)d\lambda \tag{6.108}$$

where the integral path C_0 is indicated in Figure 6-7. Substitution of (6.108) into (6.87) yields the lateral wave in air region as follows:

$$E_{2z}^{(l)} = \frac{Idl}{4\pi i\omega\varepsilon_0\varepsilon_{hc2}} \int_{C_0} \frac{1}{2}\left[A_2(\lambda)e^{V_2(z+D_1)} + B_2(\lambda)e^{-V_2(z+D_1)}\right]\frac{\lambda^3}{V_2}H_0^{(2)}(\lambda\rho)d\lambda \quad (6.109)$$

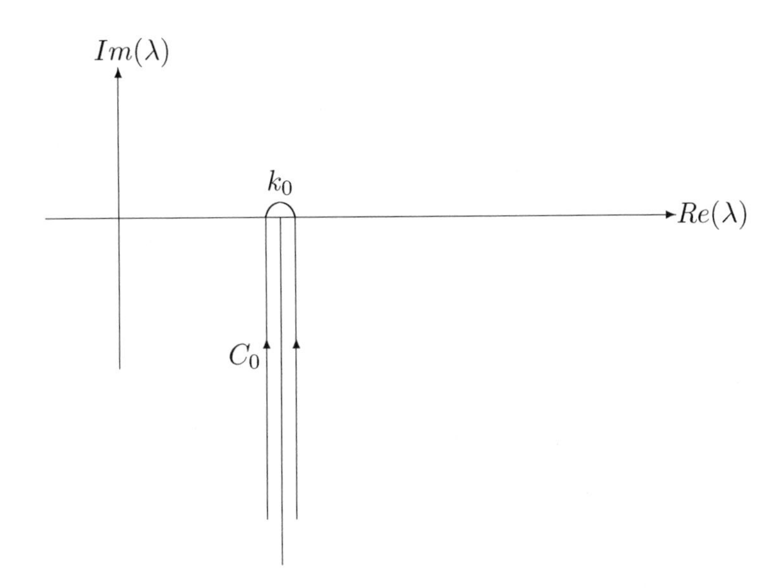

Figure 6-7 Deformed branch for lateral wave

Expending $A_2(\lambda)$ and $B_2(\lambda)$ into Taylor series, the integral in (6.109) yields an approximate form as follows:

$$E_{2z}^{(l)} = E_0\ f(p)\ G_s(h)\ G_s(z) \quad (6.110)$$

with

$$E_0 = \frac{-i\omega\mu_0 Idl}{2\pi\rho}e^{-ik_0\rho} \quad (6.111)$$

$$f(p) = \frac{-2}{2p}$$

$$p = -ik_0\rho\frac{\left[\triangle_2 + \frac{K_1}{\eta}\ tanh(V_1D_1)\right]^2}{1 - tanh^2(V_1D_1)}\Bigg|_{\lambda=k_0}$$

$$\triangle_2 = Z_2/\eta$$

$$G_s(h) = \frac{1}{\varepsilon_{vc2}}\frac{e^{V_2h}\left[1 + R_g e^{-V_2(2D_2+2h)}\right]}{1 + R_g e^{-V_2D_2}}\Bigg|_{\lambda=k_0}$$

$$G_s(z) = \frac{1}{\varepsilon_{vc2}} \frac{e^{V_2(D_1+z)}\left[1 + R_g e^{-V_2(2D_2+2D_1+2z)}\right]}{1 + R_g e^{-V_2 D_2}}\Bigg|_{\lambda=k_0} \tag{6.112}$$

Obviously, $E_{2z}^{(l)}$ is a lateral wave, the amplitude of $E_{2z}^{(l)}$ is inversely proportional to the square of the distance ρ, and the dominant phase shift of $E_{2z}^{(l)}$ is $-ik_0\rho$, which means that the dominant path of the lateral wave is in the air.

In (6.110), E_0 is twice the field of a vertical dipole in free space. $f(p)$ is the Sommerfeld attenuation function, which accounts for the imperfect conductivity of the ground and the presence of two slabs (foliage and trunks), and where

$$Z_2 = K_2 \frac{K_g + K_2\ tanh(V_2 D_2)}{K_2 + K_g\ tanh(V_2 D_2)} \tag{6.113}$$

is the surface impedance on the upper foliage-trunks interface and $\triangle_2 = Z_2/\eta$ is the so-called normalized surface impedance, which accounts for the impedance conductivity of the ground and the trunks slab.

$G_s(h)$ is the Height-Gain for the dipole height h and $G_s(z)$ is the Height-Gain for the observer height $(D_1 + h)$. The Height-Gain accounts for the influence of the reflected wave from the ground on the lateral wave. As we expected both $G_s(h)$ and $G_s(z)$ have the same formula since both the source and the observer are located in the same slab.

6.4.4 Lateral wave of a vertical dipole located in foliage

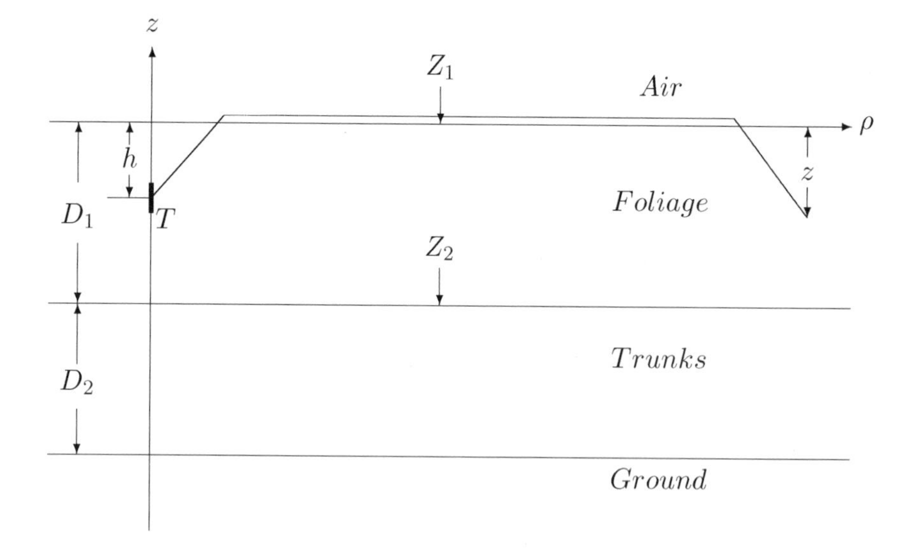

Figure 6-8 Geometry of the lateral wave

In both the transmitting dipole and the observer located in foliage layer, the lateral wave may be obtained by the same procedure as last subsection, namely,

$$E_{2z}^{(l)} \approx E_0 \; f(p) \; G_s(h) \; G_s(z) \tag{6.114}$$

with

$$\begin{aligned}
E_0 &= \frac{-i\omega\mu_0 I dl}{2\pi\rho} e^{-ik_0\rho} \\
f(p) &= -\frac{1}{2} \\
p &= -ik_0\rho\Delta_1^2/2 \\
\Delta_1 &= \frac{Z_1}{\eta} \\
G_s(h) &= \frac{1}{\varepsilon_{vc1}} \frac{e^{V_1 h}\left[1 + R_2 e^{-V_1(2D_1+2h)}\right]}{1 + R_g e^{-V_1 D_1}}\Bigg|_{\lambda=k_0} \\
G_s(z) &= \frac{1}{\varepsilon_{vc1}} \frac{e^{V_1 z}\left[1 + R_g e^{-V_1(2D_1+2z)}\right]}{1 + R_g e^{-V_1 D_1}}\Bigg|_{\lambda=k_0} \\
R_2 &= \frac{K_1 - Z_2}{K_1 + Z_2} \\
Z_2 &= K_2 \frac{K_g + K_2 \; tanh(V_2 D_2)}{K_2 + K_g \; tanh(V_2 D_2)} \\
Z_1 &= K_1 \frac{Z_2 + K_1 \; tanh(V_1 D_1)}{K_1 + Z_2 \; tanh(V_1 D_1)}
\end{aligned}$$

and h and z are the embedded height of the transmitting dipole and the observation point below the air-foliage interface as shown in Figure 6-8, and both h and z are negative.

6.4.5 Numerical results for the frequency range of $f < 200MHz$

The parameters of the forest in UHF obtained by R. H. Lang, A. Schneider, S. Seker, and F. J. Altman [28] is used to calculate the propagation loss of the lateral wave, the direct wave and the reflected wave as following discussion.

To simplify the analysis, one can make some assumptions as follows:

(1) Assume the distance, from the transmitting antenna (T) to the receiving antenna R, $\rho >> (D_1 + D_2)$ (the height of the forest), then the radiowave can be viewed as propagating in parallel to the forest floor. In this case, there is no depolarization of the electromagnetic wave in the forward scattering direction by vertical trunks (from R. H. Lang, A. Schneider, S. Seker, and F. J. Altman [28] pp.4-6).

(2) Assume the branches and leaves in each tree are distributed uniformly in

azimuth. In this case, there is no depolarization in the forward scattering direction by branches or leaves (from R. H. Lang, A. Schneider, S. Seker, and F. J. Altman [28] pp.4-11 and pp. 4-14).

In these assumption above, the ensemble of each forest component (trunks, branches or leaves) can be represented as a uniaxial anisotropic slab, with an effective tensor permittivity in each slab $\overline{\overline{\varepsilon}}_i$ as follows:

$$\overline{\overline{\varepsilon}}_i = \begin{bmatrix} \varepsilon_{hci} & & \\ & \varepsilon_{hci} & \\ & & \varepsilon_{vci} \end{bmatrix} \qquad (i = 1, 2) \tag{6.115}$$

where

$$\begin{aligned} \varepsilon_{hci} &= \varepsilon_{hi} + \sigma_{hi}/i\omega\varepsilon_0 \qquad (i = 1, 2) \\ \varepsilon_{vci} &= \varepsilon_{vi} + \sigma_{vi}/i\omega\varepsilon_0 \qquad (i = 1, 2) \end{aligned} \tag{6.116}$$

and σ_{hi} and ε_{hi} are the horizontal conductivity and relativity, respectively.

After we found the electric field E_{2z} for the lateral wave,the direct wave and the reflected wave, respectively, we may express the radiowave propagation loss by the definition from T. Tamir [35], as follows:

$$L = 10\ log_{10} \left[\left| \frac{2I}{dlE_{2z}} \right|^2 \frac{R(Z_T)}{R_0} \frac{R(Z_R)}{R_0} \cdot R_0^2 \right] \tag{6.117}$$

where E_{2z} may be the lateral wave, the direct wave and the reflected wave, respectively. with its propagation loss L in dB. $R(Z_T)$ and $R(Z_R)$ are the radiation resistances of the transmitting dipole and the receiving dipole, respectively, at the height Z_T and Z_R with respect to the ground, and

$$R_0 = 80 \left(\frac{Idl}{\lambda} \right)^2 \tag{6.118}$$

is the radiation resistance of a small dipole in unbounded free space (vacuum). $R(Z_T)/R_0$ and $R(Z_R)/R_0$ can be obtained from the curve given by Vogler, L. E., and J. L. Noble [36].

To compare the propagation loss of the lateral wave with the results given by David Anthony Rogers, A. J. Giarola, et al., [37] we use the same parameters as David Anthony Rogers, A. J. Giarola, et al., which are as follows:

$$\begin{aligned} &\varepsilon_{hc1} = \varepsilon_{vc1} = 1.12 - i\ 0.12 \times 10^{-3}/\omega\varepsilon_0 \\ &\varepsilon_{hc2} = \varepsilon_{vc2} = 1.03 - i\ 0.03 \times 10^{-3}/\omega\varepsilon_0 \\ &\varepsilon_{gc} = 20 - i\ 10 \times 10^{-3}/\omega\varepsilon_0 \\ &D_1 = 9m, \qquad D_2 = 1m \\ &(a)\ Z_T = Z_R = 10m, \qquad (b)\ Z_T = Z_R = 3m \end{aligned}$$

and are from R. H. Lang, A. Schneider, S. Seker, and F. J. Altman citeLang82 in the frequency range of $2MHz$ to $250MHz$.

Substitution of (6.114) and (6.118) into (6.117) gives the propagation loss of the lateral wave of a vertical dipole as given in Table 8-1 (for $Z_T = R_R = 3m$) and Table 8-2 (for $Z_T = R_R = 10m$).

Obviously, the theoretical results obtained by author L_{author} is in good agreement with the experimental results L_{Log} from David Anthony Rogers, A. J. Giarola, et al.

[37]. From which we understand that the theoretical parameters of forest in UHF from R. H. Lang, A. Schneider, S. Seker, and F. J. Altman citeLang82 is in good agreement with the experiment data.

Table 8-1 Table 8-1 $(Z_T = Z_R = 3\,m)$

$f(MHz)$	2	5	10	20	50	100	200	250
$L_{author}(dB)$	59	79	92	98	104	121	131	127
$L_{Tamir(dB)*}$	70	77	86	98	109	120	130	130
$\triangle L$	-11	+2	+6	0	-5	1	+1	-3

L_{Tamir} is the experimental data used by Tamir [25] and David Anthony Rogers, A. J. Giarola, et al.

[37] except the data 70,

L_{author} is author's results.

Table 8-1 Table 8-2 $(Z_T = Z_R = 10\,m)$

$f(MHz)$	2	5	10	20	50	100	200	250
$L_{author}(dB)$	54.42	75	86	91	95	111	135	133
$L_{Tamir(dB)*}$	57.5	72	82	88	97	107	123	127
$\triangle L$	-3.08	+3	+4	3	-2	4	+12	+6

L_{Tamir} is the experimental data used by Tamir [25] and David Anthony Rogers, A. J. Giarola, et al., [37],

L_{author} is author's results.

6.4.6 Numerical results for the frequency range of $f < 2000MHz$

The theoretical parameters of the forest from R. H. Lang, A. Schneider, S. Seker, and F. J. Altman (Lang 82) [28] in the frequency range of $200MHz$ to $2000MHz$ has been used by author to calculate the wireless wave propagation loss in the forest.

According to (Lang 82) [28], the mean (or the coherent) field satisfies Maxwell's equations "in the mean" and the ensemble of discrete scatters (such as the trunks, the branches or the leaves) can be replaced by an equivalent continuous medium described by an effective dyadic permittivity $\overline{\overline{\varepsilon}}$, or alternatively, by an effective dyadic susceptibility in the form of

$$\overline{\overline{\varepsilon}} = \overline{\overline{I}} + \overline{\overline{X}} \tag{6.119}$$

Generally, $\overline{\overline{\varepsilon}}$ is in form of

$$\overline{\overline{X}} = \sum X_{\alpha\beta} \overline{\alpha}^0 \overline{\beta}^0, \quad \overline{\alpha}^0, \overline{\beta}^0 \ will\ be\ \{\overline{h}^0, \overline{V}^0, \overline{i}^0\} \tag{6.120}$$

namely, $\overline{\overline{X}}$is possesses nine components, where $\overline{h}^0$, $\overline{V}^0$, $\overline{i}^0$ are the unit vector in the direction of horizontal, vertical and propagation, respectively.

Under the assumptions we used before, that $\rho >> (D_1 + D_2)$ and the branches and leaves in each tree distribute uniform in azimuth, $\overline{\overline{X}}$ becomes very simple, namely

$$\overline{\overline{X}} = \begin{bmatrix} X_{hh} & & \\ & X_{hh} & \\ & & X_{vv} \end{bmatrix} \quad (i = 1, 2) \tag{6.121}$$

since

$$X_{hv} = X_{vh} = 0$$

or alternatively, the ensemble of each forest component (trunks, branches or leaves) can be replaced as an uniaxial anisotropic slab with an effective dyadic permittivity in form of

$$\overline{\overline{\varepsilon}}_i = \begin{bmatrix} \overline{\overline{\varepsilon}}_{hci} & & \\ & \overline{\overline{\varepsilon}}_{hci} & \\ & & \overline{\overline{\varepsilon}}_{vci} \end{bmatrix} \quad (i = 1, 2) \tag{6.122}$$

In trunks slab

$$\begin{aligned} \varepsilon_{hc2} &= 1 + X_{hhT} \\ \varepsilon_{vc2} &= 1 + X_{vvT} \end{aligned} \tag{6.123}$$

In foliage slab

$$\begin{aligned} \varepsilon_{hc1} &= 1 = X_{hhB} + X_{hhL} \\ \varepsilon_{vc1} &= 1 + X_{vvB} + X_{vvL} \end{aligned} \tag{6.124}$$

where $\overline{\overline{X}}_T$, $\overline{\overline{X}}_B$ and $\overline{\overline{X}}_L$ are the effective dyadic susceptibility of trunks, branches and leaves, respectively, and all of therm have the same form as (6.121).

The calculation of $\overline{\overline{X}}_T$, $\overline{\overline{X}}_B$ and $\overline{\overline{X}}_L$ is involved in a lot of complicated formulae, fortunately, those calculation has been completed by (Lang 82) [28] and the calculation results in Figure 4-1, Figure 4-2, Figure 4-3 and Figure 4-4 in (Lang 82) [28]. Therefore, the parameters $\overline{\overline{X}}_T$, $\overline{\overline{X}}_B$ and $\overline{\overline{X}}_L$ may be obtained from those figures above.

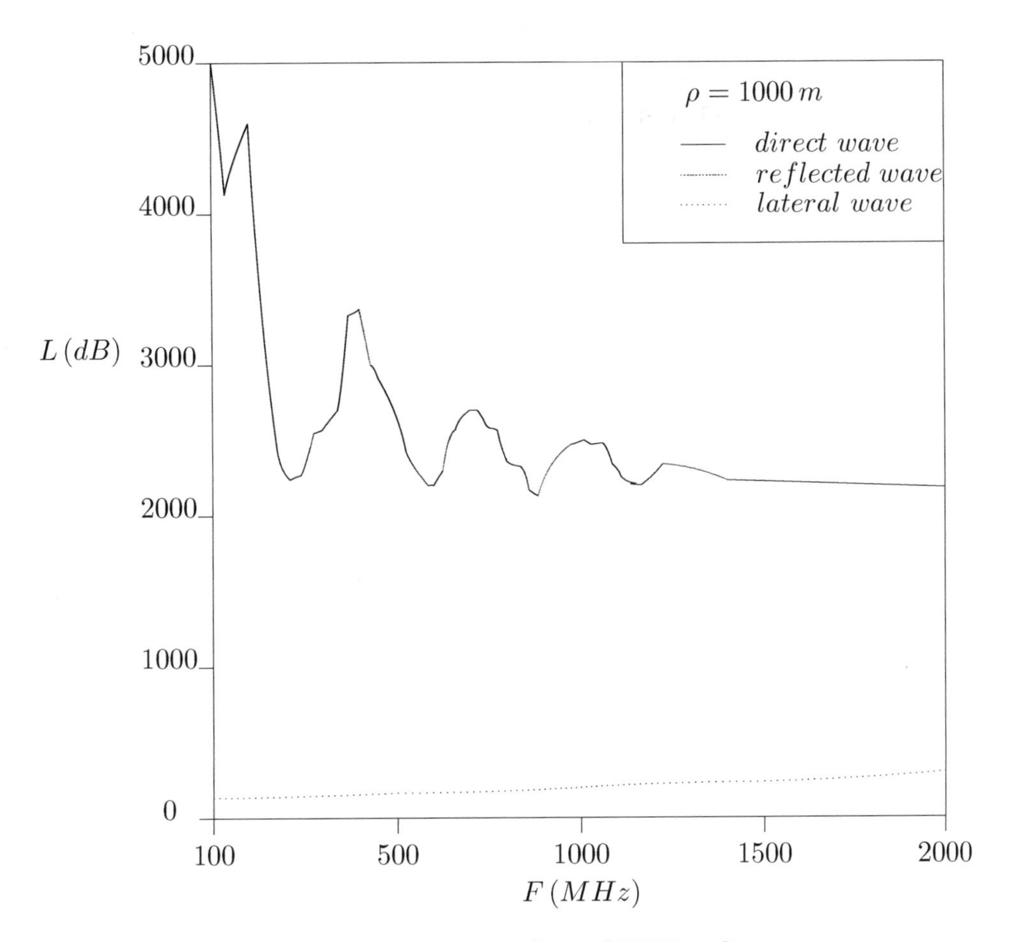

Figure 6-9 Propagation loss of UHF radiowave

The geometry parameters of the trunks, the branches and the leaves are:

$$a_T = 10.0cm, \quad \rho_T = 1/m^2 \qquad (for trunks)$$
$$a_B = 1.0cm, \quad l = 1.0cm, \quad \rho_B = /m^3, \quad \theta_B = 45^0 \qquad (for branches)$$
$$a_L = 5.0cm, \quad t = 1.0mm, \quad \rho_L = 200/m^3, \quad \theta = 30^0 \qquad (for leaves)$$

where $a_i, (i = T, B, l)$ is the ratio of each tree, each branch and each leave, respectively, l is the length of each branch, t is the thickness of each leave, $\rho_i, (i = T, B, L)$ is the probability density of the trunk, the branches and the leaves, respectively. And $\theta_B = 45^0$ means all branches have the same polar angle 45^0, $\theta = 30^0$ means that the probability density function of leaves is assumed to be uniform in azimuth and uniform in elevation angle over the range $0 - 30^0$ degrees.

The permittivity of the air is ε_0. The relative permittivity and the conductivity of the ground are

$$\varepsilon_g = 20, \qquad \sigma_g = 0.01 mho/m$$

and the free space susceptibility μ_0 is everywhere.

The geometry parameters of the forest model in Figure 6-3 are

$$D_1 = 6m, \quad D_2 = 4m, \quad h = z + D_1 = -1m, \rho = 1000m$$

Substitution of ε_{hc_1}, ε_{vc_1}, ε_{hc_2}, ε_{vc_2}, ε_{gc}, ε_0, μ_0, D_1, D_2, h and $(z + D_1)$ into (6.98), (6.100), (6.110) and then into (6.117), one can obtains the propagation loss of the direct wave, reflected wave and the lateral wave in UHF $(f = 200 \sim 2000MHz)$ as indicated in Figure 6-9.

Obviously, the propagation loss of the lateral wave is much lower than that of the direct wave and the reflected wave when the distance is 1000 m. In this case, the propagation path in the air region is the dominate part for the lateral wave, however, the dominate propagation path is in the forest region for the direct wave and the reflected wave. Which is valid at least for frequency $f < 2000MHz$ and the distance $\rho = 100m$. If the distance $\rho = 50m$, however, the propagation loss of the lateral wave will be close to that of the direct wave and the reflected wave since the propagation path in the air region is not the dominate part for the lateral wave.

6.5 Variational method in anisotropic medium

Now we will discuss the operator theory in anisotropic medium, in which, we will deduce a variational formula for the propagation constant in a waveguide with any transverse cross section filled with lossy anisotropic dielectric, as indicated in Figure 6-10, by means of the operator theory.

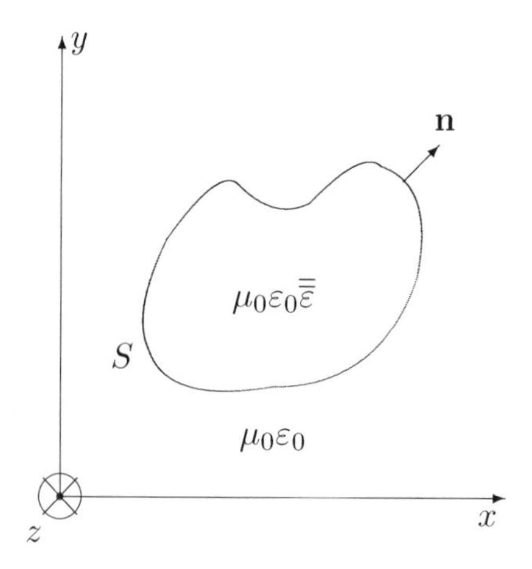

Figure 6-10 The transverse cross section of non-regular dielectric waveguide

6.5.1 The self-adjointness of the operator

To find the variational formula for the propagation constant, we'd better to discuss the self-adjointness of the operator first.

To this end, a lossy non-regular dielectric waveguide is shown in Figure 6-10, in which, the relative dielectric constant $\bar{\bar{\varepsilon}}$ is a complex tensor as follows:

$$\bar{\bar{\varepsilon}} = \begin{bmatrix} \varepsilon_{xx} & \varepsilon_{xy} & \varepsilon_{xz} \\ \varepsilon_{yx} & \varepsilon_{yy} & \varepsilon_{yz} \\ \varepsilon_{zx} & \varepsilon_{zy} & \varepsilon_{zz} \end{bmatrix} \tag{6.125}$$

and the field in the waveguide satisfies the Maxwell equations

$$-\nabla \times \vec{\mathcal{E}} = i\omega\mu_0\vec{\mathcal{H}} \tag{6.126}$$

$$\nabla \times \vec{\mathcal{H}} = i\omega\varepsilon_0\bar{\bar{\varepsilon}} \cdot \vec{\mathcal{E}} \tag{6.127}$$

To set up the variational formula, we have to introduce an adjoint waveguide as a supplementary quantity of the original wave-guide. From the mathematic point of view, the variational principle is based on the generalized interaction between the original system and the adjoint system. From the physical point of view, the relationship between the original waveguide and the adjoint waveguide is based on the fact, that the modes are mutually orthogonal to each other in real waveguide.

In the adjoint waveguide, the Maxwell equations of the fields $\vec{\mathcal{E}}^a, \vec{\mathcal{H}}^a$ are

$$-\nabla \times \vec{\mathcal{E}}^a = i\,\omega\,\mu_0\,\vec{\mathcal{H}}^a \tag{6.128}$$

$$\nabla \times \vec{\mathcal{H}}^a = i\,\omega\,\varepsilon_0\,\bar{\bar{\varepsilon}}^{\,a} \cdot \vec{\mathcal{E}}^a \tag{6.129}$$

Where $\bar{\bar{\varepsilon}}^{\,a}$ is the relative dielectric constant of the filled medium in the adjoint waveguide and is a complex tensor. Obviously, from (6.126) to (6.129), the fields $\vec{\mathcal{E}}, \vec{\mathcal{H}}$ in the original waveguide and the fields $\vec{\mathcal{E}}^a, \vec{\mathcal{H}}^a$ in the adjoint waveguide are all complex numbers. In which the time factor $\exp(i\omega t)$ is omitted.

It is known, from Figure 6-10, that the tensor $\bar{\bar{\varepsilon}}(x, y)$ is the function of x and y only, since it is independent of z. That means that the waveguide is a two-dimensional structure and the relationship along z is $\exp(-\gamma z)$, i.e.

$$\vec{\mathcal{E}}(x, y, z) = \mathbf{E}(x, y)\exp(-\gamma z) \tag{6.130}$$

$$\vec{\mathcal{E}}^a(x, y, z) = \mathbf{E}^a(x, y)\exp(-\gamma^a z) \tag{6.131}$$

$$\vec{\mathcal{H}}(x, y, z) = \mathbf{H}(x, y)\exp(-\gamma z) \tag{6.132}$$

$$\vec{\mathcal{H}}^a(x, y, z) = \mathbf{H}^a(x, y)\exp(-\gamma^a z) \tag{6.133}$$

Where γ and γ^a are the propagation constants along the original waveguide and the adjoint waveguide, respectively. They are complex numbers as well, namely

$$\gamma = \alpha + i\,\beta \tag{6.134}$$

Then form (6.127),

$$\vec{\mathcal{E}} = \bar{\bar{\varepsilon}}^{\,-1} \cdot \frac{\nabla \times \vec{\mathcal{H}}}{i\omega\varepsilon_0} \tag{6.135}$$

Substituting (6.135) into (6.126), we have

$$\mathcal{L}\vec{\mathcal{H}} \equiv \nabla \times (\overline{\overline{\varepsilon}}^{-1} \cdot \nabla \times \vec{\mathcal{H}}) - \omega^2 \mu_0 \varepsilon_0 \vec{\mathcal{H}} = 0 \tag{6.136}$$

Similarly, from (6.128) and (6.129), we have

$$\mathcal{L}^a \vec{\mathcal{H}}^a \equiv \nabla \times (\overline{\overline{\varepsilon}}^{a-1} \cdot \nabla \times \vec{\mathcal{H}}^a) - \omega^2 \mu_0 \varepsilon_0 \vec{\mathcal{H}}^a = 0 \tag{6.137}$$

Where

$$\mathcal{L} = (\nabla \times \overline{\overline{\varepsilon}}^{-1} \cdot \nabla \times) \ - \omega^2 \mu_0 \varepsilon_0 \tag{6.138}$$

$$\mathcal{L}^a = (\nabla \times \overline{\overline{\varepsilon}}^{a-1} \cdot \nabla \times) \ - \omega^2 \mu_0 \varepsilon_0 \tag{6.139}$$

Obviously, the operator $\mathcal{L}$ will become $\mathcal{L}^a$ provided that $\overline{\overline{\varepsilon}}$ is replaced by $\overline{\overline{\varepsilon}}^a$.

Now the definition of the inner product is

$$< \mathbf{A}, \mathbf{B} >= \int\int_S \mathbf{A} \cdot \mathbf{B}\, dxdy \tag{6.140}$$

Where S is the transverse cross section of the waveguide, $\mathbf{A}$ is the field in the original waveguide or its linear transformation, $\mathbf{B}$ is the field in the adjoint waveguide or its linear transformation. It follows that the complex inner product is

$$< \mathbf{A}, \mathbf{B}^* >= \int\int_S \mathbf{A} \cdot \mathbf{B}^*\, dxdy \tag{6.141}$$

Since the integral of (6.140) and (6.141) are over the transverse cross section of the waveguide, $< \mathbf{A}, \mathbf{B} >$ or $< \mathbf{A}, \mathbf{B}^* >$ must be the function of x and y only. Which will come to true provided that $\exp\{\gamma z\}$ and $\exp\{\gamma^a z\}$ cancel each other out, where $\exp\{\gamma z\}$ is in the original waveguide and $\exp\{\gamma^a z\}$ is in the adjoint waveguide. To this end, it is required that

$$\gamma^a = -\gamma \quad (For\ the\ real\ inner\ product\ (6.140)) \tag{6.142}$$

$$\gamma^a = \gamma^* \quad (For\ the\ complex\ inner\ product\ (6.141)) \tag{6.143}$$

Which may be used to set up the relationship between the field f in the original waveguide and the field f^a in the adjoint waveguide.

For the non-adjoint operator:

The relationship between f and f^a is

$$< \mathcal{L}^a f^a, f >=< f^a, \mathcal{L} f > \tag{6.144}$$

Where the operators $\mathcal{L}$ and $\mathcal{L}^a$ satisfy the boundary conditions as follows:
(1) On the electric wall $\quad \mathbf{n} \times \vec{\mathcal{E}} = \mathbf{n} \times \vec{\mathcal{E}}^a = 0$
(2) On the magnetic wall $\quad \mathbf{n} \times \vec{\mathcal{H}} = \mathbf{n} \times \vec{\mathcal{H}}^a = 0$
(3) Radiation condition (for the unbounded system)

(6.145)

Where, $\mathbf{n}$ is the normal unit vector on the wall S of the waveguide as shown in Figure 6-10.

The conditions (1) and (2) are the nature boundary conditions. In condition (1), it is assumed that there is some physical symmetry surface or the perfect conductor boundary in the waveguide. If we take the lossy of the waveguide wall into account, the waveguide wall may be considered to be a lossy dielectric extended to infinity. In this case, we have to use the condition (3). The condition (2) is used for the calculation only, and may be used for the magnetic symmetry surface alone.

For the self-adjoint operator:

We have

$$< \mathcal{L} f^a, f > = < f^a, \mathcal{L} f > \tag{6.146}$$

Compared with (6.144) we understand that this implies that

$$\mathcal{L}^a = \mathcal{L} \tag{6.147}$$

or $\mathcal{L}$ is a self-adjoint operator. In this case,, from (6.138) and (6.139), we have

$$\bar{\bar{\varepsilon}}^a = \bar{\bar{\varepsilon}} \tag{6.148}$$

Which implies that, for the case of self-adjoint, the geometry shape of the transverse cross section for the original waveguide is the same as that for the adjoint waveguide. And then, from (6.136) and (6.137), we have

$$\begin{aligned} \mathbf{H}(x, y) &= \mathbf{H}^a(x, y) \\ \mathbf{E}(x, y) &= \mathbf{E}^a(x, y) \end{aligned} \tag{6.149}$$

Which implies that, for the case of self-adjoint, the adjoint fields are the same as the original fields, such that the order of the matrix will be reduced to the half. Consequently, we hope that the calculation is in the case of self-adjoint. However, payment for the gain is as follows:

(1) The medium could be lossy, however the tenor of the medium must be symmetry, namely, it is required that

$$\bar{\bar{\varepsilon}} = \bar{\bar{\varepsilon}}^T \tag{6.150}$$

and then the waveguide is reciprocal.

(2) Or the tenor $\bar{\bar{\varepsilon}}$ is asymmetry, however the tenor is Hermitian (i.e. the tenor is not lossy), namely, it is required that

$$\bar{\bar{\varepsilon}} = (\bar{\bar{\varepsilon}}^T)^* \tag{6.151}$$

Where $(\bar{\bar{\varepsilon}}^T)^*$ is the complex transpose of $\bar{\bar{\varepsilon}}$.

Now (6.150) is in correspondence with the real inner product, (6.151) is in correspondence with the complex inner product.

In this section, we are going to analyze a lossy dielectric waveguide. For the lossy dielectric waveguide, the $\overline{\overline{\varepsilon}}$ is a lossy one and the waveguide is reciprocal, the tenor of the medium must be symmetry, and then the inner product will be real.

Finally, we will discuss that which components of the filed should be contained in the variational formula. Four situations may be considered:

(1) The variational formula contains E and H (6 components).

(2) The variational formula contains E (3 components) alone.

(3) The variational formula contains H (3 components) alone.

(4) The variational formula contains E_z, H_z (2 components) only.

First, (1) and (2) are not suitable since E is discontinuous when E cross-over the boundary between two dielectrics and then the problem will become complicated since we have to take the discontinuity into account. However, H and E_z, H_z are continuous in the whole transverse cross section. In this case, it seems like that we'd better choice (4) since the number of the components is minimum. However, it is provable that for a normal symmetry tensor $\overline{\overline{\varepsilon}}$, the variational formula for (4) will give a properties equation with non-standard form. Fortunately, (3) will give a properties equation with standard form such that it is easy to obtain the solution from the properties equation.

6.5.2 The variational formula

First, (6.136) and (6.137) are able to be written as

$$\mathcal{L}\, f = 0 \tag{6.152}$$

$$\mathcal{L}^a\, f^a = 0 \tag{6.153}$$

Where f and f^a are the fields in the dominant waveguide and the adjoint waveguide, respectively. And $\mathcal{L}$ and $\mathcal{L}^a$ are the operators of the dominant waveguide and the adjoint waveguide, respectively.

To deduce the stable formula for the propagation constant, we have to consider the simultaneous solution of (6.152) and (6.153). According to the generalized interaction principle [2] the first order variational δI of the functional I, is able to be expressed as

$$\delta I = <\delta\, f^a,\, \mathcal{L}\, f> + <\mathcal{L}^a\, f^a,\, \delta\, f> \tag{6.154}$$

Where $\delta\, f$ and $\delta\, f^a$ are the variations of f and f^a, respectively.

Taking the integral of (6.154) with respect to the unknown fields f and f^a gives the functional

$$I = <f^a,\, \mathcal{L}\, f> \tag{6.155}$$

And then from (6.144), we have

$$I = <\mathcal{L}^a\, f^a,\, f> \tag{6.156}$$

[2] V. H Rumsey, "Reaction Concept in electromagnetic theory", Phys. Rev., Vol. 94, pp. 1483-1491, June. 1954.

For the self-adjoint situation

$$I =< \mathcal{L} f^a, f > \tag{6.157}$$

For the non-adjoint situation:

Now we will deduce the non-adjoint functional first, and the self-adjoint functional is a special situation of the non-adjoint functional only.

In order to make all of the stable quantities to be contained in the functional I, we may use the functional formula

$$I(f, f^a) = 0 \tag{6.158}$$

for a passive medium. As we mentioned before, we have made a decision to make use of the real inner product as well as that the functional formula contains H alone. In this case, substituting (6.137) into (6.156) and (6.158), we have

$$< \nabla \times (\overline{\overline{\varepsilon}}^{a-1} \cdot \nabla \times \vec{\mathcal{H}}^a), \vec{\mathcal{H}} > -\omega^2 \mu_0 \varepsilon_0 < \vec{\mathcal{H}}^a, \vec{\mathcal{H}} >= 0 \tag{6.159}$$

Consider the symmetry of the curl operation $\nabla\times$, i.e.

$$< \nabla \times (\overline{\overline{\varepsilon}}^{a-1} \cdot \nabla \times \vec{\mathcal{H}}^a), \vec{\mathcal{H}} >=< (\overline{\overline{\varepsilon}}^{a-1} \cdot \nabla \times \vec{\mathcal{H}}^a), \nabla \times \vec{\mathcal{H}} >$$

we have

$$< (\overline{\overline{\varepsilon}}^{a-1} \cdot \nabla \times \vec{\mathcal{H}}^a), \nabla \times \vec{\mathcal{H}} > -\omega^2 \mu_0 \varepsilon_0 < \vec{\mathcal{H}}^a, \vec{\mathcal{H}} >= 0 \tag{6.160}$$

This is the first functional formula.

Similarly, Substituting (6.136) into (6.155) and (6.158), considering the symmetry of the curl operation again, we have another functional formula as follows

$$< \nabla \times \vec{\mathcal{H}}^a, \overline{\overline{\varepsilon}}^{-1} \cdot \nabla \times \vec{\mathcal{H}} > -\omega^2 \mu_0 \varepsilon_0 < \vec{\mathcal{H}}^a, \vec{\mathcal{H}} >= 0 \tag{6.161}$$

From (6.160), it is easy to obtain the variational formula for the frequency ω as follows:

$$\omega^2 = \frac{< \overline{\overline{\varepsilon}}^{a-1} \cdot \nabla \times \vec{\mathcal{H}}^a, \nabla \times \vec{\mathcal{H}} >}{\mu_0 \varepsilon_0 < \vec{\mathcal{H}}^a, \vec{\mathcal{H}} >} \tag{6.162}$$

Note, this variational formula is for the non-adjoint situation, and then it is able to be used for any dielectric tensor. Unfortunately, the frequency from (6.162) is a complex value. For a time harmonic field, it is able to make the imaginary part as smaller as possible by means of the successive substitution. We have to do the same thing for the self-adjoint situation.

Now, to deduce the variational formula for the propagation constant from (6.160) and (6.161), we have to decompose the operator $\nabla\times$ into the sum of the transversal operator and the longitudinal operator as follows:

$$\nabla\times \equiv \nabla_t \times + \nabla_z\times \tag{6.163}$$

Where

$$\nabla_t = \frac{\partial}{\partial x}\mathbf{x}^0 + \frac{\partial}{\partial y}\mathbf{y}^0, \qquad \nabla_z = \frac{\partial}{\partial z}z^0 \tag{6.164}$$

and $\dfrac{\partial}{\partial z} = -\gamma$. Substituting (6.163) into (6.160), considering that the real inner product has to satisfy $\gamma^a = -\gamma$, we have

$$< \overline{\overline{\varepsilon}}^{a-1} \cdot \nabla_t \times \vec{\mathcal{H}}^a, \nabla_t \times \vec{\mathcal{H}} > - \gamma < \overline{\overline{\varepsilon}}^{a-1} \cdot \nabla_t \times \vec{\mathcal{H}}^a, \mathbf{z}^0 \times \vec{\mathcal{H}} >$$

$$+ \gamma < \overline{\overline{\varepsilon}}^{a-1} \cdot \mathbf{z}^0 \times \vec{\mathcal{H}}^a, \nabla_t \times \vec{\mathcal{H}} > - \gamma^2 < \overline{\overline{\varepsilon}}^{a-1} \cdot \mathbf{z}^0 \times \vec{\mathcal{H}}^a, \mathbf{z}^0 \times \vec{\mathcal{H}} >$$

$$-\omega^2 \mu_0 \varepsilon_0 < \vec{\mathcal{H}}^a, \vec{\mathcal{H}} > = 0 \qquad (6.165)$$

This is the non-adjoint variational formula for the propagation constant. Obviously, this is a properties equation with non-standard form as follows:

$$\gamma^2 \, Ax + \gamma \, Bx + Cx = 0 \qquad (6.166)$$

For the self-adjoint situation:

$$\overline{\overline{\varepsilon}}^a = \overline{\overline{\varepsilon}}$$

$$\mathbf{H}(x, y) = \mathbf{H}^a(x, y)$$

And then in (6.165) the γ terms cancel each other out. And (6.165) becomes

$$\gamma^2 = \frac{< \overline{\overline{\varepsilon}}^{-1} \cdot \nabla_t \times \mathbf{H}, \nabla_t \times \mathbf{H} > -\omega^2 \mu_0 \varepsilon_0 < \mathbf{H}, \mathbf{H} >}{< \overline{\overline{\varepsilon}}^{-1} \cdot \mathbf{z}^0 \times \mathbf{H}, \mathbf{z}^0 \times \mathbf{H} >} \qquad (6.167)$$

This properties equation is a standard form as

$$\gamma^2 \, Ax = Cx \qquad (6.168)$$

The integral form of (6.167) is

$$\gamma^2 = \frac{\int \int_S \overline{\overline{\varepsilon}}^{-1} \cdot (\nabla_t \times \mathbf{H}) \cdot (\nabla_t \times \mathbf{H}) dxdy - \omega^2 \mu_0 \varepsilon_0 \int \int_S \mathbf{H} \cdot \mathbf{H} dxdy}{\int \int_S \overline{\overline{\varepsilon}}^{-1} \cdot (\mathbf{z}^0 \times \mathbf{H}) \cdot (\mathbf{z}^0 \times \mathbf{H}) dxdy} \qquad (6.169)$$

Now we will prove (6.167) is stable. To this end, rewrite (6.167) as

$$< \overline{\overline{\varepsilon}}^{-1} \cdot \nabla_t \times \mathbf{H}, \nabla_t \times \mathbf{H} > -\omega^2 \mu_0 \varepsilon_0 < \mathbf{H}, \mathbf{H} >$$

$$= \gamma^2 < \overline{\overline{\varepsilon}}^{-1} \cdot \mathbf{z}^0 \times \mathbf{H}, \mathbf{z}^0 \times \mathbf{H} > \qquad (6.170)$$

and then to prove that when there exists a variation $\delta \mathbf{H}$ of $\mathbf{H}$ in (6.170), the first order variation $\delta \gamma^2$ of γ^2 in (6.170) should be zero, i.e. $\delta \gamma^2 = 0$.

To this end, when $\mathbf{H}$ is replaced by $\mathbf{H} + \delta \mathbf{H}$, γ^2 will be replaced by $\gamma^2 + \delta \gamma^2$. Substitution of those in (6.170) gives us

$$-(\gamma^2 + \delta \gamma^2) < \overline{\overline{\varepsilon}}^{-1} \cdot \mathbf{z}^0 \times (\mathbf{H} + \delta \mathbf{H}), \mathbf{z}^0 \times (\mathbf{H} + \delta \mathbf{H}) >$$

$$-\omega^2\mu_0\varepsilon_0 < \mathbf{H}+\delta\mathbf{H}, \mathbf{H}+\delta\mathbf{H} >$$

$$+ < \overline{\overline{\varepsilon}}^{-1}\cdot\nabla_t\times(\mathbf{H}+\delta\mathbf{H}), \nabla_t\times(\mathbf{H}+\delta\mathbf{H}) >= 0 \tag{6.171}$$

Then, subtracting (6.170) from (6.171) , neglecting the second order of γ^2, we have

$$< \delta\mathbf{H},\ \nabla\times(\overline{\overline{\varepsilon}}^{-1}\cdot\nabla\times\mathbf{H}) - \omega^2\mu_0\varepsilon_0\mathbf{H} > -\delta\gamma^2 < \overline{\overline{\varepsilon}}^{-1}\cdot\mathbf{z}^0\times\mathbf{H}, \mathbf{z}^0\times\mathbf{H} >= 0 \tag{6.172}$$

In which, from (6.136), the first term is zero. Considering that $< \overline{\overline{\varepsilon}}^{-1}\cdot\mathbf{z}^0\times\mathbf{H}, \mathbf{z}^0\times\mathbf{H} >\neq 0$ in general situation, we have $\delta\gamma^2 = 0$. This implies that (6.167) is stable.

Therefore, (6.167) is able to be used for finding γ, the propagation constant of the passive uniform dielectric waveguide with dielectric constant $\overline{\overline{\varepsilon}}$ and any transverse cross section. The error is better than the trial function **H** by one order. The condition to make use of this formula is that the relative dielectric tensor $\overline{\overline{\varepsilon}}$ has to be symmetry (complex) tensor, and it is deduced under the self-adjoint condition and real inner product.

The variational formula (6.165) is for the non-adjoint situation. In this case, we can use the real inner product or the complex inner product.

6.6 Appendix Hertz Vector

In a linear isotropic homogeneous medium, both the electric field and the magnetic field may be expressed with the same Hertz vector but different operator.

1. Electric Hertz vector $\mathbf{\Pi}$

In a linear isotropic homogeneous medium, we have $\nabla \cdot \mathbf{H} = 0$, which means that $\mathbf{H}$ is a curl field, and then it may be expressed by means of the curl of a vector Π as follows:

$$\mathbf{H} = i\omega\varepsilon\nabla \times \Pi \tag{6.173}$$

Substituting (6.173) into the Faraday's law

$$\nabla \times \mathbf{E} = -i\omega\mu\mathbf{H} \tag{6.174}$$

considering that $\nabla \times \nabla\phi \equiv 0$, we have

$$\mathbf{E} = k^2\mathbf{\Pi} - \nabla\phi \tag{6.175}$$

Substituting $\mathbf{E}$ and $\mathbf{H}$ into the Ampere's law

$$\nabla \times \mathbf{H} = i\omega\varepsilon\mathbf{E} + \mathbf{J} \tag{6.176}$$

we have

$$\nabla \times \nabla \times \mathbf{\Pi} = k^2\mathbf{\Pi} - \nabla\phi - i\frac{\mathbf{J}}{\omega\varepsilon} \tag{6.177}$$

Where ϕ is an arbitrary scale function and may be taken as

$$\phi = -\nabla \cdot \mathbf{\Pi}$$

And then (6.177) becomes

$$\nabla\nabla \cdot \mathbf{\Pi} - \nabla \times \nabla \times \mathbf{\Pi} + k^2\mathbf{\Pi} = i\frac{\mathbf{J}}{\omega\varepsilon} \tag{6.178}$$

In this case, (6.175) becomes

$$\mathbf{E} = \nabla\nabla \cdot \mathbf{\Pi} + k^2\,\mathbf{\Pi} \tag{6.179}$$

and (6.173) becomes

$$\mathbf{H} = i\omega\varepsilon\nabla \times \mathbf{\Pi} \tag{6.180}$$

In this case, the problem becomes to find the unique unknown $\mathbf{\Pi}$, which may be obtained by means of the equation (6.178) and the boundary conditions.

In infinite space, the solution of (6.178) is

$$\mathbf{\Pi} = \frac{-i}{4\pi\omega\varepsilon}\int \frac{\mathbf{J}\exp(-ikr)}{r}dr$$

which implies that, in general situation, $\mathbf{\Pi}$ is caused by the external current source $\mathbf{J}$. Consequently, $\mathbf{\Pi}$ is called the electric Hertz vector.

2. Magnetic Hertz vector $\mathbf{\Psi}$

In free source space, we have $\nabla \cdot \mathbf{E} = 0$ and we may expressed the field as

$$\mathbf{E} = -i\omega\mu\nabla \times \mathbf{\Psi} \tag{6.181}$$

$$\mathbf{H} = \nabla\nabla \cdot \mathbf{\Psi} + k^2\,\mathbf{\Psi} \tag{6.182}$$

Where Ψ satisfies

$$\nabla\nabla \cdot \mathbf{\Psi} - \nabla \times \nabla \times \mathbf{\Psi} + k^2\,\mathbf{\Psi} = 0 \tag{6.183}$$

In which, (6.181) is from $\nabla \cdot \mathbf{E} = 0$ directly. Substituting (6.181) into the Ampere's law (6.176), considering that $\mathbf{J} = 0$, and then considering that ψ is an arbitrary scale function and may be taken as follows

$$\psi = -\nabla \cdot \mathbf{\Psi}$$

we have (6.182). And then substituting (6.181) and (6.182) into the Faraday's law (6.174), we have (6.183).

Note, in free source space, all of $\mathbf{E}$ in (6.181) and $\mathbf{H}$ in (6.182) as well as $\mathbf{E}$ in (6.178) and $\mathbf{H}$ in (6.179) satisfy the Maxwell equations. Considering that the medium is a linear medium, these two solutions may be superposed to

$$\mathbf{E} = (\nabla\nabla \cdot + k^2)\mathbf{\Pi} - i\omega\mu\nabla \times \mathbf{\Psi} \tag{6.184}$$

$$\mathbf{H} = i\omega\varepsilon\nabla \times \mathbf{\Pi} + (\nabla\nabla \cdot + k^2)\mathbf{\Psi} \tag{6.185}$$

Obviously, this $\mathbf{E}$ and $\mathbf{H}$ still satisfy the Maxwell equations in linear isotropic homogeneous medium. Where $\mathbf{\Pi}$ satisfies (6.178) and $\mathbf{\Psi}$ satisfies (6.183).

Suppose that there exists an imaginary magneto fluid, then we have to add a term of magneto-fluid density in the right-hand side of (6.183). And then $\mathbf{\Psi}$ may be considered to be the magnetic potential caused by the magneto fluid. Consequently, $\mathbf{\Psi}$ is called the magnetic Hertz vector.

Chapter 7

Nonlinear wave in optical fiber comm. system

7.1 Introduction

Till now, we consider that all of the media in the electromagnetic field are linear media, in which:

(1) $\mathbf{J} = \sigma \mathbf{E}$ for the conducting medium,

(2) $\mathbf{D} = \varepsilon \mathbf{E}$ for the dielectric medium,

(3) $\mathbf{B} = \mu \mathbf{H}$ for the magnetic medium.

and σ, ε, μ are all constants. In this case, the wave equations are the linear wave equations, which leads to the propagation property of the linear waves in the linear medium.

Actually, the linear wave is an approximation solution under the following conditions:

(1) the power of the source is small and;

(2) the space of the field is big enough,

such that the amplitude of the wave is small enough that the non-linear part of the field is ignorable compared with the linear part. In this case, the waves are propagating without any distortion since the wave of different frequency is propagating in an uniform velocity. However, one situation is special that the waves are propagating in different velocity for different frequency. This medium is said to be the dispersion medium. In this case, the duration of the pulse will be extended when the wave of the pulse is propagating in the dispersion medium.

Nowadays, the powerful laser device and the single-mode optical fiber have been used in the world in these several ten years. And the field in the single-mode optical fiber is no long so small that the nonlinearity in this medium has to be considered, in other words, the nonlinearity in the medium has to be taken into account. In this case, the phenomena in this medium may be described by the nonlinear wave equations. Although the general theory and the stability of the nonlinear wave equations are still the research problem, the study of some nonlinear wave equations have resulted in some analytic solutions already.

Especially in the single-mode fiber , the diameter of the core in a single-mode fiber is $10\mu m$ only. Under the high power transmission, the power density is very large in the core. In this case, a lot of interesting phenomena such as the stimulated Raman scattering, the stimulated Brillouin scattering , the self-focus phenomenon, the Kerr effect, the four-photon mixture , the nonlinear absorption and the soliton transmission etc. will occur in the single-mode fiber.

Those phenomena are able to be described by the following nonlinear wave equations:[42]

(1) Nonlinear Schrödinger equation.

$$i\frac{\partial\varphi}{\partial\tau}+\frac{\omega''}{2}\frac{\partial^2\varphi}{\partial\xi^2}+Q|\varphi|^2\varphi=0 \tag{7.1}$$

(2) KdV equation.

$$\frac{\partial u}{\partial t}+\alpha\, u\frac{\partial u}{\partial x}+\mu\frac{\partial^3 u}{\partial x^3}=0 \tag{7.2}$$

(3) Burgers equation

$$\frac{\partial u}{\partial t}+\alpha\, u\frac{\partial u}{\partial x}-\mu\frac{\partial^2 u}{\partial x^2}=0 \qquad (\mu\geq 0) \tag{7.3}$$

(4) Nonlinear coupling wave equation [63],[64],[66]

$$\begin{aligned}\frac{\partial P_p}{\partial z}+\frac{1}{V_p}\frac{\partial P_p}{\partial t}&=-\frac{g_p}{KA}P_pP_s-\alpha_pP_p\\ \frac{\partial P_s}{\partial z}+\frac{1}{V_s}\frac{\partial P_s}{\partial t}&=\frac{g_s}{KA}P_pP_s-\alpha_sP_p\end{aligned} \tag{7.4}$$

or [67]

$$\begin{aligned}\frac{\partial Q^*}{\partial t}+V_{ph}\frac{\partial Q^*}{\partial t}+PQ^*&=-K_1K_sE_L^*\\ \frac{\partial E_s}{\partial z}-\frac{1}{V_s}\frac{\partial E_s}{\partial t}-\frac{\alpha}{2}E_s&=K_3E_LQ^*\\ \frac{\partial E_l}{\partial z}+\frac{1}{V_L}\frac{\partial E_L}{\partial t}+\frac{\alpha}{2}E_L&=K_2E_sQ\end{aligned} \tag{7.5}$$

In this chapter, we will discuss a lot of interesting nonlinear phenomena.

First, it is the same as the nonlinearity in the circuit, that the nonlinearity in the field will cause the new frequencies. Secondary, from the wave point of view, when the light intensity is lager enough, we have to take the nonlinear term in the refraction index of the single-mode fiber into account, i.e. in the single-mode fiber,

$$n=n_0+\frac{n^2}{2n_0}\,|\mathbf{E}|^2 \tag{7.6}$$

And then the field in the single-mode fiber has to satisfy the nonlinear wave equation

$$\nabla^2\mathbf{E}-\left(\frac{n_0}{c}\right)^2\mathbf{E}_{tt}=\left(\frac{n_2^2}{c}\right)^2(|\mathbf{E}|^2\,E)_{tt} \tag{7.7}$$

Where $(n_0/c)^2\mathbf{E}_{tt}$ is the linear term. $(n_2^2/c)^2(|\mathbf{E}|^2E)_{tt}$ is the nonlinear term. Similar to the circuit, this nonlinearity will cause the new frequencies. Obviously, the nonlinear term is proportional to the cube of the field E. Therefore, the nonlinearity in the single-mode fiber is caused only when the optical intensity is lager enough.

For instance, if the single-mode fiber is under an injection of the powerful pumping light, it will cause the vibration of the molecule in the single-mode fiber . In this case, the pumping wave (ω_p) and the vibration wave of the molecule (Ω) will cause the new frequencies such as the Stocks wave (ω_s) and the anti-Stocks wave (ω_a) etc. due to the effect of the nonlinear term in the right-hand side of (7.7). Where

$$\omega_s = \omega_p - \Omega$$

$$\omega_a = \omega_p + \Omega$$

The phenomenon of the Stocks wave caused by the pumping wave is called the Stimulated Raman Scattering . Of cause, the explanation of the Stimulated Raman Scattering by means of the wave point of view is not completed. We may give the explanation from the quantum point of view. Namely, when the pumping source (f_p) is injected into a single-mode fiber , it will cause the electrons on the external layer of the molecule transit from the lower energy level E_1 to the high energy level E_2. In this case, the electrons obtain the required energy $\triangle E = E_2 - E_1$ from the pumping source. This energy is able to be denoted by $\triangle E = h\nu$. Where ν is the wave number and h is the Planck's constant. And ν is related to the width of the two energy levels $\triangle E$. In this process, the required energy $\triangle E$ is able to be considered to produce a photo of the stocks wave $f_s = f_p - \nu$ while a pumping photon with energy hf_p is annihilated. This is the explanation from the quantum point of view for the Stimulated Raman Scattering. In this case, the interaction between the power of the pumping wave P_p and the power of the Stocks wave P_s is able to be expressed as (7.4).

The stimulated Brillouin scattering in a single-mode fiber is caused by the interaction between the optical fiber mode and the acoustic wave under an injection of a narrow pumping source. The acoustic wave Q may occurs under the condition that a narrow pumping light is injected into a single-mode fiber. And the Stocks wave $\omega_s = \omega_L - \Omega_a$ is produced via the nonlinear term in (7.7). In which ω_L is the frequency of the pumping light. Ω_a is the frequency of the acoustic wave. In this case, the interaction between the pumping wave E_L, acoustic wave Q and the Stocks wave E_s in the single-mode fiber is able to be described by the coupling wave equation (7.5).

As well known, the optical wave velocity in a single-mode fiber is related to the refraction index n and then is related to the light intensity under the injection of the powerful light $|\mathbf{E}|^2$ as described in (7.6). And then it will cause the distortion of the waveform for the incident wave, namely the wave will be squeezed some where and loose some where in the single-mode fiber. Which is said to be the self-focus phenomenon. The equation to describe this phenomenon is said to be the Schrödinger equation (7.1). Where φ represents the field intensity, $Q|\varphi|^2\varphi$ is caused by the nonlinear refraction index in (7.6).

Another nonlinear phenomenon is said to be the Soliton transmission . This is a transmission process, in which, the shape of a pulse signal is holden in the whole transmission process. As well known, the dispersion in a single-mode fiber will cause the pulse broadening. However, the nonlinearity in the single-mode fiber will cause the pulse narrowing. Under a certain condition, if the broadening caused by the dispersion is equal to the narrowing caused by the nonlinearity, then the two effects will be in balance. And then Soliton transmission occurs in the single-mode fiber . The equation to describe the Soliton transmission is said to be the KdV equation (7.2). In which, the second term denotes the nonlinearity effect and the third term denotes the dispersion effect.

The second term in Burgers equation (7.3) denotes the nonlinearity effect in the medium and the third term denotes the dissipation in the medium. That means that the nonlinearity will cause the pulse narrowing and the dissipation will cause the pulse smoothing. In other words, the dissipation in the medium will relieve the nonlinearity effect in the medium.

All of the nonlinearity effects mentioned above will be explained in the following section. However, we would like concentrate on the analytical approach of those nonlinear wave equations. In which,

(1) The characteristic method is normally used to find the solution of the nonlinear hyperbolic wave equation.

(2) The perturbation approach is used to find the far field of the nonlinear wave.

(3) And the inverse scattering method is used to find the solution of the initial value problem for the KdV equation and the Schrödinger equation.

The analytical approach of those nonlinear equations above have been discussed in a book "Nonlinear Waves" by the authors Tosiya Taniuti and Katsunobu Nishihara [38] in the area of physics and applied mathematics applied to particle physics, plasma physics, ecology and elsewhere.

Today, we will discuss the nonlinear waves in optical fiber transmission system based on the analytical technology given by Tosiya Taniuti and Katsunobu Nishihara [38] and author's study[63],[64],[66].

7.2 The contour method

The contour method or the characteristic curve method is a good idea. According to this idea, a nonlinear wave equation is able to be equivalent to two linear equations, which can be used not only for the linear wave equations, the nonlinear wave equations in this Chapter but also for the linear coupling wave equations in Chapter 9. Therefore, we are going to spend some time to explain some basic idea of the characteristic curve method so as to help reader to handle the key points. Fortunately, Tosiya Taniuti and Katsunobu Nishihara in [38] presents some examples, which can help us to understand the key points of the contour method or the characteristic curve method.

7.2.1 The contour method

The start point of the discussion about the nonlinear wave is by means of the conception of the linear wave.

1. Wave velocity of the linear wave:

For linear wave, the wave velocity

$$v = \frac{dx}{dt} = contant \tag{7.8}$$

that means that the linear wave is propagation along x direction with a constant velocity v no meter the wave amplitude u is large or small. Which is just the normal case for the optical fiber transmission system. In this case, the waveform of the pulse modulation signal is invariant in the whole transmission system. Which can be reached by using a small optical signal inside the optical single mode fiber.

For nonlinear wave, the wave velocity may be in the form of

$$v = \frac{dx}{dt} = \alpha u \tag{7.9}$$

that means that the wave velocity v of the nonlinear wave depends on the wave amplitude u, the larger the wave amplitude u the faster the wave velocity v. Which is different from the linear wave.

2. The moving trace of the linear wave:

For the linear wave, the moving trace of the wave can be obtained by integral the equation (7.8) as follows:

$$x - vt = x_i = contant, \quad (i = 1, 2, \cdots) \tag{7.10}$$

where x_i is the initial position of the wave amplitude u_i. Which can be indicated in Figure 7-1.

Obviously,

1. The moving traces $x - vt = x_i$ are parallel lines,

2. The wave amplitude u_i on each trace is invariant along the whole trace $x - vt = x_i$, which can be expressed as

$$\frac{du(x,t)}{dt} = 0 \tag{7.11}$$

where $\frac{d}{dt}$ is the directional derivative along the trace $x - vt = x_i$. In other words, the derivative of the wave amplitude $u(x,t)$ along the trace $x - vt = x_i$ is zero, namely,

$$du = \frac{\partial u}{\partial t}dt + \frac{\partial u}{\partial x}dx = 0 \tag{7.12}$$

where, from (7.9), the wave velocity is $v = \frac{dx}{dt}$, and then we have, from (7.12),

$$\frac{\partial u}{\partial t} + v\frac{\partial u}{\partial x} = 0 \tag{7.13}$$

Which is the wave equation for the linea wave. And the solution of (7.13) is

$$u(x,t) = f(x_i) = f(x - vt) \tag{7.14}$$

Therefore, the wave equation (7.13) may be equivalent to two equations as follows:

$$\frac{dx}{dt} = v = constant \tag{7.15}$$

$$\frac{du(x,t)}{dt} = 0 \tag{7.16}$$

where $\frac{d}{dt}$ in (7.16) is the directional derivative along the line

$$\varphi(x,t) = x - vt = \xi \tag{7.17}$$

which is obtained by the integral of (7.15). And then (7.16) denotes that the wave amplitude u is invariant on the line $\varphi(x,t) = x - vt = \xi$. Therefore, from physical point of view , each trace $\varphi(x,t) = x - vt = \xi$ is so called *a contour* , and from mathematic point of view, each trace is so called a characteristic curve . And this method is so called *the contour method* or *the characteristic curve method* .

The terms: "the characteristic curve method" or "the contour method"?

Perhaps people would like to make use of the term "the characteristic curve method" since this is the common term. However, for a student, some times it is not so easy to understand that "the characteristic curve" actually is a special wave moving trace, on which the wave amplitude is invariant. If we point out that "the characteristic curve" actually is a contour, namely the wave amplitude is invariant on a contour, that is easy to be understood and then it is easy to be used to discuss the moving trace of the nonlinear wave. Therefore, the author would like to give the students both of them: "the contour method" is easy to be understood, "the characteristic curve method" is the common term.

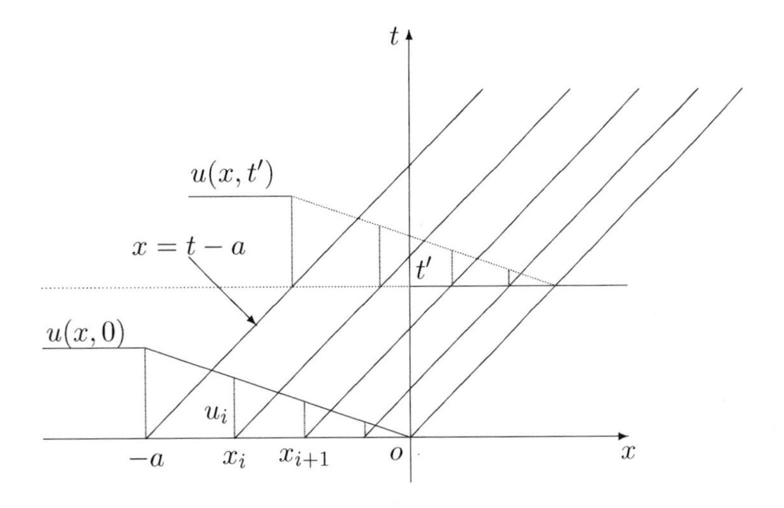

Figure 7-1 The moving trace of the linear wave

3. The moving trace of the nonlinear wave:

When we discuss the the KdV equation (7.2) and the Burgers equation (7.3), we understand that the first two terms of these two nonlinear wave equations are in form of

$$\frac{\partial u}{\partial t} + \alpha u \frac{\partial u}{\partial x} = 0 \tag{7.18}$$

In which, the wave velocity v is

$$\frac{dx}{dt} = v = \alpha u \tag{7.19}$$

where α is a constant and the integral of (7.19) gives

$$\varphi(x, t) = \xi(constant) \tag{7.20}$$

which is a family of contours (or a family of characteristic curves).

Now we will following the contour method of linear wave to discuss the nonlinear wave equation (7.18). An example shown us that a family of contours (or a family of characteristic curve) $\varphi(x, t) = \xi(constant)$ is indicated in Figure 7-2. In which,

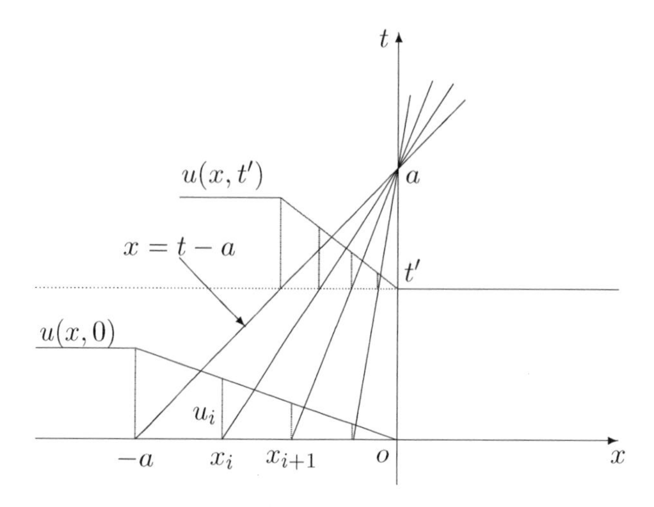

Figure 7-2 The moving trace of the nonlinear wave

1. The moving traces of the nonlinear wave are not parallel lines since the wave velocity depends on the wave amplitude. The "higher" the wave amplitude u_i the faster the wave velocity v_i, and then the moving traces are going to a intersection point a. This is a different from the moving traces of the linear wave.

2. However, the wave amplitude u_i is still invariant on each moving trace $\varphi(x,t) = \xi(constant)$, and then the directional derivative of u along each moving trace $\varphi(x,t) = \xi(constant)$ is still equal to zero, which can be expressed as

$$\frac{du(x,t)}{dt} = 0 \tag{7.21}$$

Where $\frac{d}{dt}$ denotes the directional derivative along the moving trace $\varphi(x,t) = \xi(constant)$.

4. The contour method (or the characteristic curve method):

And then, the nonlinear wave equation (7.18)

$$\frac{\partial u}{\partial t} + \alpha u \frac{\partial u}{\partial x} = 0 \tag{7.22}$$

may be equivalent to two linear equations as follows

$$v = \frac{dx}{dt} = \alpha u \tag{7.23}$$

$$\frac{du(x,t)}{dt} = 0 \tag{7.24}$$

Where the $\frac{d}{dt}$ in (7.24) is the directional derivative along the contour (or the characteristic curve) $\varphi(x,t) = \xi$. This is so called the contour method or (the characteristic curve method) for the nonlinear wave equation. In other words the contour method (or the characteristic curve method) can make the nonlinear equation (7.22) to be equivalent to two linear equations (7.23) and (7.24).

The discussion above is from physical point of view. Now we may prove the equation (7.24) in terms of the mathematic point of view as follows.

[Proof of (7.24)]:

Equation (7.20) tell us that $\varphi(x,t)$ is a constant along the contour (or the characteristic curve) $\varphi(x,t) = \xi(constant)$, and then

$$d\varphi = \frac{\partial \varphi}{\partial t}dt + \frac{\partial \varphi}{\partial x}dx = 0 \tag{7.25}$$

on the contour (or the characteristic curve). Substituting $\frac{dx}{dt} = v(x,t)$ into (7.25), we have

$$\frac{\partial \varphi}{\partial t} + v(x,t)\frac{\partial \varphi}{\partial x} = 0 \tag{7.26}$$

on the contour (or characteristic curve). In addition, we have

$$du = \frac{\partial u}{\partial t}dt + \frac{\partial u}{\partial x}dx \tag{7.27}$$

on the contour (or characteristic curve). Substituting $\frac{dx}{dt} = v(x,t)$ into (7.27), we have

$$\begin{aligned}\frac{du}{dt} &= \frac{\partial u}{\partial t} + v\frac{\partial u}{\partial x} = \frac{\partial u}{\partial \varphi}\frac{\partial \varphi}{\partial t} + v\frac{\partial u}{\partial \varphi}\frac{\partial \varphi}{\partial x} \\ &= \left(\frac{\partial \varphi}{\partial t} + v\frac{\partial \varphi}{\partial x}\right)\frac{\partial u}{\partial \varphi} = 0\end{aligned}$$

and then (7.24) is proved. Where use of (7.23) and (7.26) have been made.

Again, the nonlinear wave equation (7.22) is the first two terms of the KdV equation and the Schrödinger equation. In which, the second term is the nonlinear term. This term denotes that the wave velocity is a linear function of the wave amplitude u. The higher the wave amplitude u the faster the wave velocity v. And then the higher wave amplitude will overtake the lower wave amplitude in the single-mode fiber. This is the so called the phenomenon of "the wave overtaking" for the nonlinear wave. The intersection point a in Figure 7-2 has shown us the phenomenon of "the wave overtaking" already. Which will make the distortion for the going nonlinear wave in the single-mode fiber transmission system.

Now we will discuss the solution of a nonlinear wave equation by example.

Example 1 Find the solution for the following nonlinear wave equation

$$\begin{cases} \dfrac{\partial u}{\partial t} + u\dfrac{\partial u}{\partial x} = 0 \\ \\ u|_{t=0} = u(x,0) \end{cases} \tag{7.28}$$

Obviously, this equation may be equivalent to

$$\frac{dx}{dt} = u \tag{7.29}$$

$$\frac{du}{dt} = 0 \tag{7.30}$$

Where $\frac{d}{dt}$ is the derivative along the contour (or the characteristic curve), and the contour (or the characteristic curve) equation may be obtained by the integral of (7.29) as follows:

$$x = u(x,0)t + \varphi(x,t) \tag{7.31}$$

where the initial condition $u|_{t=0} = u(x,0)$ has been considered and $\varphi(x,t)$ is the integral constant of (7.29). Now if

$$u(x,0) = -x/a, \quad -a \le x \le 0 \tag{7.32}$$

the contour (or the characteristic curve) (7.31) is shown in Figure 7-2.

Where, from (7.31) and (7.32), the contours (or the characteristic curves) are all of the straight lines with different slope. When $u(x,t)$ move from the initial value $u(x,0)$ to $u(x,t')$, the wave amplitude on each contour (or characteristic curve) is invariant. In which the higher the wave amplitude u, the faster the wave velocity $v = u$. And then the phenomenon of "the wave overtaking" occurs and all of the contour (or the characteristic curves) come to a point $(x = 0, t = a)$. This point contains all of the waves with different velocity u, where u is the wave amplitude, that means that this point is a multi-value point and then it is non-determinate. Therefore, this point is nonsense for the nonlinear equation. This is one of the important feature of the nonlinear wave equation and is different from the linear wave equation. Therefore when we try to obtain the solution of the nonlinear wave equation, we should take this point into account.

Summery, from all of the discussion above, the wave amplitude u of the nonlinear wave equation is invariant on each contour (or on each characteristic curve). This is the key point of the contour method (or the characteristic curve method). The contours (or the characteristic curves) of the nonlinear wave equation (7.28) are straight lines going to a point $(x = 0, t = a)$. Except this point, the solution of the nonlinear wave equation can be obtained by means of the following discussion.

As we mentioned before, the nonlinear wave equation (7.28) may be equivalent to (7.29) and (7.30). Now the integral constant of (7.29) is

$$\varphi(x,t) = x - u(x,0)t$$

and then the integral of (7.30) give us

$$u = u(\varphi) = u[x - u(x,0)t] \tag{7.33}$$

Where φ is the integral constant from the integral of (7.29). Therefore this result satisfies the nonlinear wave equation as well as the initial condition in (7.28). The solution (7.33) is valid for all of the point (x,t) except the point $(x = 0, t = a)$.

7.2.2 The application of the contour method

1. **For the linear wave equation**

Example 2: Find the solution of the well known linear wave equation

$$\frac{\partial^2 u}{\partial t^2} - \frac{\partial^2 u}{\partial x^2} = 0$$

This linear equation is able to be expanded to

$$\Big(\frac{\partial}{\partial t} + \frac{\partial}{\partial x}\Big)\Big(\frac{\partial}{\partial t} - \frac{\partial}{\partial x}\Big)u = 0 \tag{7.34}$$

Let

$$\varphi \equiv \frac{\partial u}{\partial t}, \quad \psi \equiv \frac{\partial u}{\partial x} \tag{7.35}$$

then (7.34) becomes two first order equations

$$\Big(\frac{\partial}{\partial t} + \frac{\partial}{\partial x}\Big)(\psi - \phi) = 0 \tag{7.36}$$

$$\Big(\frac{\partial}{\partial t} - \frac{\partial}{\partial x}\Big)(\psi + \phi) = 0 \tag{7.37}$$

Obviously, the contour equation (or the characteristic curve equation) is that the wave velocity

$$v = \frac{dx}{dt} = \pm 1 \tag{7.38}$$

Where, the positive sign is corresponding to (7.36) and the negative sign is corresponding to (7.37). That means that $(\psi - \phi)$ is invariant on the contour (or the characteristic curve)

$$x - t = \xi \tag{7.39}$$

and $(\psi + \phi)$ is invariant on the contour (or the characteristic curve)

$$x + t = \eta \tag{7.40}$$

These can be written as

$$\psi - \phi = r(\xi) \tag{7.41}$$

$$\psi + \phi = S(\eta) \tag{7.42}$$

Where, ξ, η are the integral constants, and $r(\xi)$, $S(\eta)$ represent the function of ξ and η, respectively. Both of them are said to be *the Reman invariants*, which means that,

1. The wave amplitude $\psi - \phi$ is invariant on the contour $\xi = x - t$,
2. The wave amplitude $\psi - \phi$ is invariant on the contour $\eta = x + t$.

Given the initial conditions

$$\phi(x, 0) = \phi_0(x), \qquad \psi(x, 0) = \psi_0(x) \tag{7.43}$$

we have, from (7.41) and (7.42),

$$r(\xi) = \psi_0(\xi) - \phi_0(\eta)$$
$$S(\eta) = \psi_0(\eta) + \phi_0(\eta)$$

And then, from (7.41) and (7.42), we have

$$\begin{aligned}\phi &= -\frac{1}{2}[r(\xi) - S(\eta)] \\ &= -\frac{1}{2}[\psi_0(\xi) - \phi_0(\xi) - \psi_0(\eta) - \phi_0(\eta)]\end{aligned}$$

$$\begin{aligned}\psi &= \frac{1}{2}[r(\xi) + S(\eta)] \\ &= \frac{1}{2}[\psi_0(\xi) - \phi_0(\xi) + \psi_0(\eta) + \phi_0(\eta)]\end{aligned} \tag{7.44}$$

Since ϕ, ψ are the function of ξ and η, then from (7.35), u will be the function of ξ and η either, i.e.

$$u = f(\xi) + g(\eta) \tag{7.45}$$

Where f and g are unknown functions and can be found as follows.

Now, from (7.35), (7.45), (7.39) and (7.40), we have

$$\begin{aligned}\psi &= \frac{\partial u}{\partial x} = \frac{\partial u}{\partial \xi}\frac{\partial \xi}{\partial x} + \frac{\partial u}{\partial \eta}\frac{\partial \eta}{\partial x} = \frac{\partial u}{\partial \xi} + \frac{\partial u}{\partial \eta} \\ \phi &= \frac{\partial u}{\partial t} = \frac{\partial u}{\partial \xi}\frac{\partial \xi}{\partial t} + \frac{\partial u}{\partial \eta}\frac{\partial \eta}{\partial t} = -\frac{\partial u}{\partial \xi} + \frac{\partial u}{\partial \eta}\end{aligned} \tag{7.46}$$

Substituting (7.45) into (7.46) and (7.44) , rearranging the result, we have

$$\begin{aligned}\frac{\partial u}{\partial \xi} &= f'(\xi) = \frac{1}{2}\left[\psi_0(\xi) - \phi_0(\xi)\right] \\ \frac{\partial u}{\partial \eta} &= g'(\eta) = \frac{1}{2}\left[\psi_0(\eta) + \phi_0(\eta)\right]\end{aligned} \tag{7.47}$$

Considering (7.35) again, we have $\psi_0(\xi) = \bar{u}_0(\xi)$, $\psi_0(\eta) = \bar{u}_0(\eta)$ [1]. And then the integral of (7.47) is

$$f(\xi) = \frac{1}{2}u_0(\xi) - \frac{1}{2}\int_0^{x-t} \phi_0(\xi)d\xi$$

[1] Considering (7.35) we have

$$\psi_0(x) = \frac{du_0(x)}{dx}$$

which is equivalent to

$$\psi_0(\xi) = \bar{u}_0(\xi)$$

$$g(\eta) = \frac{1}{2}u_0(\eta) + \frac{1}{2}\int_0^{x+t} \phi_0(\eta)d\eta \tag{7.48}$$

Substituting (7.48) into (7.45) we obtain the following solution, which satisfies the initial conditions $u(x,0) = u_0(x)$, and $\left.\frac{\partial u}{\partial t}\right|_{t=0} = \phi_0(x)$ as follows

$$u = f(\xi) + g(\eta) = \frac{1}{2}[u_0(x-t) + u_0(x+t)] + \frac{1}{2}\int_{x-t}^{x+t} \phi_0(\eta)d\eta \tag{7.49}$$

This is the well known D' Alembert's solution.

2. **For the nonlinear wave equation**

Example 3: Find the solution of the following nonlinear wave equation by means of the contour method (or the characteristic curve method)

$$\frac{\partial u}{\partial t} + v(u)\frac{\partial u}{\partial x} = 0 \tag{7.50}$$

Obviously, the wave velocity is

$$\frac{dx}{dt} = v(u) \tag{7.51}$$

and then, the integral of (7.51) is

$$x = v(u)t + \varphi(x,t) \tag{7.52}$$

where $\varphi(x,t)$ is an integral constant. In this case, according to the contour method (or the characteristic method) the wave amplitude u is invariant on the contour (or on the characteristic curve) as follows

$$\varphi(x,t) = x - v(u)t = constant \tag{7.53}$$

and then the solution of $u(x,t)$ is

$$u(x,t) = f[\varphi(x,t)] = f[x - v(u)t] \tag{7.54}$$

Example 4: Find the solution of the following nonlinear wave equation by means of the contour method (or the characteristic curve method):

$$\frac{\partial u}{\partial t} + v(u)\frac{\partial u}{\partial x} = g(u) \tag{7.55}$$

Obviously, the wave velocity is

$$\frac{dx}{dt} = v(u) \tag{7.56}$$

and then, the integral of (7.56) is

$$x = v(u)t + \varphi(x,t) \tag{7.57}$$

where

$$\varphi(x,t) = x - v(u)t = \xi \tag{7.58}$$

is the integral constant.

Now (7.57) and

$$t = t \tag{7.59}$$

can be viewed as a parameter transformation and then we have

$$\frac{d}{dt} = \frac{\partial}{\partial x}\frac{\partial x}{\partial t} + \frac{\partial}{\partial t}\frac{\partial t}{\partial t} = v(u)\frac{\partial}{\partial x} + \frac{\partial}{\partial t} \tag{7.60}$$

and then, the equation (7.55) become

$$\frac{du}{dt} = g(u) \tag{7.61}$$

Obviously, considering that the wave amplitude $u(x,t)$ is invariant on the contour (or the characteristic curve) $\varphi(x,t) = x - v(u)t = \xi$, the solution of (7.61) is

$$u(x,t) = \int g(u)dt + f[\varphi(x,t)] = \int g(u)dt + f[x - v(u)t] \tag{7.62}$$

Example 5: Find the solution of the following nonlinear wave equation by means of the contour method (or the characteristic method)

$$\frac{\partial u}{\partial t} + \sum_{i=1}^{n} v_i(u)\frac{\partial u}{\partial x_i} = 0 \tag{7.63}$$

Obviously, the wave velocity is

$$\frac{dx}{dt} = v_i(u), \qquad (i = 1, 2, \cdots, n) \tag{7.64}$$

and then, the integral of (7.64) is

$$x = v_i(u)t + \varphi_i(x,t), \qquad (i = 1, 2, \cdots, n) \tag{7.65}$$

where

$$\varphi_i(x,t) = x - v_i(u)t = \xi, \qquad (i = 1, 2, \cdots, n) \tag{7.66}$$

is the integral constant, and the wave amplitude $u_i(x,t)$ is invariant on the contour (or the characteristic curve) $\varphi_i(x,t) = x - v_i(u)t = \xi$, namely

$$\frac{du_i}{dt} = 0, \qquad (i = 1, 2, \cdots, n) \tag{7.67}$$

where $\frac{d}{dt}$ is the derivative along the contour (or the characteristic curve) $\varphi_i(x,t) = x - v_i(u)t = \xi_i$.

Obviously, the wave amplitude u is consist of $u_i(x,t)$, $(i = 1, 2, \cdots, n)$, each $u_i(x,t)$ is invariant on each contour (or on each characteristic curve) $\varphi_i(x,t) = x - v_i(u)t = \xi$, and then u in (7.63) has to be expressed as a column matrix form U, namely,

$$U = (u_1, u_2, \cdots, u_n)^T \tag{7.68}$$

And then the nonlinear wave equation (7.63) may be expressed as a matrix form as follows

$$U_t + A(U)U_x = 0 \tag{7.69}$$

Where $A(U)$ is a $n\times n$ matrix, the elements in which are the functions of $u_1, u_2, \cdots, u_n$, x and t. And the n eigenvalues of A, $v_1, v_2, \cdots, v_n$, are the velocity of the n sub-wave, respectively. Which can be expressed as

$$\frac{dx}{dt} = v_i \qquad (i = 1, 2, \cdots, n) \tag{7.70}$$

The integral of (7.70) is a set of the contour (or the characteristic curve), which is

$$\varphi_i(x,t) = \xi_i \qquad (i = 1, 2, \cdots, n), \tag{7.71}$$

where ξ_i is the integral constant.

Considering (7.70), the derivative of (7.71) along the contour (or the characteristic curve) is

$$\varphi_{it} + v_i\, \varphi_{ix} = 0 \qquad (i = 1, 2, \cdots, n) \tag{7.72}$$

where the ith sub-wave u_i in (7.69) is invariant along the contour (or the characteristic curve) φ_i, and the solution of (7.69) should be

$$U = U(\varphi_i) \qquad (i = 1, 2, \cdots, n) \tag{7.73}$$

Substituting (7.73) into (7.69) gives

$$(\varphi_{it} + A\varphi_{ix})\frac{dU}{d\varphi_i} = 0 \qquad (i = 1, 2, \cdots, n) \tag{7.74}$$

Substituting (7.72) into (7.74), we have

$$(A - v_i\, I)\frac{dU}{d\varphi_i} = 0 \qquad (i = 1, 2, \cdots, n)$$

Where I is an unit matrix. This denotes that $\frac{dU}{d\varphi_i}$ is proportional to the right eigenvector R_i. Namely

$$\frac{dU}{d\varphi_i} \propto R_i \qquad (i = 1, 2, \cdots, n) \tag{7.75}$$

Assume that $R_i = [r_1^{(i)}, r_2^{(i)}, \cdots, r_n^{(i)}]^T$, and we may set $r_1^{(i)} = 1$ as the normalization. Then from the explicit formula of (7.75)

$$\begin{pmatrix} \frac{du_1}{d\varphi_i} \\ \vdots \\ \frac{du_n}{d\varphi_i} \end{pmatrix} \propto \begin{pmatrix} r_1^{(i)} \\ \vdots \\ \vdots \\ r_n^{(i)} \end{pmatrix}$$

and $r_1^{(i)} = 1$, we have

$$\varphi_i = u_1 \tag{7.76}$$

And then it is able to assume that

$$\frac{dU}{du_1} = R_i$$

Namely

$$\begin{pmatrix} \frac{du_1}{du_1} \\ \frac{du_2}{du_1} \\ \vdots \\ \frac{du_n}{du_1} \end{pmatrix} = \begin{pmatrix} r_1^{(i)} \\ r_2^{(i)} \\ \vdots \\ \vdots \\ r_n^{(i)} \end{pmatrix} \tag{7.77}$$

Since R_i is a definite function of $u_1, u_2, \cdots, u_n$, (7.77) is able to be viewed as an ordinary differential equation system of $n-1$ variables $u_2, \cdots, u_n$ with respect to an independent variable u_1. And then the solutions

$$u_k = u_k(u_1) \qquad (k = 1, 2, \cdots, n) \tag{7.78}$$

are able to be obtained.

Finally, from (7.76) and (7.72), it is known that u_1 satisfies

$$u_{1t} + v_i(u_1)\, u_{1x} = 0 \tag{7.79}$$

This equation is equivalent to the following two ordinary differential equations

$$\frac{dx}{dt} = v_i(u_1)$$

$$\frac{du_1}{dt} = 0$$

From these two equations as well as the initial conditions, it is able to obtain the solution of u_1. And then substituting the result into (7.78), we may obtain the solution of $u_k (k = 1, 2, \cdots, n)$.

7.3 The perturbation method for far field

——Nonlinear phenomena in single-mode fiber

When we study the nonlinearity in a long-haul optical fiber communication system, the wave in a single-mode fiber is a small amplitude wave and the nonlinear wave propagation in a single-mode fiber is able to be described approximately by the Burgers equation, the KdV equation or the Schrödinger equation. All of those are the far field phenomena with respect to the source. The far field phenomena is most important in the design of a long-haul optical fiber transmission system. Therefore, we have to spend more time to discuss the Burgers equation, the KdV equation and the Schrödinger equation.

Before those discussion, we would like to introduce some conception about the perturbation method and the G-M transformation method to deal with the first two terms nonlinear wave equation from the Burgers equation and the KdV equation since those are very useful for the discussion of the the Burgers equation, the KdV equation and the Schrödinger equation.

7.3.1 The perturbation method

The perturbation method has been used for the linear wave equation analysis. Now we will introduce the perturbation method for the nonlinear wave equation analysis under some conditions.

As we known, the Burgers equation can be modified to

$$u_t + \frac{\alpha}{2}(u^2)_x - \mu(u)_{xx} = 0 \tag{7.80}$$

The KdV equation may be modified to

$$u_t + \frac{\alpha}{2}(u^2)_x + \mu(u)_{xx} = 0 \tag{7.81}$$

For far field, $|x| \to \infty$, u is decaying in exponential law and then the nonlinear term $(u^2)_x$ in the Burgers equation and the KdV equation is smaller than the other two linear terms (since they are the first order of u). In this case, it is possible to find the solution of the Burgers equation and the KdV equation by means of the perturbation method, namely, the solution u can be expended into a power series of ε.

Usually, the perturbation method has to deal with three problems:

1. The solution u is able to be expanded to a constant $u^{(0)}$ with a power series of ε, in which the ε is related with the equation or the boundary condition,

2. Substitution of the series u into the equation and the boundary condition will give us a series equation, in which, the coefficient of each power of ε should be equal to zero. From which, we are able to find the first order solution u_1, then the second order solution u_2 until u_n step by step,

3. Finally, we have to check the convergence of the series to get the final solution

u.

This procedure is also good for finding the solution of the nonlinear wave equation under the condition of $|x| \to \infty$ as following discussion.

As we mentioned in (7.69), the nonlinear parabolic equation with n elements

$$U_t + A(U)U_x = 0$$

is able to be resulted in finding the n number of nonlinear equations

$$u_t + v_j(u)u_x = 0 \qquad (j = 1, 2, \cdots, n) \tag{7.82}$$

where the unknown u_1 in (7.79) has been written as u in (7.82). When the wave amplitude u is small, the solution u is able to be obtained by means of the perturbation method. Namely, let ε be a small parameter, then the solution u is able to be expanded to a constant u_0 with a power series of ε, i.e.

$$u = u_0 + \varepsilon u_1 + \varepsilon^2 u_2 + \cdots \tag{7.83}$$

Substituting (7.83) into (7.82), considering that the coefficient of each power of ε should be equal to zero, we have

$$O(\varepsilon): \qquad u_{1t} + v_0 u_{1x} = 0 \tag{7.84}$$

$$O(\varepsilon^2): \qquad u_{2t} + v_0 u_{2x} = -v_{u0} u_1 u_{1x} \tag{7.85}$$

Where $v_0 \equiv v(u_0)$, $v_{u0} \equiv \left.\dfrac{dv}{du}\right|_{u=u_0}$, and j has been omitted.

Before the further discussion, we have to introduce a G-M transformation method, which is a powerful method to deal with the nonlinear wave equation.

7.3.2 G-M transformation method

Now a transformation of

$$\xi = x - v_0 t, \qquad \tau = t \tag{7.86}$$

is helpful to transform the nonlinear equations (7.84) into an original differential equation as follows:

$$u_{1t} = 0 \tag{7.87}$$

And then the solution of this equation is

$$u_1 = f(\xi) \tag{7.88}$$

In this case, the nonlinear equation(7.85) becomes

$$u_{2t} = -v_{u0} u_1 u_{1x} = -v_{u0} f f_\xi \tag{7.89}$$

Therefore

$$u_2 = -v_{u0} f f_\xi t \tag{7.90}$$

Where, use of

$$\frac{\partial}{\partial t} = \frac{\partial}{\partial \tau}\frac{\partial \tau}{\partial t} + \frac{\partial}{\partial \xi}\frac{\partial \xi}{\partial t} = \frac{\partial}{\partial \tau} - v_0 \frac{\partial}{\partial \xi}$$
$$\frac{\partial}{\partial x} = \frac{\partial}{\partial \xi} \tag{7.91}$$

has been made in driving (7.87) and (7.89).

Now we should point out that the solution (7.90) is nonsense when the time t reach to an order of ε^{-1}. This is because that in this case, $|\varepsilon^2 u_2| \approx |\varepsilon u_1|$, and then the series expansion (7.83) is divergence.

To solve this problem, we may introduce a new transformation

$$\xi = x - v_{0j}t, \qquad \tau = \varepsilon t \tag{7.92}$$

to prolong the time scale. In this case, the coefficient of $O(\varepsilon)$ power is

$$u_{1\tau} + v_{u0}u_1u_{1\xi} = 0 \tag{7.93}$$

In which use of the transformation (7.92) has been made.

From physical point of view, the new transformation (7.92) can prolong the time scale so as in new coordinates system (ξ, τ) the intersection point $a(x, t)$ (or the nonsense point) in Figure 7-2 will be moved far away from the source area and then the contours (or the characteristic curves) of the nonlinear wave are approximately in parallel to each other.

For instance, before the new transformation (7.92), the phase velocity of the nonlinear wave equation (7.82) is able to be expanded to

$$\frac{dx}{dt} = v_j = v_{j0} + \left(\frac{\partial v_j}{\partial u}\right)\Big|_{u=u_0} \cdot \varepsilon u_1 + \cdots \tag{7.94}$$

In which, use of (7.83) has been made. And $v_{j0} \equiv v_j(u_0)$ and $\left(\partial v_j/\partial u\right)\Big|_{u=u_0}$ are constants. That means that the difference of the phase velocity between the nonlinear wave and the linear wave is an order of $O(\varepsilon)$.

After the new transformation (7.92), the phase velocity of the nonlinear wave equation (7.93)

$$\frac{d\xi}{d\tau} = \frac{dx}{d(\varepsilon t)} - \frac{v_{j0}}{\varepsilon} \approx \left(\frac{\partial v_j}{\partial u}\right)\Big|_{u=u_0} u_1 \tag{7.95}$$

is a constant. And then the contours (or the characteristic curves) of the nonlinear wave become parallel lines. Where use of (7.92) and (7.94) has been made in driving (7.95).

In other words, the new transformation (7.92) made the new time (τ) ε times smaller than the original time (t). Therefore, the variation of τ is slower than the variation of t or the variation of the time τ become more fine than t before the crossed point t_B. And then the contours (or the characteristic curves) of the nonlinear wave look like in parallel before the crossed point t_B.

If you read it carefully you may find that (7.93) is invariant under the following scale transformation:

$$u_1 \rightarrow \varepsilon u_1, \quad \xi \rightarrow \varepsilon^{-\gamma}\xi, \quad \tau \rightarrow \varepsilon^{-(\gamma+1)}\tau$$

Where, in (7.93), $u_{1\tau}$ is going toward the order of $\varepsilon^{\gamma+2}$, $u_1 u_\xi$ is going toward the order of $\varepsilon^{\gamma+2}$, and then (7.93) is invariant. In other words, it is able to make the transformation of (7.93) by means of

$$\xi = \varepsilon^{\gamma}(x - v_0 t), \qquad \tau = \varepsilon^{\gamma+1} t \tag{7.96}$$

This transformation is said to be *the G-M transformation* , where γ is an arbitrary constant. That means, in G-M transformation, ξ is been enlarged $\varepsilon^{-\gamma}$ times and t is been enlarged $\varepsilon^{-(\gamma+1)}$ times, so that, the variation of the curve in the coordinates of ξ and τ will be more fine than before. And (7.92) is only a special case with $\gamma = 0$. The key point of those transformation is that the enlargement of t is ε^{-1} times lager than that of ξ. In this case the expansion of the perturbation method (7.83) will be convergent.

G-M transformation performs two functions:

Therefore, the G-M transformation performs two functions:

1. The transformation of

$$\xi = \varepsilon^{\gamma}(x - v_0 t) \tag{7.97}$$

may transform a nonlinear equation into an ordinary differential equation,

2. The transformation of

$$\tau = \varepsilon^{\gamma+1} t \tag{7.98}$$

may prolong the time scale so as the intersection point $t_B(x,t)$ will be moved to far away from the source region and then the contours (or the characteristic curves) looks like in parallel. In this case, the solution of the perturbation method is good in the region before the intersection point and after the source region.

Note, since γ is an arbitrary constant, and then the solution of (7.82) is not unique. To make γ to be unique, it is necessary to add third term based on (7.82). Burgers equation is just an example.

In fact, in a long-haul optical fiber transmission system,

1. The Burgers equation has taken the dissipation in the single-mode fiber into account and the dissipation in the single-mode fiber will make the wave burst smooth.

2. The KdV equation has taken the dispersion into account, and the balance of the nonlinear effect and the dispersion effect will make a soliton transmission in a long-haul optical fiber system, which means that a pulse signal in a single-mode fiber will be holden in the optical fiber transmission system.

3. The Schrödinger equation is caused by the nonlinear refraction index of the

single-mode fiber, which will result in the self-focus in the single-mode fiber.

Now we are going to discuss one by one as follows.

G-M transformation method and the contour method:

The G-M transformation method is based on the contour method (or the characteristic method).

In fact, from (7.17), the first G-M transformation

$$\xi = x - vt, \quad t = t \tag{7.99}$$

is just the contour (or the characteristic curve) equation of the nonlinear wave equation

$$\frac{\partial u}{\partial t} + \alpha u \frac{\partial u}{\partial x} = 0 \tag{7.100}$$

And (7.99) is from the integral of the equation

$$\frac{dx}{dt} = v = \alpha u \tag{7.101}$$

and then, we have proved in (7.27) that

$$\frac{du}{dt} = \frac{\partial u}{\partial t} + \alpha u \frac{\partial u}{\partial x} \tag{7.102}$$

Which means that the transformation of (7.99) definitely lead to the equation (7.102).

7.3.3 Burgers equation and the weak dissipation transmission

As well known, in a long-haul optical fiber communication system, the single-mode fiber is a weak dissipation system with the loss $L < 0.3 - 0.5dB/km$. In this case, the far field in a single-mode fiber can be described by *the Burgers equation* in form of

$$\frac{\partial u}{\partial t} + \alpha u \frac{\partial u}{\partial x} - \mu \frac{\partial^2 u}{\partial x^2} = 0 \qquad (\mu > 0) \tag{7.103}$$

$$when \quad x \to \pm\infty \qquad u \to u \pm \infty \tag{7.104}$$

Where the second term in (7.103) is the nonlinear term caused by the nonlinearity in an optical single mode fiber and the third term is the dissipation term caused by the loss in the optical single mode fiber with μ being positive.

If $\mu = 0$, (7.103) becomes a nonlinear parabolic equation. As we mentioned before, the nonlinear parabolic equation displays the smooth feature only in a finite time. The characteristic curves of the nonlinear parabolic equation will be intercross to each other, and then the wave become discontinuous when the time reach to $t = t_B$, which means that the nonlinear term (second term) in (7.103) will make the wave burst. And now the third term will make the wave smooth, which is equivalent to make $t_B \to \infty$. The solution of (7.103) will show us this conclusion.

Parameters transformation

Now the equation (7.103) is able to be transformed into an ordinary differential equation via the following parameters transformation, i.e. let

$$u = u(\xi), \qquad \xi = (x - vt)$$

then (7.103) becomes

$$-vu_\xi + uu_\xi - \mu u_{\xi\xi} = 0 \tag{7.105}$$

Where u_ξ represents the partial derivative of u with respect to ξ and $\alpha = 1$ has been set for the simplification. Let

$$\bar{u} = u - v$$

then (7.105) becomes

$$\bar{u}\bar{u}_\xi - \mu \bar{u}_{\xi\xi} = 0 \tag{7.106}$$

Taking the integral of (7.106) with respect to ξ from ∞ to $\bar{u}$, considering the boundary condition (7.104), we have

$$\frac{1}{2}(\bar{u}^2 - \bar{u}_\infty^2) = \mu \bar{u}_\xi$$

And then

$$\int_\infty^{\bar{u}(\xi)} \frac{d\bar{u}}{\bar{u}^2 - \bar{u}_\infty^2} = \frac{\xi}{2\mu} \tag{7.107}$$

The integral (7.107) is in form of [2]

$$\int \frac{dx}{x^2 - 1} = -tanh^{-1} x$$

where $x = \dfrac{\bar{u}}{\bar{u}_\infty}$, and then the solution of the integral (7.107) is

$$-\frac{1}{\bar{u}_\infty} \tanh^{-1}\left(\frac{\bar{u}}{\bar{u}_\infty}\right) = \frac{\xi}{2\mu}$$

And then

$$\bar{u} = \bar{u}_\infty \tanh\left(-\frac{\bar{u}_\infty}{2\mu}\,\xi\right) \tag{7.108}$$

This solution is shown in Figure 7-3.

---

[2] I.S. Gradshteyn/I.M. Ryzhik, "Table of Integrals, and products" pp.63, 2.143,3

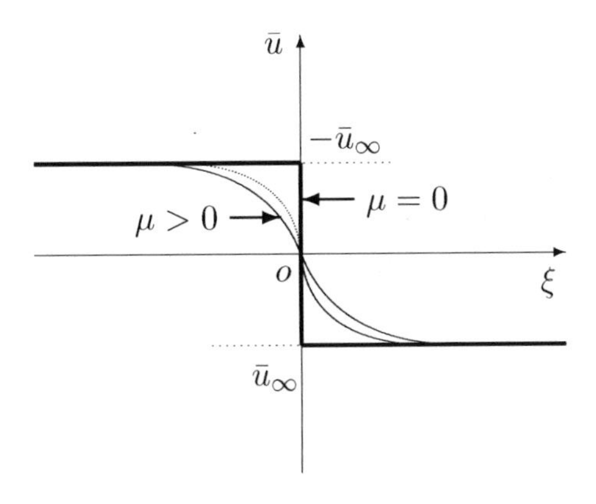

Figure 7-3 The solution of the Burgers equation

Obviously, the solution of the Burgers equation is a smooth continuous solution, the wave form is an exponential form along the ξ axis. And the width of the exponential section is $2\mu/|\bar{u}_\infty|$.

However, when $\mu = 0$, the stable solution (7.108) of the Burgers equation becomes a step function with the discontinuity point at $\xi = 0$. In this case, the Burgers equation (7.103) becomes a nonlinear parabolic equation (7.93). Which means that the behavior of the solution of the nonlinear parabolic equation presents a burst point at $\xi = 0$ when $\mu = 0$, and then the third term (dissipation term) of the Burgers equation made it smooth.

Now we will discuss the $G - M$ transformation of the Burgers equation. It is able to see that the Burgers equation is invariant under the following scalar transformation, i.e. when

$$u \to \varepsilon u, \quad \xi \to \varepsilon^{-1}\xi, \quad t \to \varepsilon^{-2}t$$

u_t approaches to the order of ε^3, uu_ξ approaches to the order of ε^3, $u_{\xi\xi}$ approaches to the order of ε^3, and then (7.103) is invariant. That means that the $G - M$ transformation of the Burgers equation is

$$\xi = \varepsilon(x - vt), \quad \tau = \varepsilon^2 t$$

Since the $G - M$ transformation of the Burgers equation is unique and then the solution is unique either.

7.3.4 KdV equation and the soliton transmission

As well known, in a long-haul optical fiber communication system, the nonlinearity in a single-mode fiber will make the pulse narrowing in the transmission. However the dispersion in a single-mode fiber will make the pulse broadening in the transmission. If the narrowing effect and the broadening effect of the pulse are in balance in the transmission, the pulse shape will be holden in the transmission. This transmission is said to be the soliton transmission. The equation to describe this phenomenon is called *the KdV equation*, i.e.

$$\frac{\partial u}{\partial t} + u\frac{\partial u}{\partial x} + \mu\frac{\partial^3 u}{\partial x^3} = 0 \tag{7.109}$$

Where μ is a constant and it can be positive or negative. If $\mu > 0$, the solution is a soliton. The process to obtain the solution is similar to that of the Burgers equation. Now we will use it to discuss the KdV equation for far field in a long-haul optical fiber communication system.

Let

$$\xi = x - vt, \quad u = u(\xi)$$

then (7.109) becomes

$$-vu_\xi + uu_\xi + \mu u_{\xi\xi\xi} = 0 \tag{7.110}$$

Where u_ξ represents the derivative of u with respect to ξ. Let

$$\bar{u} = u - v$$

then (7.110) becomes an original differential equation

$$\bar{u}\bar{u}_\xi + \mu\bar{u}_{\xi\xi\xi} = 0 \tag{7.111}$$

Taking an integral of (7.111) with respect to ξ from ∞ to ξ we have

$$\frac{1}{2}(\bar{u}^2 - \bar{u}_\infty^2) + \mu\bar{u}_{\xi\xi} = 0 \tag{7.112}$$

Where $\lim_{\xi\to\infty} \bar{u} = \bar{u}_\infty = constant$.

Multiplied both sides of (7.112) by $\bar{u}_\xi$, and then taking the integral from ∞ to ξ, in which

$$\int_\infty^\xi (\bar{u}^2 - \bar{u}_\infty^2)\bar{u}_\xi d\xi = \int_\infty^\xi (\bar{u}^2 - \bar{u}_\infty^2)d\bar{u} = \left[\frac{1}{3}\bar{u}^3 - \bar{u}_\infty^2\bar{u}\right]\Big|_\infty^\xi = \frac{1}{3}\bar{u}^3 - \bar{u}_\infty^2\bar{u} + \frac{2}{3}\bar{u}_\infty^3$$

and

$$\int_\infty^\xi \bar{u}_{\xi\xi}\bar{u}_\xi d\xi = \int_\infty^\xi \frac{\partial}{\partial\xi}(\bar{u}_\xi)\bar{u}_\xi d\xi = \frac{1}{2}\int_\infty^\xi \bar{u}_\xi^2 d\bar{u}_\xi = \frac{1}{2}\bar{u}_\xi^2\Big|_\infty^\xi = \frac{1}{2}\bar{u}_\xi^2$$

and then we have

$$\frac{1}{6}\bar{u}^3 - \frac{1}{2}\bar{u}_\infty^2\bar{u} + \frac{1}{3}\bar{u}_\infty^3 + \frac{\mu}{2}\bar{u}_\xi^2 = 0$$

Where

$$\frac{1}{6}\bar{u}^3 - \frac{1}{2}\bar{u}_\infty^2\bar{u} = \frac{1}{6}(\bar{u} - \bar{u}_\infty)^2(\bar{u} - \bar{u}_\infty + 3\bar{u}_\infty)$$

Taking the variables separation of (7.113) and then integral, we have

$$\int \frac{d\bar{u}}{(\bar{u} - \bar{u}_\infty)[-1/3(\bar{u} - \bar{u}_\infty) - \bar{u}_\infty]^{1/2}} = \frac{\xi}{\sqrt{\mu}} + C \tag{7.113}$$

Let $C = 0,\ \bar{u}_\infty \leq 0$, the integral (7.113) is in form of [3]

$$\int \frac{dx}{x\sqrt{z}} = \frac{1}{\sqrt{a}} \ln \frac{\sqrt{z} - \sqrt{a}}{\sqrt{z} + \sqrt{a}} \tag{7.114}$$

Where

$$\begin{aligned} x &= (\bar{u} - \bar{u}_\infty), \\ b &= -\frac{1}{3}, \\ a &= -\bar{u}_\infty \\ z &= a + bx = -\bar{u}_\infty - \frac{1}{3}(\bar{u} - \bar{u}_\infty) \end{aligned}$$

and then the integral (7.113 becomes

$$\frac{1}{\sqrt{-\bar{u}_\infty}} \ln \frac{[-1/3(\bar{u} - \bar{u}_\infty) - \bar{u}_\infty]^{1/2} - [-\bar{u}_\infty]^{1/2}}{[-1/3(\bar{u} - \bar{u}_\infty) - \bar{u}_\infty]^{1/2} + [-\bar{u}_\infty]^{1/2}} = \frac{\xi}{\sqrt{\mu}}$$

After rearranged, we have

$$\bar{u} = \bar{u}_\infty - 3\bar{u}_\infty sech^2[1/2(-\bar{u}_\infty/\mu)^{1/2}\xi] \tag{7.115}$$

Where $\bar{u} = u - v,\ \bar{u}_\infty = u_\infty - v,\ \xi = x - vt$. Let $-3\bar{u}_\infty = \varepsilon$, then (7.115) becomes

$$u = u_\infty + \varepsilon\, sech^2\left\{\left(\frac{\varepsilon}{12\mu}\right)^{1/2}\left[x - \left(u_\infty + \frac{\varepsilon}{3}\right)t\right]\right\} \tag{7.116}$$

This is the soliton solution of the KdV equation . This solution leads us to some important conclusions as follows:

(1) If the amplitude of the wave envelope $\varepsilon = constant$, then the soliton will be transmission forward with a constant velocity of $v = u_\infty + \varepsilon/3$ and then the soliton will preserves its identity.

(2) The effect of the dispersion factor μ is to broaden the soliton since the width of the waveform of the soliton (7.116) is proportional to $\sqrt{\mu}$.

(3) If $u_\infty = \lim_{x\to\pm\infty} u = 0$, then the solution of the soliton becomes

$$u = \varepsilon\, sech^2\left[\left(\frac{\varepsilon}{12\mu}\right)^{1/2}\left(x - \frac{\varepsilon}{3}t\right)\right] \tag{7.117}$$

In this case, the propagation velocity of the soliton $v = \varepsilon/3$ is proportional to the amplitude ε, namely, the higher the wave amplitude ε the higher the wave velocity. This phenomena is valid either for (7.116).

[3] I.S. Gradshteyn/I.M. Ryzhik, "Table of Integrals, and products." 1980. pp.73, 2.224,5

7.3.5 Exponential law and Hirota method

T. Taniuti and K. Nishihara in [38] has discussed the Hirota method in the area of physics and applied mathematics applied to particle physics, plasma physics and ecology. Now we will discuss the multi-soliton transmission in a long-haul optical fiber communication system in terms of the Hirota method.

Exponent law of nonlinear wave equation:

As we mentioned before, the solution (7.117) of the KdV equation (7.109) is in the form of

$$u \approx sech^2\xi = \left(\frac{1}{e^{\xi}+e^{-\xi}}\right)^2, \quad \xi = \gamma(x - vt) \tag{7.118}$$

In which, when $x \to -\infty, u \propto e^{-2\xi}$. In this case, u is decaying in exponential law. This is an important feather of the nonlinear wave equation and can be used to analyze the far field of the nonlinear wave equations such as the Burgers equation and the KdV equation.

Hirota method:

Now we will discuss the KdV equation

$$u_t + 6uu_x + \mu u_{xxx} = 0 \tag{7.119}$$

by means of the Hirota method.

Hirota notices that when $|x| \to \infty$, the solution of u in (7.118) is decaying in exponential law $u \propto e^{-2\xi}$, and then the accuracy solution of u at $x \to -\infty$ is a quotient of the polynomial of e^{θ} in form of

$$u = G/F \tag{7.120}$$

Where, G and F are unknown polynomial of e^{θ} and can be obtained by substitution of (7.120) into (7.119).

Step 1: Function transformation:

To this end, making a function transformation of

$$u = \phi_x \tag{7.121}$$

In this case, the KdV equation (7.119) becomes

$$\phi_t - 3(\phi_x)^2 + \phi_{xxx} = 0 \tag{7.122}$$

In which, it is reasonable to consider that $\phi \to 0$ when $x \to -\infty$.

Now, from (7.120), suppose that

$$\phi = g/f \tag{7.123}$$

Substitution of (7.123) into (7.122) we have

$$(g_t f - g f_t)/f^2 - 3(g_x f - g f_x)^2/f^4$$

$$+(g_{xxx} f - 3g_{xx} f_x - 3g_x f_{xx} - g f_{xxx})/f^2$$

$$+6(f g_x f_x^2 + f g f_{xx} f_x - g f_x^3)/f^4 = 0 \tag{7.124}$$

This is a coupling equation between g and f.

Step 2: Decoupling equation:

For the sake of the simplification, introduce a decoupling equation

$$g = -2f_x \tag{7.125}$$

Substitution of (7.125) into (7.124) we found that 6 terms $f_x^2 - f_x^2$, $12ff_xf_{xx} - 12ff_xf_{xx}$, and $12ff_x^2f_{xx} - 12ff_x^2f_{xx}$ have been cancelled and then (7.124) becomes

$$-f(f_t + f_{xxx})_x + f_x(f_t + f_{xxx}) + 3(f_{xxx}f_x - f_{xx}^2) = 0 \tag{7.126}$$

Step 3: Perturbation method:

Considering , from (7.122), that $\phi \to 0$ when $x \to -\infty$, and then the nonlinear terms $(\phi_x)^2 = u^2$ in (7.122) is smaller that two linear terms, therefore, f is able to be linearly expanded to

$$f = f_0 + \varepsilon f_1 + \varepsilon^2 f_2 + \cdots \tag{7.127}$$

Substitution of (7.127) into (7.126), we have

$$f_0 = constant \tag{7.128}$$

$$\left(\frac{\partial}{\partial t} + \frac{\partial^3}{\partial x^3}\right) f_{1x} = 0 \tag{7.129}$$

$$\left(\frac{\partial}{\partial t} + \frac{\partial^3}{\partial x^3}\right) f_{2x} = 3(f_{1xxx}f_{1x} - f_{1xx}^2) \tag{7.130}$$

$$\left(\frac{\partial}{\partial t} + \frac{\partial^3}{\partial x^3}\right) f_{3x} = -\left[f_1\left(\frac{\partial}{\partial t} + \frac{\partial^3}{\partial x^3}\right) f_{2x} - f_{1x}\left(\frac{\partial}{\partial t} + \frac{\partial^3}{\partial x^3}\right) f_2\right]$$

$$+3(f_{2xxx}f_{1x} + f_{1xxx}f_{2x} - 2f_{1xx}f_{2xx}) \tag{7.131}$$

Considering that (7.124) still holds if f times a constant, and then we may consider that $f_0 = 1$ without loosing the generality of the solution. In this case:

For a single soliton:

From (7.129), it is reasonable to consider that

$$f_1 = e^{\theta} \qquad (\theta = px - \Omega t + \delta, \quad \Omega = p^3) \tag{7.132}$$

such that

$$\Big(\frac{\partial}{\partial t} + \frac{\partial^3}{\partial x^3}\Big) f_{1x} = 0 \tag{7.133}$$

Substitution of (7.132) into (7.130) we have

$$\Big(\frac{\partial}{\partial t} + \frac{\partial^3}{\partial x^3}\Big) f_{2x} = 0 \tag{7.134}$$

Obviously, this equation is the same as (7.133) and then the solution f_2 is the same as f_1, therefore it is unnecessary. And then we may consider that $f_2 = 0$. Substituting it into (7.131), we found that the equation of f_3 is the same as (7.134) and then we may consider that $f_3 = 0$. The same process will show us that $f_n = 0$ $(n = 4, 5, \cdots)$. Therefore

$$f = 1 + \varepsilon e^{\theta} \tag{7.135}$$

Let $\varepsilon = 1$, then from (7.135), (7.125) and (7.123) we have

$$\phi = \frac{-2pe^{\theta}}{1 + e^{\theta}} \tag{7.136}$$

Substitution of (7.136) into (7.121), we have

$$u = \phi_x = -\frac{2p^2 e^{\theta}}{(1 + e^{\theta})^2} = -\frac{p^2}{2} sech^2(\theta/2)$$

For double solitons:

From (7.129), it is reasonable to consider that

$$f_1 = e^{\theta_1} + e^{\theta_2} \qquad (\theta_i = p_i x - \Omega_i t + \delta_i, \ i = 1, 2) \tag{7.137}$$

such that

$$\Big(\frac{\partial}{\partial t} + \frac{\partial^3}{\partial x^3}\Big) f_{1x} = 0 \tag{7.138}$$

Substituting (7.137) into (7.130), we have

$$\Big(\frac{\partial}{\partial t} + \frac{\partial^3}{\partial x^3}\Big) f_2 = 3\,[p_1^3 p_2 + p_1 p_2^3 - 2p_1^2 p_2^2]e^{\theta_1 + \theta_2} \tag{7.139}$$

which means that the solution of f_2 should be

$$f_2 = Ae^{\theta_1 + \theta_2} \tag{7.140}$$

Substitution of (7.140) into (7.139), we have

$$f_2 = \frac{(p_1 - p_2)^2}{(p_1 + p_2)^2} e^{\theta_1 + \theta_2} \tag{7.141}$$

and then (7.131) becomes

$$\left(\frac{\partial}{\partial t} + \frac{\partial^3}{\partial x^3}\right) f_{3x} = 0 \tag{7.142}$$

This is the same as (7.129), that means that this f_3 is unnecessary. The same process will show us that $f_n = 0$ ($n = 4, 5, \cdots$). Therefore

$$f = 1 + \varepsilon f_2 + \varepsilon^2 f_2 \tag{7.143}$$

Then the final result of u at $x \to -\infty$ will be

$$u = \phi_x = (g/f)_x = (-2f_x/f)_x$$

in which use of (7.125), (7.127), (7.137), (7.141) and (7.143) have been made.

If you are interesting in three solitons solution you may find it in [38].

7.3.6 Schrödinger equation and the self-focus

T. Taniuti and K. Nishihara in [38] has discussed the Schrödinger equation in the area of physics and applied mathematics applied to particle physics, plasma physics and ecology. Now we will discuss the Schrödinger equation and the self-focus in a long-haul optical fiber communication system.

As well known, the single-mode optical fiber is a linear medium in the normal transmission and its refraction index $n = n_0$. However, if the intensity of the transmission light is strong enough, the nonlinear refraction index in a single-mode fiber has to be taken into account. In this case, the refraction index is

$$n = n_0 + \Delta n = n_0 + \left(\frac{n_2}{2n_0}\right)^2 |\mathbf{E}|^2 \tag{7.144}$$

Where the second term is the nonlinear refraction index proportional to the square of the electric field $\mathbf{E}$.

The nonlinear refraction index is caused by such as the electrostrictive effect . This is because that the single-mode optical fiber is made of crystal, and the variation of the pressure and the associated variation of the density caused by the strong electric field will cause the variation of the refraction index.

Assumes that n_0 and n_2 are real constants, or in other words, if we only consider the real number party of n_0 and n_2, the Maxwell equation is

$$\nabla^2 \mathbf{E} - (n_0/c)^2 \mathbf{E}_{tt} = (n_2/c)^2 (|\mathbf{E}|^2 \mathbf{E})_{tt} \tag{7.145}$$

Where $n_0 >> n_2$.

Step 1: Make a guess of the solution:

Since $n_0 >> n_2$, it is reasonable to consider that the solution of (7.145) is a quasi-plane wave with a modulation amplitude. Since the modulation amplitude is caused by the small nonlinearity and then it should be a slow variation function of the coordinates, i.e. the solution of (7.145) is able to be expressed as

$$\mathbf{E}(r,t) = \varepsilon\varphi(\xi,\eta,\zeta)\,\mathbf{i}\,e^{i(kz-\omega t)} \tag{7.146}$$

Where

$$\xi = \varepsilon x, \quad \eta = \varepsilon y, \quad \zeta = \varepsilon^2 z$$

That means that the modulation amplitude φ is a slow variation function, in which the variation along the propagation direction (ζ) is slower than that along the transverse direction (ξ, η).

Substitution of (7.146) into (7.145), we have

$$\Big[\Big(\varepsilon^4\frac{\partial^2}{\partial\zeta^2} + 2\,i\,k\varepsilon^2\frac{\partial}{\partial\zeta} - k^2\Big) + \varepsilon^2\Big(\frac{\partial^2}{\partial\xi^2} + \frac{\partial^2}{\partial\eta^2}\Big) + \Big(\frac{\omega n_0}{c}\Big)^2\Big]\varepsilon\varphi =$$

$$-\Big(\frac{\omega n_2}{c}\Big)^2|\varphi|^2\varphi\varepsilon^3 \tag{7.147}$$

Considering that the coefficient of the first order of ε in both side should be equal to each other, we have the dispersion equation

$$k^2 = (\omega n_0/c)^2$$

And the coefficient equation of the third order of ε gives the Shcrödinger equation

$$i\varphi_\zeta + (1/2k)\nabla_t^2\varphi + (k/2)(n_2/n_0)^2|\varphi|^2\varphi = 0 \tag{7.148}$$

Where

$$\nabla_t^2 = \frac{\partial^2}{\partial\xi^2} + \frac{\partial^2}{\partial\eta^2}$$

Step 2: Shcrödinger equation:

Now we will consider two real situations:

(1) For a planar waveguide, $\partial/\partial\eta = 0$, $\nabla_t^2 = \partial^2/\partial\xi^2$.

(2) For a single-mode fiber, to find the accuracy solution of the Schrödinger equation in the cylindric coordinate system is still a difficult problem for the time being. Therefore, we have to ignore the variation of the field along φ in the single-mode fiber. In this case, $\partial/\partial\varphi = 0$, $\nabla_t^2 = \partial^2/\partial r^2$.

For these two situations, the Shcrödinger equation (7.148) is reduced to

$$i\varphi_\tau + (\omega''/2)\varphi_{\xi\xi} + Q|\varphi|^2\varphi = 0 \tag{7.149}$$

Where, from (7.148) to (7.149),

$$\zeta \to \tau, \quad (1/2k) \to \omega''/2, \quad \nabla_t^2 = \partial^2/\partial\xi^2$$

and the sign of Q is the same as $\omega''/2$, meanwhile, when $|\xi| \to \infty$, $\varphi \to 0$. Therefore (7.149) is good for the true situation.

Step 3: The analytical solution of the Schrödinger equation

Since (7.149) is a complex equation, namely, it consists of a real part and a imaginary part, and then φ should contain an amplitude part and a phase part, i.e.

$$\varphi = \Phi(\xi, \tau)e^{i\theta(\xi,\tau)} \tag{7.150}$$

Substitution of (7.150) into (7.149), rearranging the result, separating the real part and the imaginary part, we have

$$(\omega''/2)\Phi_{\xi\xi} - (\omega''/2)\Phi\theta_\xi^2 - \Phi\theta_\tau + Q\Phi^3 = 0 \tag{7.151}$$

$$(\omega''/2)\Phi\theta_{\xi\xi} + \omega''\Phi_\xi\theta_\xi + \Phi_\tau = 0 \tag{7.152}$$

Step 4: Parameter transformation:[4]

Let

$$\theta(\xi, \tau) = \theta(\xi - v_c\tau), \quad \Phi(\xi, \tau) = \Phi(\xi - v_a\tau)$$

Where v_c and v_a are constants. And then (7.151) and (7.152) become

$$(\omega''/2)\Phi_{\xi\xi} - (\omega''/2)\Phi\theta_\xi^2 + v_c\Phi\theta_\xi + Q\Phi^3 = 0 \tag{7.153}$$

$$(\omega''/2)\Phi\theta_{\xi\xi} + \omega''\Phi_\xi\theta_\xi - v_a\Phi_\xi = 0 \tag{7.154}$$

Multiplying (7.154) by Φ and then integral, we have

$$\Phi^2(\omega''\theta_\xi - v_a) = C_1 \tag{7.155}$$

Considering that when $|\xi| \to \infty$, $\varphi \to 0$, we have $C_1 = 0$ and then

$$\theta_\xi = v_a/\omega'' \tag{7.156}$$

$$\theta = (v_a/\omega'')(\xi - v_c\tau) + C_2 \tag{7.157}$$

Substituting (7.156) into (7.153), we have

$$\omega''^2\Phi_{\xi\xi} - (v_a^2 - 2v_cv_a)\Phi + 2\omega''Q\Phi^3 = 0 \tag{7.158}$$

---

[4] This is from the contour method (or the characteristic curve method).

Multiplying (7.158) by Φ_ξ, and then integral, meanwhile, considering that when $|\xi| \to \infty$, $\varphi \to 0$ we have

$$\omega''^2\Phi_\xi^2 - (v_a^2 - 2v_c v_a)\Phi^2 + \omega'' Q\Phi^4 = 0 \tag{7.159}$$

From (7.159), we will able to obtain

$$\int \frac{d\Phi}{\sqrt{P(\Phi)}} = \xi - v_a\tau \tag{7.160}$$

Where

$$P(\Phi) = (Q/\omega'')[\Phi^2(\Phi_0^2 - \Phi^2)]$$

$$\Phi_0 = \left[\frac{v_a^2 - 2v_a v_c}{\omega'' Q}\right]^{1/2} \tag{7.161}$$

The left-hand side of (7.160) is an integral formula in form of [5]

$$\int \frac{dx}{x\sqrt{R}} = -\frac{1}{\sqrt{a}}\frac{2a + bx + 2\sqrt{aR}}{x} \quad (a > 0) \tag{7.162}$$

where

$$R = a + bx + cx^2$$
$$x = \phi$$
$$a = \phi_0^2, \quad b = 0, \quad c = -1$$

And then, from (7.162) we have

$$\int \frac{d\Phi}{\sqrt{P(\Phi)}} = -\frac{\omega''}{\sqrt{\omega'' Q}}\frac{1}{\Phi_0}\ln\frac{\Phi_0 + \sqrt{\Phi_0^2 - \Phi^2}}{\Phi} + C_3 \tag{7.163}$$

Substituting (7.163) into (7.160), we have

$$\frac{\Phi_0 + \sqrt{\Phi_0^2 - \Phi^2}}{\Phi} = \exp\Big[-\Phi_0\sqrt{\frac{Q}{\omega''}}(\xi - v_a\tau - C_3)\Big]$$

And then

$$\Phi = \Phi_0 sech\Big[\Phi_0\sqrt{\frac{Q}{\omega''}}(\xi - v_a\tau - C_3)\Big] \tag{7.164}$$

Where the contribution of C_3 for the solution is a constant and it is reasonable to assume that $C_3 = 0$. Substituting (7.157) and (7.164) into (7.150), considering that

[5] I. S. Gradshteyn/I. M. Ryzhik, "Tables of Integrals, Series, and Products." 1980. pp. 84, 2.266. TI (137)

the contribution of C_2 for the solution of φ is a constant phase only, it is reasonable to assume that $C_2 = 0$. In this case, from (7.164), the amplitude slow function is

$$\varphi = \Phi_0 sech\left[\Phi_0\sqrt{\frac{Q}{\omega''}}(\xi - v_a\tau)\right]\exp\left[i\frac{v_a}{\omega''}(\xi - v_c\tau)\right] \quad (7.165)$$

This is the analytical solution of the Schrödinger equation (7.149).

Now we will discuss the feature of this solution as follows:

(1) The wave envelope denoted by a parabolic function holds in the transmission. Therefore, this wave is said to be the envelope soliton.

(2) In the phase term denoted by a exponential function, v_a is proportional to the wave amplitude Φ_0 as given in (7.161). Which means that the phase delay of the wave depends on the wave amplitude in the transmission, which will results in the focus. This focus is caused by the transmission medium. Therefore this focus is said to be the self-focus phenomenon.

(3) (7.161) and (7.165) show us that there are two independent parameters Φ_0 and v_c in the solution of the nonlinear Schrödinger equation, and then the wave amplitude Φ_0 and the wave velocity v_c are able to be given independently.

Summery: In the process to obtain the result, the key points are:
(1) The function transformation (7.150)

$$\varphi = \Phi(\xi, \tau)e^{i\theta(\xi,\tau)}$$

(2) Decoupling (7.155)

$$\Phi^2(\omega''\theta_\xi - v_a) = C_1 = 0$$

7.3.7 Nonlinear coupling wave equations and Stimulated Raman Scattering

This subsection is going to discuss the nonlinear coupling wave equations and Stimulated Raman Scattering in a long-haul optical fiber communication system studied by author and his students [63][64][65][66].

1. The root of the nonlinear coupling wave equations

In the nonlinear dispersion medium such as an optical single mode fiber, the Maxwell equation is able to be simplified to

$$\nabla^2\mathbf{E} - \left(\frac{1}{c^2}\right)(\mathbf{E} + 4\pi\mathbf{P})_{tt} = \left(\frac{4\pi}{c^2}\right)\mathbf{P}_{tt}^{(NL)} \quad (7.166)$$

Where $\mathbf{P}$ and $\mathbf{P}^{(NL)}$ denotes the linear polarization vector and the nonlinear polarization vector, respectively. In normal situation, they are able to be expressed as

$$\mathbf{P}(\mathbf{r}, t) = \int_{-\infty}^{\infty} d^3\mathbf{r}' dt' \chi^{(1)}(\mathbf{r} - \mathbf{r}', t - t') \cdot \mathbf{E}(\mathbf{r}', t')$$

$$\begin{aligned}
\mathbf{P}^{(NL)}(\mathbf{r},t) &= \int_{-\infty}^{\infty} d^3\mathbf{r}'dt'd^3\mathbf{r}''dt''\chi^{(2)}(r-\mathbf{r}'.t-t';\mathbf{r}-\mathbf{r}'',t-t'') \\
&: \mathbf{E}(\mathbf{r}',t')\mathbf{E}(\mathbf{r}'',t'') + \int_{-\infty}^{\infty} d^3\mathbf{r}'dt'd^3\mathbf{r}''dt''d^3\mathbf{r}'''dt''' \\
&\chi^{(3)}(\mathbf{r}-\mathbf{r}',t-t';\mathbf{r}-\mathbf{r}'',t-t'';\mathbf{r}-\mathbf{r}''',t-t''') \\
&\vdots\, \mathbf{E}(\mathbf{r}',t')\mathbf{E}(\mathbf{r}'',t'')\mathbf{E}(\mathbf{r}''',t''') \qquad (7.167)
\end{aligned}$$

For the sake of simplification, it is reasonable to assume that the input signal in a single-mode fiber has an uniform distribution of linear polarization states over the transverse cross section. And the direction of the polarization is x, the propagation direction is z. Then

$$P = \chi^{(1)}E(r,\theta,z;t) \qquad (7.168)$$

$$P^{(NL)} = \chi^{(3)}E^3(r,\theta,z;t) \qquad (7.169)$$

Note, the single-mode fiber is made of the crystal and the crystal is an antisymmetrical material. Therefore, the second order of electrical susceptibility $\chi^{(2)} = 0$.

Substituting (7.168) and (7.169) into (7.166), we have

$$\frac{d^2}{dz^2}E(r,\theta,z;t) - \left(\frac{n_L^2}{c}\right)^2\frac{d^2}{dt^2}E(r,\theta,z;t) = \frac{16\pi\chi^{(3)}}{c^2}\frac{d^2}{dt^2}[E^3(r,\theta,z;t)] \qquad (7.170)$$

Where $\frac{d}{dx} = \frac{d}{dy} = 0$ since it is assumed that the input signal in single-mode fiber has an uniform distribution of linear polarization states over the transverse cross section.

It is easy to understand that the linear polarization vector $\mathbf{P}$ will cause the wave dispersion, and the nonlinear polarization vector $\mathbf{P}^{(\mathbf{NL})}$ will cause the new frequency components. And then the field in the single-mode fiber is able to be expressed as

$$\begin{aligned}
|E(r,\theta,z;t)| &= \sum_m \psi_m(r,\theta)\,|A_m(z,t)|\,\cos(k_m z - \omega_m t + \phi_m) \\
&+ \sum_{m'} \psi_{m'}(r,\theta)\,|A_{m'}(z,t)|\,\cos(k_{m'} z - \omega_{m'} t + \phi_{m'})
\end{aligned} \qquad (7.171)$$

Where m denotes the wave propagating in the forward direction with a frequency ω_m, and m' denotes the wave propagating in the backward direction with another frequency $\omega_{m'}$. Both of them belong to the single transverse waves in a single-mode fiber .

In the nonlinear fiber optics, the Stimulated Raman Scattering (SRS) is an important phenomenon since *the Distributed Raman Amplifier* is one amplification scheme that can provide a broad and relatively flat gain profile over a wider wavelength range than EDFA (*Erbium Doped Fiber Amplifier*) amplification techniques. The latter has been conventionally used in optical communication systems. Therefore, we would like to concentrate to analyze the Stimulated Raman Scattering as follows.

Assume that the power of the pump wave (p) injected into the single-mode fiber is beyond *the threshold power*, and then the pump wave will stimulate the first order of stocks wave (s) and the first order of anti-stocks wave (a), however, it is not large enough to stimulate the higher order of stocks waves. In this case, $m = p, s, a; \quad m' = s, a.$

Next, we may consider that the field amplitude

$$F_m(z,t) = N_m A_m(z,t) \tag{7.172}$$

so that the power of the field will be

$$P_m(z,t) = |F_m(z,t)|^2$$

And then the normalization condition gives

$$N_m^2 = \frac{n_{mL}\, c}{8\pi} \int_0^{\infty} \int_0^{2\pi} \psi_m^2(r,\theta) r dr d\theta \tag{7.173}$$

Where n_{mL} is the linear refraction index at the frequency ω_m.

Substituting (7.172) into (7.171), we have

$$\begin{aligned} E(r,\theta,z;t) \quad = \quad & \frac{1}{2}\sum_m \psi_m(r,\theta)\Big[\frac{F_m(z,t)}{N_m} e^{i(k_m z - \omega_m t)} \\ & + \frac{F_m^*(z,t)}{N_m} e^{-i(k_m z - \omega_m t)}\Big] \\ & + \frac{1}{2}\sum_{m'} \psi_{m'}(r,\theta)\Big\{\frac{F_{m'}(z,t)}{N_{m'}} e^{i[k_{m'}(L-z) - \omega_{m'} t]} \\ & + \frac{F_{m'}^*(z,t)}{N_{m'}} e^{-i[k_{m'}(L-z) - \omega_{m'} t]}\Big\} \end{aligned} \tag{7.174}$$

Where L is the length of the single-mode fiber from the inject point of the pump wave to the end point of the single-mode fiber.

For the sake of simplification, we may consider the forward waves alone without considering the backward waves for the time been, i.e. let $m' = 0$ in (7.174). And then substituting (7.174) into (7.170), collect the same frequency terms in both sides, we may obtain the coupling wave equation for the Stocks wave as following:

$$\frac{ik_s\psi_s}{N_s}\left(\frac{\partial F_s}{\partial z} + \frac{n_{sL}}{c}\frac{\partial F_s}{\partial t}\right)$$

$$-\frac{16\pi\omega_s^2}{c}\Big[\chi^{(3)}(-\omega_s,-\omega_p,\omega_p,-\omega_s)\frac{6\psi_p^2\psi_s}{8N_pN_s}|F_p|^2F_s$$

$$+\chi^{(3)}(-\omega_s,-\omega_p,\omega_p,-\omega_s)\frac{3\psi_p^2\psi_a}{8N_p^2N_a}|F_a|^2F_a^*e^{i(2k_p-k_a-kc)z}$$

$$+\chi^{(3)}(-\omega_s,-\omega_s,\omega_s,-\omega_s)\frac{3\psi_s^3}{8N_s^3}|F_s|^2F_s$$

$$+\chi^{(3)}(-\omega_s,-\omega_a,\omega_a,-\omega_s)\frac{6\psi_s\psi_a^2}{8N_sN_a^2}|F_a|^2F_s\Big] \tag{7.175}$$

Of course, we understand that the forward waves contains not only the Stocks wave F_s but also the pumping wave F_p and the anti-Stocks wave F_a. And the coupling wave equations for the pumping wave F_p or the anti-Stocks wave F_a is similar to (7.175). Therefore, we will concentrate to deal with the equation (7.175) first.

Now, to eliminate the transverse distribution function ψ_a and ψ_p etc. from (7.175), it is able to multiply (7.175) by ψ_m and then take a integral over the transverse cross section, and then make a definition

$$<ijkl>=\frac{\int_0^{2\pi}\int_0^{\infty}\psi_i\psi_j\psi_k\psi_l\, rdrd\theta}{N_iN_jN_kN_l}=A_e^{-1} \tag{7.176}$$

we have

$$\frac{\partial F_s}{\partial z}+\frac{n_{sL}}{c}\frac{\partial F_s}{\partial t}$$

$$=i\frac{3\omega_s}{4}\Big[2\chi^{(3)}(-\omega_s,-\omega_p,\omega_p,-\omega_s)<spps>|F_p|^2F_s$$

$$+\chi^{(3)}(-\omega_s,-\omega_p,\omega_p,-\omega_s)<spap>|F_p|^2F_a^*e^{i(\Delta kz+2\phi_p)}$$

$$+\chi^{(3)}(-\omega_s,-\omega_s,\omega_s,-\omega_s)<ssss>|F_s|^2F_s$$

$$+2\chi^{(3)}(-\omega_s,-\omega_a,\omega_a,-\omega_s)<saas>|F_a|^2F_s\Big] \tag{7.177}$$

Similarly, we have

$$\frac{\partial F_p}{\partial z}+\frac{n_{pL}}{c}\frac{\partial F_p}{\partial t}$$

$$=i\frac{3\omega_p}{4}\Big[\chi^{(3)}(-\omega_p,-\omega_p,\omega_p,-\omega_p)<pppp>|F_p|^2F_p$$

$$+2\chi^{(3)}(-\omega_p,-\omega_s,\omega_s,-\omega_p) < pssp > |F_s|^2F_p$$

$$+2\chi^{(3)}(-\omega_p,-\omega_a,\omega_a,-\omega_p) < paap > |F_a|^2F_p\Big] \quad (7.178)$$

and

$$\frac{\partial F_a}{\partial z}+\frac{n_{aL}}{c}\frac{\partial F_a}{\partial t}$$

$$= i\frac{3\omega_a}{4}\Big[2\chi^{(3)}(-\omega_a,-\omega_p,\omega_p,-\omega_a) < appa > |F_p|^2F_a$$

$$+\chi^{(3)}(-\omega_a,-\omega_p,\omega_s,-\omega_p) < apsp > |F_p|^2F_s^*e^{i(\Delta kz+2\phi_p)}$$

$$+\chi^{(3)}(-\omega_a,-\omega_a,\omega_a,-\omega_a) < aaaa > |F_a|^2F_a$$

$$+2\chi^{(3)}(-\omega_a,-\omega_s,\omega_s,-\omega_a) < assa > |F_s|^2F_a\Big] \quad (7.179)$$

Where

$$\Delta k = 2k_p - k_s - k_a$$

$$2\omega_p = \omega_s + \omega_a$$

And in $\chi^{(3)}(-\omega_i,-\omega_j,\omega_k,-\omega_l)$, $(i,j,k,l,=p,s,a)$, we have

$$-\omega_i = -\omega_j + \omega_k - \omega_l$$

$$\mathbf{Im}\Big[\chi^{(3)}(-\omega_i,-\omega_j,\omega_k,-\omega_l)\Big] = -\mathbf{Im}\Big[\chi^{(3)}(\omega_i,\omega_j,-\omega_k,\omega_l)\Big]$$

It is worth to point out, from (7.176) to (7.178), that the amplitude coupling between the pumping wave F_p, Stocks wave F_s and the anti-Stocks wave F_a is caused by the imaginary part of $\chi^{(3)}$, which results in the Stimulated Raman Scattering, and the phase coupling between the pumping wave F_p, Stocks wave F_s and the anti-Stocks wave F_a is caused by the real part of $\chi^{(3)}$, which results in the self-phase modulation phenomenon.

2. **Stimulated Raman Scattering (SRS)**

From wave point of view, when the pumping power is beyond a threshold value in a single-mode fiber, it will stimulate the forward Stock wave and the anti-Stocks wave. This phenomenon is said to be *the Stimulated Raman Scattering.*

Step 1: Nonlinear coupling wave equation:

The nonlinear coupling wave equation described the Stimulated Raman Scattering in a single-mode fiber is able to be obtained from (7.178) and (7.177) as follows:

$$\frac{\partial F_p}{\partial z}+\frac{n_{pL}}{c}\frac{\partial F_p}{\partial t}=-\frac{3\omega_p}{2}|\mathbf{Im}[\chi^{(3)}(-\omega_p,-\omega_s,\omega_s,-\omega_p)]|<pssp>|F_s|^2F_p \quad (7.180)$$

$$\frac{\partial F_s}{\partial z}+\frac{n_{sL}}{c}\frac{\partial F_s}{\partial t}=\frac{3\omega_s}{2}|\mathbf{Im}[\chi^{(3)}(-\omega_s,-\omega_p,\omega_p,-\omega_s)]|<spps>|F_p|^2F_s \quad (7.181)$$

Where F_p and F_s represent the amplitude factor of the forward pumping wave and the forward Stocks wave, respectively.

Multiplying (7.180) and (7.181) by F_p^* and F_s^*, respectively, considering that $F_p\cdot F_p^*=P_P$, $F_s\cdot F_s^*=P_s$, we have the coupling wave equations of the Stimulated Raman Scattering as follows:

$$\frac{\partial P_p}{\partial z}+\frac{1}{v_p}\frac{\partial P_p}{\partial t}=-\frac{g_p}{KA_e}P_sP_p-\alpha_pP_p \quad (7.182)$$

$$\frac{\partial P_s}{\partial z}+\frac{1}{v_s}\frac{\partial P_s}{\partial t}=\frac{g_s}{KA_e}P_pP_s-\alpha_sP_s \quad (7.183)$$

Where

$$g_p=3\omega_p|\mathbf{Im}[\chi^{(3)}(-\omega_p,-\omega_s,\omega_s,-\omega_p)]|$$

$$g_s=3\omega_s|\mathbf{Im}[\chi^{(3)}(-\omega_s,-\omega_p,\omega_p,-\omega_s)]|$$

are the gain coefficients of the pumping wave and the Stocks wave, respectively. And

$$v_p=c/n_{pL},\quad v_s=c/n_{sL}$$

are the propagation velocity of the pumping wave and the Stocks wave, respectively. K is the polarization factor. For a single-mode fiber with a holding polarization, $K=1$. For a single-mode fiber with non-holding polarization, $K=2$, which is the worst case. Therefore, $1\le K\le 2$. The definition of A_e is in (7.176).

Step 2: Parameter transformation:

Now we will find the solution of (7.182) and (7.183). Normally, the single-mode fiber is a low dissipation fiber. In this case, it is reasonable to consider that the linear attenuation coefficient for both the pumping wave and the Stocks wave are approximately the same, therefore, it is reasonable to assume that $\alpha_p=\alpha_s=\alpha$. In this case, set a parameter transformation as follows [6]

$$z'=z,\quad t_p'=t-z/v_p,\quad t_s'=t-z/v_s \quad (7.184)$$

[6] This is from the contour method (or the characteristic curve method).

And then, in (7.182) and (7.183),

$$\frac{\partial}{\partial z} = \frac{\partial}{\partial z'}\frac{\partial z'}{\partial z} + \frac{\partial}{\partial t'}\frac{\partial t'}{\partial z} = \frac{\partial}{\partial z'} + \frac{\partial}{\partial t'}\Big(-\frac{1}{v}\Big)$$

$$\frac{1}{v}\frac{\partial}{\partial t} = \frac{1}{v}\Big(\frac{\partial}{\partial z'}\frac{\partial z'}{\partial t} + \frac{\partial}{\partial t'}\frac{\partial t'}{\partial t}\Big) = \frac{1}{v}\Big(\frac{\partial}{\partial t'}\Big)$$

Therefore,

$$\frac{\partial}{\partial z} + \frac{1}{v}\frac{\partial}{\partial t} = \frac{\partial}{\partial z'}$$

And then (7.182) and (7.183) become

$$\frac{\partial \tilde{P}_p}{\partial z'} = -\frac{g_p}{KA_e}\tilde{P}_p\tilde{P}_s - \alpha\tilde{P}_p \tag{7.185}$$

$$\frac{\partial \tilde{P}_s}{\partial z'} = \frac{g_s}{KA_e}\tilde{P}_p\tilde{P}_s - \alpha\tilde{P}_s \tag{7.186}$$

Where

$$\tilde{P}_p = P_p(z', t_p'), \quad \tilde{P}_s = P_s(z', t_s')$$

And the boundary conditions are

$$\tilde{P}_p(z', t_p')|_{z'=0} = P_p(0, t_p')$$

$$\tilde{P}_s(z', t_s')|_{z'=0} = P_s(0, t_s')$$

Obviously, the parameter transformation (7.184) transforms the partial differential equations (7.182) and (7.183) into the ordinary differential equations (7.185) and (7.186).

Step 3: Find the solution:

To fine the solution of (7.185) and (7.186), suppose that

$$\tilde{P}_p(z', t_p') = C_p(z', t_p')e^{-\alpha z'} \tag{7.187}$$

$$\tilde{P}_s(z', t_s') = C_s(z', t_s')e^{-\alpha z'} \tag{7.188}$$

Substituting (7.187) and (7.188) into (7.185) and (7.186), we have

$$\frac{\partial C_p(z', t_p')}{\partial z'} = -\tilde{g}_p C_p(z', t_p')\, C_s(z', t_s')\, e^{-\alpha z'} \tag{7.189}$$

$$\frac{\partial C_s(z',t'_s)}{\partial z'} = \tilde{g}_s C_p(z',t'_p)\, C_s(z',t'_s)\, e^{-\alpha z'} \tag{7.190}$$

Where $\tilde{g}_p = g_p/KA_e$, $\tilde{g}_s = g_s/KA_e$. And the boundary conditions are

$$\begin{aligned} C_p(z',t'_p)|_{z'=0} &= P_p(0,t'_p) \\ C_s(z',t'_s)|_{z'=0} &= P_s(0,t'_s) \end{aligned} \tag{7.191}$$

From (7.189) and (7.190), we have

$$\frac{\partial}{\partial z'}(\tilde{g}_s C_p + \tilde{g}_p C_s) = 0 \tag{7.192}$$

Therefore,

$$\tilde{g}_s C_p + \tilde{g}_p C_s = G(t') \tag{7.193}$$

Where $G(t')$ is an integral constant in the integral of (7.192) with respect to z', and then $G(t')$ is independent of z'. Therefore, we may consider that

$$G(t') = G(t')|_{z'=0} = [\tilde{g}_s C_p(z',t'_p) + \tilde{g}_p C_s(z',t'_s)]_{z'=0}$$

and then

$$G(t') = \tilde{g}_s P_p(0,t'_p) + \tilde{g}_p P_s(0,t'_s) \tag{7.194}$$

Substituting (7.194) into (7.193), we have

$$\begin{aligned} C_p &= \frac{G(t')}{\tilde{g}_s} - \frac{\tilde{g}_p}{\tilde{g}_s} C_s \\ C_s &= \frac{G(t')}{\tilde{g}_p} - \frac{\tilde{g}_s}{\tilde{g}_p} C_p \end{aligned} \tag{7.195}$$

Substituting (7.195) into (7.189) and (7.190), we have

$$\begin{aligned} \frac{\partial C_p}{\partial z'} &= -\,G(t')\, C_p\, e^{-\alpha z'} + \tilde{g}_s\, C_p^2\, e^{-\alpha z'} \\ \frac{\partial C_s}{\partial z'} &= G(t')\, C_s\, e^{-\alpha z'} - \tilde{g}_p\, C_s^2\, e^{-\alpha z'} \end{aligned}$$

Or

$$\begin{aligned} \frac{\partial}{\partial z'}\Big(\frac{1}{C_p}\Big) - G(t')\Big(\frac{1}{C_p}\Big)e^{-\alpha z'} &= -\tilde{g}_s\, e^{-\alpha z'} \\ \frac{\partial}{\partial z'}\Big(\frac{1}{C_s}\Big) - G(t')\Big(\frac{1}{C_s}\Big)e^{-\alpha z'} &= \tilde{g}_p\, e^{-\alpha z'} \end{aligned} \tag{7.196}$$

These are an ordinary differential equations with the form of

$$Y' + P(x)Y = Q(x)$$

and the solution is

$$Y = \exp\Big\{-\int Pdx\Big\}\Big(\int Q\exp\Big\{\int Pdx\Big\}dx + C\Big)$$

Following this equation, considering the boundary conditions (7.191) and (7.194), we have

$$C_p(z', t_p') = \frac{G(t')}{\tilde{g}_s}\frac{1}{1+H(z',t')} \tag{7.197}$$

$$C_s(z', t_s') = \frac{G(t')}{\tilde{g}_p}\frac{H(z',t')}{1+H(z',t')} \tag{7.198}$$

Where

$$H(z',t') = \frac{\omega_p}{\omega_s}\frac{P_s(0,t_s')}{P_p(0,t_p')}\exp[G(t')L_e] \tag{7.199}$$

$$L_e = [1 - e^{-\alpha z'}]/\alpha$$

$$G(t') = \tilde{g}_s[P_p(0,t_p') + (\omega_p/\omega_s)P_s(0,t_s')]$$

$$t_p' = t - \frac{n_{pL}}{c}z, \qquad t_s' = t - \frac{n_{sL}}{c}z \tag{7.200}$$

Substituting (7.197) and (7.200) into (7.187) and (7.188), considering, from (7.184), that $z' = z$, we have

$$P_p\Big(z, t - \frac{z}{v_p}\Big) = P_0(0,t')\exp(-\alpha z)\frac{1}{1+H} \tag{7.201}$$

$$P_s\Big(z, t - \frac{z}{v_s}\Big) = \Big(\frac{\omega_s}{\omega_p}\Big)P_0(0,t')\exp(-\alpha z)\frac{1}{1+H} \tag{7.202}$$

Where

$$H = H(z',t')$$

$$P_0(0,t') = P_p\Big(0, t - \frac{z}{v_p}\Big) + \frac{\omega_p}{\omega_s}P_s\Big(0, t - \frac{z}{v_s}\Big) \tag{7.203}$$

(7.201) and (7.202) are *the transient-state solution* of the coupling wave equations (7.182) and (7.183) under the condition of $\alpha_p = \alpha_s = \alpha$.

Now we will consider two situations:

(1) **Quasi-steady-state solution:**

For a pulse signal in a single-mode fiber, if the duration of the pulse signal is larger than the time delay [7] for several times, the dispersion in the single-mode fiber is able to be ignored, in this case, the transient-state solution is said to be *the quasi-steady-state solution.* And then,

$$v_p = v_s = v, \qquad \alpha_p = \alpha_s = \alpha$$

and the Stimulated Raman Scattering coupling wave equations (7.182) and (7.183) become

$$\frac{\partial P_p}{\partial z} + \frac{1}{v}\frac{\partial P_p}{\partial t} = -\frac{g_p}{KA_e}P_pP_s - \alpha P_p \tag{7.204}$$

$$\frac{\partial P_s}{\partial z} + \frac{1}{v}\frac{\partial P_s}{\partial t} = \frac{g_s}{KA_e}P_sP_p - \alpha P_s \tag{7.205}$$

The solutions of (7.204) and (7.205) can be obtained by substituting the following transformation

$$t' = t - \frac{z}{v_p} = t - \frac{z}{v_s} = t - \frac{z}{v} \tag{7.206}$$

into (7.201) and (7.202) to get

$$P_p\left(z,\, t - \frac{z}{v}\right) = P_0\left(0,\, t - \frac{z}{v}\right)\exp(-\alpha z)\frac{1}{1+H} \tag{7.207}$$

$$P_S\left(z,\, t - \frac{z}{v}\right) = \frac{\omega_s}{\omega_p}P_0\left(0,\, t - \frac{z}{v}\right)\exp(-\alpha z)\frac{H}{1+H} \tag{7.208}$$

Where $P_0\left(0,\, t - \frac{z}{v}\right)$, H are able to be obtained by substituting (7.206) into (7.203) and (7.199).

(2) **Steady-state solution:**

For a continuous wave in a single-mode fiber, the solution of the Stimulated Raman Scattering is said to be *the steady-state solution.* In this case, the coupling wave equations are

$$\frac{\partial P_p}{\partial z} = -\frac{g_p}{KA_e}P_pP_s - \alpha P_p \tag{7.209}$$

$$\frac{\partial P_s}{\partial z} = \frac{g_s}{KA_e}P_pP_s - \alpha P_s \tag{7.210}$$

[7] The time delay is caused by the wave travelling the total length L of the single-mode fiber.

Where the time factor $\exp(i\omega t)$ has been considered in (7.174). And the $\partial/\partial t$ term is absent since the wave amplitude is invariant along the single-mode fiber and then it is said to be *the steady-state solution.* The solution is able to be obtained by substituting

$$t = z/v \tag{7.211}$$

into (7.207) and (7.208) to get

$$P_p(z) = P_0(0)\exp(-\alpha z)\frac{1}{1+H} \tag{7.212}$$

$$P_s(z) = \frac{\omega_s}{\omega_p}P_0(0)\exp(-\alpha z)\frac{H}{1+H} \tag{7.213}$$

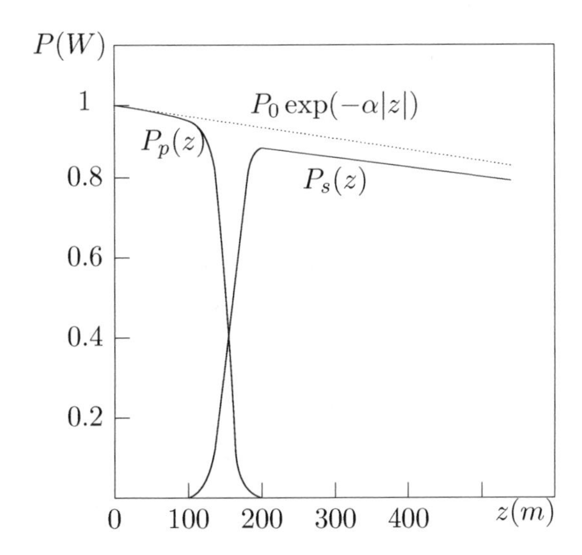

Figure 7-4 The steady-state solution of SRS

Where H is able to be obtained by substituting (7.211) into (7.199). $P_0(0)$ is able to be obtained by substituting (7.211) into (7.203) to get

$$P_0(0) = P_p(0) + \frac{\omega_p}{\omega_s}P_s(0)$$

The steady-state solution (7.212) and (7.213) describes the transformation process between the pumping power $P_p(z)$ and the Stocks power $P_s(z)$ as shown in Figure 7-4.

Problem:

1. Starting from (7.170) to prove that:

(1) The quasi-steady-state equations are (7.204) (7.205) and the solutions are (7.207) and (7.208).

(2) The steady-state equations are (7.209), (7.210) and the solutions are (7.212) and (7.213). [**Hint:** You may follow the process of finding the transient-state equations and the solutions].

2. Starting from (7.170) to prove that the steady-state coupling wave equations of the pumping power P_p, the forward Stocks power P_F and the anti-Stocks power P_B are

$$\frac{\partial P_p}{\partial z} = -\frac{g_p}{KA_e}(P_F + P_B)P_p$$

$$\frac{\partial P_F}{\partial z} = \frac{g_s}{KA_e}P_pP_F$$

$$\frac{\partial P_B}{\partial z} = -\frac{g_s}{KA_e}P_pP_B$$

respectively. [**Hint:** By substituting (7.174) into (7.170)]. Where $g_s = g_p(\omega_s/\omega_p)$. And the solutions of these equations are

$$P_F(z) = \frac{\omega_s}{\omega_p}\frac{P_i + Wtanh(g_sUz/2KA_e)}{1 - Vtanh[g_sU(L-z)/2KA_e]}$$

$$P_B(z) = \frac{\omega_s}{\omega_p}\frac{P_i + Wtanh(g_sU(L-z)/2KA_e)}{1 - Vtanh[g_sU(L-z)/2KA_e]}$$

$$P_P(z) = P_0 - \frac{P_i + Wtanh(g_sUz/2KA_e)}{1 - Vtanh[g_sUz/2KA_e]}$$

$$+\frac{P_i + Wtanh(g_sU(L-z)/2KA_e)}{1 - Vtanh[g_sU(L-z)/2KA_e]}$$

Where

$$P_0 = P_{p0} + (\omega_p/\omega_s)(P_{qs} + RP_{FL}) - (\omega_P/\omega_s)P_{FL}$$

$$P_i = (\omega_p/\omega_s)(P_{qs} + RP_{FL})$$

$$U = [4(\omega_p/\omega_s)^2P_{FL}(P_{qs} + RP_{FL}) + P_0^2]^{1/2}$$

$$V = [P_0 - 2(\omega_p/\omega_s)(P_{qs} + RP_{FL})]$$

$$W = (U - P_0V)/2$$

P_{qs} is the noise power caused by the Stimulated Raman Scattering. R is the reflection coefficient at the end surface of the single-mode fiber . P_{FL} is the forward wave power at the end surface of the single-mode fiber . L is the length of the single-mode fiber .

3. **Stimulated Brillouin Scattering (SBS)**

The Stimulated Raman Scattering is caused by the interaction between the optical fiber mode and the molecular variation mode under an injection of a large pumping power.

The feature of the Stimulated Raman Scattering is that the bandwidth of the Raman scattering is very wide (about $300cm^{-1}$) and the Raman frequency shift is large (about $400cm^{-1}$).

The Stimulated Brillouin Scattering is caused by the interaction between the optical fiber mode and the acoustic wave under an injection of a narrow pumping source. And the line-width of the pumping source is about 100MHz. The Brillouin frequency shift is about $1cm^{-1}$. Under this condition, the maximin gain of the Stimulated Brillouin Scattering is two orders large than the maximin gain of the Stimulated Raman Scattering. Since the matching condition of the wave vector, the Brillouin amplification occurs only in the backward direction. Therefore, substituting $m = p, \quad m' = s = B$ into (7.174), we have the coupling wave equations for the Stimulated Brillouin Scattering as follows:

$$\begin{aligned} \frac{\partial P_p}{\partial z} + \frac{n_{pL}}{c}\frac{\partial P_p}{\partial t} &= \frac{g_p}{KA_e}P_B P_p - \alpha_p P_p \\ \frac{\partial P_B}{\partial z} - \frac{n_{sL}}{c}\frac{\partial P_B}{\partial t} &= -\frac{g_s}{KA_e}P_p P_B + \alpha_s P_B \end{aligned} \tag{7.214}$$

Where

$$g_p = 3\omega_p \left| Im[\chi^{(3)}(-\omega_p, -\omega_s, \omega_s, -\omega_p)] \right|$$

$$g_s = 3\omega_s \left| Im[\chi^{(3)}(-\omega_s, -\omega_p, \omega_p, -\omega_s)] \right|$$

That means that the Stimulated Brillouin Scattering is also caused by the imaginary part of $\chi^{(3)}$. The process to find the solution is almost the same as that in the Stimulated Raman Scattering .

4. **Self-Focus Phenomenon:**

As well known, the phase (or phase velocity) depends on the refraction index of the medium and the nonlinear refraction index in single-mode fiber depends on the light intensity. Consequently, if the medium is a single-mode fiber and the optical fiber can only support a single mode propagation, the real part of $\chi^{(3)}$ in the single-mode fiber will cause the variation of the phase (or phase velocity) with the variation of the light intensity, and then forms the self-focus phenomenon as we mentioned in subsection 9.3.5.

In this case, suppose there exist a pumping wave P_p , a Stocks wave P_s and an anti-Stocks wave P_a in the Stimulated Raman Scattering, the self-focus phenomenon will forms a phase modulation on each wave. The coupling wave equation to describe this phenomenon is able to be obtained by substituting (7.174) into (7.170) and taking the real part of $\chi^{(3)}$ as follows:

$$\frac{dP_p}{dz} = i\Big(\frac{\gamma_{pp}}{A_e}P_p + \frac{\gamma_{ps}}{A_e}P_s + \frac{\gamma_{pa}}{A_e}P_a\Big)P_p$$

$$\frac{dP_s}{dz} = i\Big(\frac{\gamma_{sp}}{A_e}P_p + \frac{\gamma_{ss}}{A_e}P_s + \frac{\gamma_{sa}}{A_e}P_a\Big)P_s$$

$$\frac{dP_a}{dz} = i\Big(\frac{\gamma_{ap}}{A_e}P_p + \frac{\gamma_{sa}}{A_e}P_s + \frac{\gamma_{aa}}{A_e}P_a\Big)P_a \tag{7.215}$$

Where

$$\gamma_{ii} = (3\omega_i/2)\, Re[\chi^{(3)}(-\omega_i, -\omega_i, \omega_i, -\omega_i)]$$

$$\gamma_{ij} = 3\omega_i\, Re[\chi^{(3)}(-\omega_i, -\omega_j, \omega_j, -\omega_i)] \tag{7.216}$$

$$\omega_{ji} = (\omega_j/\omega_i)\gamma_{ij}$$

(7.215) is the steady-state equation, in which P_p is the forward pumping wave power, P_s is the Stocks wave power and P_a is the anti-Stocks wave power. If the anti-Stocks wave is omitted the solution becomes

$$P_p(l+z) = P_p(l)e^{i\delta k_p(l)z} \tag{7.217}$$

$$P_s(l+z) = P_s(l)e^{i\delta k_s(l)z} \tag{7.218}$$

Note, this solution is under the condition of $\alpha L < 0.1$. Where L is the total length of the single-mode fiber , $l = L/z$ and

$$\delta k_i(l) = (\gamma_{ip}/A_e)P_p(l) + (\gamma_{is}/A_e)P_s(l), \quad i = p, s$$

Obviously, the phase $\delta k_i(l)z$ $(i = p, s)$ in (7.217) and (7.218) depends on the light power or the light intensity in the single-mode fiber. In other words, this is the phase modulation on each wave P_p and P_s.

7.4 Function Transformation method

7.4.1 Function Transformation method

An efficient method to find the solution of the nonlinear equation is to transform the nonlinear equation into a linear equation via the function transformation. This is said to be *the function transformation method.*

Suppose there is a liner equation

$$L[v] = 0 \tag{7.219}$$

and the solution v of the linear equation is a nonlinear equation of a variable u, i.e.

$$v = F(u) \tag{7.220}$$

Now, substituting (7.220) into (7.219), we have a nonlinear equation as follows:

$$L[F(u)] = 0. \tag{7.221}$$

If the given problem is to find the solution u of the nonlinear equation (7.221), then we may find the solution v of the linear equation (7.220) first, then go through the inverse transformation of (7.220) to obtain u as

$$u = F^{-1}[v] \tag{7.222}$$

Now the problem is: if the inverse transformation F^{-1} and the boundary condition of v are easy to be determined?.

7.4.2 The application of the function transformation method

Assume v is the solution of the heat conduction equation

$$v_t = \mu\, v_{xx} \qquad (\mu > 0) \tag{7.223}$$

Then via the Cole-Hopf transformation

$$u = -(2\mu/\alpha)(1/v)v_x \tag{7.224}$$

we found that $u(x,t)$ satisfies the Burgers equation

$$u_t + \alpha u u_x - \mu u_{xx} = 0 \tag{7.225}$$

In this case, the nonlinear equation (7.225) has been transformed into the linear equation (7.223) via the transformation of (7.224), and then to find the solution u from the nonlinear equation (7.225) becomes to find the solution v from the linear equation (7.223) and then the solution u may be obtained by (7.224) after we obtained v. Where the inverse transformation of (7.224)

$$v(x,t) = \exp\Big[-\frac{\alpha}{2\mu} \int u(x,t)dx\Big]$$

is also easy to be obtained.

7.4.3 The nonlinear equation of the Josephson junction

Now we will discuss that how to find the phase difference $\varphi(t)$ of the Josephson junction as indicated in Figure 7-5. This was a project studied by author and Dr. Jamal Deen when both of them have been in Simon Fraser University, Burnaby, Canada.

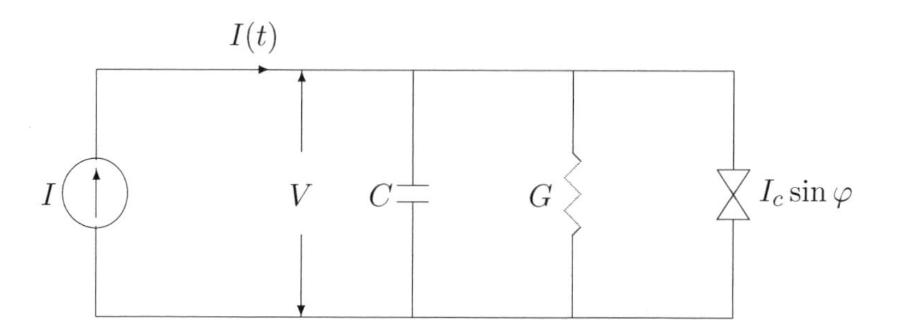

Figure 7-5 The equivalent circuit of a Josephson junction

Figure 7-5 is an equivalent circuit of a super-conducting junction or a Josephson junction. In which, $\varphi(t)$ is the phase difference across the point contact of the Josephson junction, and is related to the voltage V in form of

$$\frac{d\varphi}{dt} = \frac{2e}{\hbar}V \tag{7.226}$$

which results in

$$\frac{dV}{dt} = \frac{\hbar}{2e}\varphi_{tt} \tag{7.227}$$

Where $\varphi_{tt} = \dfrac{d^2\varphi}{dt^2}$.

And the current $I(t)$ is the sum of the following currents:

1. The capacity current I_C in form of

$$I_C = C\frac{dV}{dt} = \frac{\hbar C}{2e}\varphi_{tt} \tag{7.228}$$

where use of (7.227) been made.

2. The conductance current I_G in form of

$$I_G = GV = \frac{\hbar G}{2e}\varphi_t \tag{7.229}$$

where use of (7.226) has been made.

3. The phase-different current in form of

$$I_p = \varepsilon \cos \varphi G V = \varepsilon \cos \varphi \frac{\hbar G}{2e} \varphi_t \tag{7.230}$$

Where the phase-different linear conductance of a weakly damped superconducting quantum point contact has been obtained by A Levy Yeyati, A Martin-Rodero and J C Cuevas [41].

4. Junction current in form of

$$I_J = I_c \sin \varphi \tag{7.231}$$

And then, the equation describing the circuit in Figure 7-5 is

$$I = I_C + I_G + I_P + I_J$$

or

$$I = \frac{\hbar C}{2e} \varphi_{tt} + \frac{\hbar G}{2e} (1 + \varepsilon \cos \varphi) \varphi_t + I_c \sin \varphi \tag{7.232}$$

Obviously, this is a nonlinear equation of $\varphi(t)$ and is a complicated nonlinear equation. Now we will try to obtain the solution by the function transformation.

To this end, the equation (7.232) can be written as

$$I = a \varphi_{tt} + b(1 + \varepsilon \cos \varphi) \varphi_t + I_c \sin \varphi \tag{7.233}$$

where

$$a = \frac{\hbar C}{2e}, \quad b = \frac{\hbar G}{2e} \tag{7.234}$$

The function transformation is consist of three steps as follows:

1. **Step 1:**

Considering that the equation (7.233) contains $\cos \varphi$ and $\sin \varphi$ we may set the first function transformation in form of

$$e^{i\varphi} = x \tag{7.235}$$

so as the terms $\cos \varphi$ and $\sin \varphi$ become the functions of x as follows:

$$\sin \varphi = \frac{1}{2i} \left(x - \frac{1}{x} \right) \tag{7.236}$$

$$\cos \varphi = \frac{1}{2} \left(x + \frac{1}{x} \right) \tag{7.237}$$

Then, from (7.235), we have

$$\frac{1}{i} \text{[illegible]} \tag{7.238}$$

$$\frac{d\varphi}{dt} = \frac{1}{i}\frac{1}{x}\frac{dx}{dt}$$
$$\frac{d^2\varphi}{dt^2} = \frac{1}{i}\frac{1}{x}\frac{d^2x}{dt^2} - \frac{1}{i}\frac{1}{x^2}\left(\frac{dx}{dt}\right)^2$$

and then, substitution of (7.236), (7.237) and (7.238) into (7.233) we have

$$-i\frac{a}{x}\frac{d^2x}{dt^2} + i\frac{a}{x^2}\left(\frac{dx}{dt}\right)^2 + b\Big[1+\frac{\varepsilon}{2}\Big(x+\frac{1}{x}\Big)\Big]\Big(-i\frac{1}{x}\frac{dx}{dt}\Big) - i\frac{I_c}{2}\Big(x-\frac{1}{x}\Big) - I = 0 \quad (7.239)$$

Multiplying both sides of (7.239) by $\left(i\frac{x}{a}\right)$ we have

$$x_{tt} - \frac{1}{x}\Big(x_t\Big)^2 + A\Big[1+\frac{\varepsilon}{2}\Big(x+\frac{1}{x}\Big)\Big]x_t + B\Big(x^2-1\Big) + i\bar{I}x = 0 \quad (7.240)$$

with

$$A = \frac{b}{a}, \quad B = \frac{I_c}{2a}, \quad \bar{I} = -\frac{I}{a} \quad (7.241)$$

2. **Step 2:**

To transform the second order nonlinear equation into a first order nonlinear equation we may set the second function transformation in form of

$$x_t = z \quad (7.242)$$

and then

$$x_{tt} = \frac{dz}{dx}\frac{dx}{dt} = z\frac{dz}{dx} \quad (7.243)$$

In this case, the second order nonlinear equation (7.240) becomes a first order nonlinear equation in form of

$$z\frac{dz}{dx} - \frac{1}{x}z^2 + q(x)z + p(x) = 0 \quad (7.244)$$

where

$$q(x) = A\Big[1+\frac{\varepsilon}{2}\Big(x+\frac{1}{x}\Big)\Big] \quad (7.245)$$
$$p(x) = B\Big(x^2-1\Big) + i\bar{I}x \quad (7.246)$$

3. **Step 3:**

To combine the first term and the second term in (7.244) into one term, we may set the third function transformation in form of

$$y = \frac{z}{x} \quad (7.247)$$

and then

$$z = xy \quad (7.248)$$

$$\frac{dz}{dx} = y + x\frac{dy}{dx}$$

or

$$\frac{dz}{dx} - \frac{z}{x} = x\frac{dy}{dx} \tag{7.249}$$

Where, use of (7.247) has been made.

Substitution of (7.249) into (7.244) we have

$$z\left(x\frac{dy}{dx}\right) + q(x)z + p(x) = 0$$

or

$$x\frac{dy}{dx} + q(x) + \frac{1}{xy}p(x) = 0 \tag{7.250}$$

where use of (7.248) has been made.

And then, from (7.250), we have

$$\frac{dy}{dx} + \frac{1}{y}P(x) = Q(x) \tag{7.251}$$

with

$$P(x) = \frac{1}{x^2}p(x) = B\left[\left(1 - \frac{1}{x^2}\right) + i\frac{\bar{I}}{x}\right] \tag{7.252}$$

$$Q(x) = -\frac{1}{x}q(x) = -A\left[\frac{1}{x} + \frac{\varepsilon}{2}\left(1 + \frac{1}{x^2}\right)\right] \tag{7.253}$$

Obviously, (7.251) is a first order nonlinear equation from the second order nonlinear equation (7.233) and the nonlinear equation (7.251) is much simpler than the nonlinear equation (7.251). Even though, to find the solution of the nonlinear equation (7.251) is still not a easy duty, we have provided three steps of the function transformation (7.235),(7.242) and (7.247), which may be useful for some analysis.

Further more, if we can find the solution $y(t)$ of the nonlinear equation (7.251), then we have, from (7.248),

$$z = xy$$

Then, from (7.242), we have

$$\frac{dx}{dt} = z \tag{7.254}$$

and then, we have

$$x = \int zdt = \int yxdt$$

or

$$e^{i\varphi(t)} = \int y(x,t)xdt$$

or

$$e^{i\varphi(t)} = \int y\left(e^{i\varphi(t)}, e^{i\omega t}\right) e^{i\varphi(t)} dt \tag{7.255}$$

where use of (7.235) has been made, and the $e^{i\omega t}$ term in (7.255) is considered to be the sin signal in $I(t)$. Obviously, the solution of the equation (7.255) is a function of $e^{i\varphi(t)}$ and is not so difficult to be obtained. After we found the $e^{i\varphi(t)}$, then we will obtain the final solution $\varphi(t)$ of (7.233).

7.5 Inverse Scattering Method

Tosiya Taniuti and Katsunobu Nishihara in [38] has discuss the *Inverse Scattering Method* for the area of particle physics, plasma physics, ecology and elsewhere.

Now we will introduce the idea about the *Inverse Scattering Method* briefly, and then introduce some formulas from Tosiya Taniuti and Katsunobu Nishihara in [38] to discuss the initial value problem of nonlinear wave in a long-haul optical fiber communication system.

Suppose a power from laser is injected into a single mode fiber, the laser power at some point inside the optical single mode fiber can be tested. And then the testing results can be viewed as an initial value. In this case, we will find the solution of the initial value problem of a nonlinear wave equation by *Inverse Scattering Method* as follows:

As the following discussion, we will find that the function transformation method not only provide a way to find the solution for a nonlinear equation, but also open a way to go to the inverse scattering method, which can be used to find a solution for the initial value problem of the nonlinear equations.

7.5.1 Inverse Scattering Method

The initial value problem of the KdV equation is to find a solution $u(x,t)$ to satisfy the KdV equation

$$u_t - 6uu_x + u_{xxx} = 0 \tag{7.256}$$

and the initial value

$$u(x,0) = u_0(x), \qquad |x| < \infty,\ t \geq 0$$

where, $u(x,0)$ is the initial amplitude of the nonlinear wave in an optical single mode fiber.

Now, this problem is able to be solved with the five steps as follows:

Step 1:

The solution $u(x,t)$ of the KdV equation (7.256) is able to be transformed into the potential u of the Schrödinger equation

$$\Psi_{xx} - u\Psi = 0 \tag{7.257}$$

via a function transformation [8]

$$u = v^2 + v_x \tag{7.258}$$

and

$$v = \Psi_x/\Psi \tag{7.259}$$

Since the KdV equation is invariant under the following transformation

$$u \to u - \lambda, \qquad x \to x + 6\lambda x$$

we may consider that $u - \lambda$ is the solution of the KdV equation (7.256) as well. And then from (7.257), we have

$$\Psi_{xx} - (u - \lambda)\Psi = 0 \tag{7.260}$$

Where it is provable that

$$\lambda_t = 0 \tag{7.261}$$

that means that the u in the Schrödinger equation (7.260) is an iso-spectral potential.

Step 2:

And then, substituting the initial value $u(x,0)$ into (7.260), it is able to find the spectra of the Schrödinger equation (7.260). These spectra consist of the discrete spectra in bounded regions and the continuous spectrum in un-bounded region and result in the scattering data at $t = 0$.

Step 3:

Then, substitution of the Schrödinger equation (7.260) into the KdV equation (7.256), we will have the time evolution equation

$$R = \Psi_t + \Psi_{xxx} - 3(u + \lambda)\Psi_x = C(t)\Psi + D(t)\,\Psi \int_0^x \frac{dx}{\Psi^2} = 0 \tag{7.262}$$

Step 4:

---

[8] Namely,

$$u = v^2 + v_x = \frac{\Psi_x^2}{\Psi^2} + \frac{\Psi_{xx}\Psi - \Psi_x^2}{\Psi^2} = \frac{\Psi_{xx}}{\Psi}$$

Substituting the scattering data at $t = 0$ into the time evolution equation (7.262), we will have the scattering data at $t = t$.

Step 5:

Substituting the scattering data at $t = t$ into the G-L-M (Gel'fand-Levitan-Marchenko) equation

$$K(x, x; t) + B(x + y; t) + \int_x^\infty B(x + z; t) K(x, z; t) dz = 0 \qquad (y > x) \qquad (7.263)$$

and

$$B(x + y; t) = (2\pi)^{-1} \int_{-\infty}^{\infty} r(k, t) e^{ik(x+y)} dk + \sum_{n=1}^{N} C_n^2(t) e^{-k_n(x+y)} \qquad (7.264)$$

we will find the solution $K(x, x; t)$ from the integral equations (7.263) and (7.264), and then we will have the solution $u(x, t)$ of the KdV equation (7.256) as follows

$$u(x, t) = -2\frac{d}{dx} K(x, x; t) \qquad (7.265)$$

Where the G-L-M equation (7.263) and (7.264) are from the the Schrödinger equation (7.260).

If we read carefully, we will find that the equation in each step is a linear equation and can be solved right away. And then the initial value problem of a nonlinear equation becomes a linear equation problem in each step.

The "Inverse Scattering Method" has been discussed more in detail by Tosiya Taniuti and Katsunobu Nishihara in [38]. Here, we only introduce an idea about the "Inverse Scattering Method" briefly.

To generalize the inverse scattering method further, we have to answer two questions as follows:

(1) For a given nonlinear equation, how to find another nonlinear equation so as the iso-spectral potential of the another nonlinear equation is the solution of the original nonlinear equation?

(2) If the another nonlinear equation has been obtained, how to derive the corresponding time evolution equation?

P.D. Lax answer these two questions by means of the following theorem.

7.5.2 Lax Pair Theorem*

The following discussion is based on some material from Wikipedia, the free encyclopedia. We have make a note * as the material from Wikipedia, the free encyclopedia.

A Lax pair, developed by Peter Lax [39], [40] to discuss solitons in continuous media, is a pair of one parameter families of matrices/operators that describe certain solution of differential equations.

Definition of the Lax Pair:

In more detail, a lax pair is a pair of one parameter operators $L(t)$, $A(t)$ acting on a fixed Hilbert space such that

$$L_t = LA - AL$$

Where A depends on L in a prescribed way, and L as a function of t is a nonlinear equation $F(u)$ of u, namely

$$L_t = F(u)$$

In the pioneering paper, Lax stated a condition under which certain one families of operators $\{L(t)\}$ are iso-spectral, namely, all the $L(t)$ have the same spectrum. Therefore, the eigenvalues and the continuous spectrum of L are independent of t, namely,

$$L\Psi = \lambda\Psi$$

and

$$\lambda_t = 0$$

and the time evolution equation of the eigenfunction Ψ is

$$\Psi_t = A\Psi$$

The discussion above may be resulted in the Lax pair theorem as follows.

Lax Pair Theorem:

A nonlinear equation

$$L_t = F(u) \tag{7.266}$$

can be reformulated as the Lax Pair equation

$$L_t = AL - LA \tag{7.267}$$

Where **the Lax pair** $L(t)$ and $A(t)$ are the one parameter t matrices/operators and acting on the Hilbert space. And all the $L(t)$ have the same spectrum or the spectrum of $L(t)$ is time-independent, namely,

$$L\Psi = \lambda\Psi \tag{7.268}$$

and

$$\lambda_t = 0 \tag{7.269}$$

and the time evolution equation of the eigenfunction Ψ is

$$\Psi_t = A\Psi \tag{7.270}$$

Example:

The KdV equation

$$u_t = 6uu_x - u_{xxx} \tag{7.271}$$

can be reformulated as the Lax Pair equation

$$L_t = AL - LA \tag{7.272}$$

with

$$L = -\partial^2 + u, \quad (a\ Sturm - Liouville\ operator) \tag{7.273}$$

$$A = 4\partial^3 - 3(u\partial + \partial u) \tag{7.274}$$

Where $\partial = \dfrac{\partial}{\partial t}$. Obviously

$$L_t = u_t$$

and

$$AL - LA = (u_{xxx} - 6uu_x)$$

and then, from (7.271), L and A satisfies the Lax Pair equation (7.272), namely,

$$L_t = AL - LA$$

In this case, from the Lax Pair theorem, the eigenvalues λ given by the equation

$$L\Psi = -\Psi_{xx} + u\Psi = \lambda\Psi \tag{7.275}$$

are time-independent, and the time evolution equation of the spectrum function Ψ is

$$\Psi_t = A\Psi \tag{7.276}$$

or

$$\Psi_t + \Psi_{xxx} - 3(u + \lambda)\Psi_x = 0 \tag{7.277}$$

Where (7.275) is the Schrödinger equation. (7.277) is the time evolution equation. That means that the process to find the solution of the initial problem of the KdV equation by means the "Inverse Scattering Method" is coincident with the Lax Pair theorem.

Proof of the Lax pair Theorem:

As we mentioned before, a Lax pair is a pair of matrices/operators $L(t)$ and $A(t)$ dependent on time t and acting on a fixed Hilbert space. The eigenvalues and the continuous spectrum of L are independent of t, and the lax Pair equation is the infinitesimal form of a family of matrices $L(t)$, all having the same spectrum, by virtue of giving by

$$L(t) = g^{-1}L(0)g(t) \tag{7.278}$$

Where g can be arbitrarily complicated and satisfy

$$g^{-1}g = gg^{-1} = I \tag{7.279}$$

where I is the unit operator.

In this case, from (7.279),

$$\begin{aligned}\frac{dL}{dt} = \frac{dg^{-1}}{dt}L(0)g(t) + g^{-1}L(0)\frac{dg}{dt} &= -g^{-2}\frac{dg}{dt}L(0)g + \left(g^{-1}L(0)g\right)\left(g^{-1}\frac{dg}{dt}\right) \\ &= -\left(g^{-1}\frac{dg}{dt}\right)\left(g^{-1}L(0)g\right) + LA = LA - AL\end{aligned}$$

with

$$A = g^{-1}\frac{dg}{dt} \tag{7.280}$$

and then the Lax Pair equation (7.267) has been proved.

To sum up:

The Lax Pair Theorem said, a given nonlinear equation

$$u_t = F(u) \tag{7.281}$$

can be reformulated as the Lax Pair equation

$$L_t = AL - LA \tag{7.282}$$

as long as we can find the Lax Pair $L(t)$ and $A(t)$. In this case, the eigenvalues λ given by the equation

$$L\Psi = \lambda\Psi \tag{7.283}$$

are time-independent, and the time evolution equation of the spectrum function Ψ is

$$\Psi_t = A\Psi \tag{7.284}$$

Based on this idea, Zakharov and Shahat gave a great contribution as follows.

In the analysis of the Inverse Scattering Method for the Schrödinger equation in [38], Zakharov and Shahat set up a generalized Inverse Scattering Theory in Zakharov-Shahat system , and then lead to an Inverse Transformation Equation similar to the $G - L - M$ integral equation, so as to find the solution for the original equation. The Generalized Inverse Scattering Theory may be used to find the solutions of the initial value problem for many nonlinear wave equations, such as the KdV equation

$$u_t - 6uu_x + u_{xxx} = 0 \tag{7.285}$$

the nonlinear Schrödinger equation

$$iu_t + u_x x + Q|u|^2 u = 0 \tag{7.286}$$

the Sine-Gordon equation

$$\varphi_{tt} - \varphi_{xx} + \sin\varphi = 0 \tag{7.287}$$

and so on. If you are interesting you may read [38].

Part III

Mode Theory

Chapter 8

Mode theory in curved structures

8.1 Introduction

Nowadays, some times we have to deal with the problem with the curved structure or the twist structure in the fiber optics (such as the optical sensors) or in the microstrip components and the waveguide components. In this case, the best way is to chose the curved coordinates system since the boundary of the curved structure can match the curved coordinates.

The pure curved coordinates system is still the orthogonal coordinates system except the twist coordinates system. The twist coordinates system is a non-orthogonal coordinates system except at the axis of the twist. In other words, it is an orthogonal coordinates system at the axis of the twist alone, however, it is a non-orthogonal coordinates system if it away from the axis of the twist. In this case, we have to make use of the coordinates transformation to transform the field into a new orthogonal coordinates system. And then we can obtain the solution by means of the Maxwell equation in the new orthogonal coordinates system. To this end, we have to discuss the twist coordinates system first and the curved coordinates system is the special case of the twist coordinates system only.

8.2 The coordinates system in curved structure

8.2.1 The twist coordinates system

Suppose that there is a smooth curve in the space drawing out by the end point of a vector $\mathbf{R}(s)$ in Figure 8-1. Where s is the curve length measured from the start point of the curve, and $\mathbf{R}(s)$ is the function of s. $\mathbf{a_t}$ is the unit vector along the tangential direction of the curve. And then $\mathbf{R}(s) + \mathbf{a_t}ds = \mathbf{R}(s + ds)$. From which, we have

$$\mathbf{a_t} = \mathbf{R}' \tag{8.1}$$

Where $\mathbf{R}'$ is the derivative of $\mathbf{R}$ with respect to s.

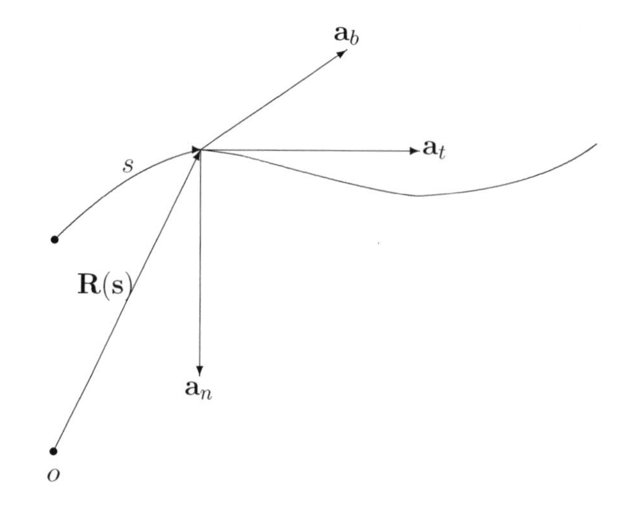

Figure 8-1 The twist coordinates system

Meanwhile, when s moves to $s + ds$, $\mathbf{a}_t$ rotates an angle $d\theta$ and then forms a plane with this rotation such that the principle normal direction at this point is located in this plane. In other words, if $\mathbf{a}_n$ is the unit vector of the curve along the principle normal direction at this point, then $\mathbf{a_n}d\theta$ will become the third side of a triangle, while another two sides of this triangle are $\mathbf{a_t}(s)$ and $\mathbf{a_t}(\mathbf{s} + \mathbf{ds})$, namely, $\mathbf{a_t}(s + ds) = \mathbf{a_t}(s) + \mathbf{a_n}d\theta$. And then $\mathbf{a_n} = \mathbf{a_t'}\dfrac{ds}{d\theta}$.

Definition: Making a definition of

$$\chi = \frac{d\theta}{ds} \tag{8.2}$$

as *the rotating rate* around the curve s, we have

$$\mathbf{a_n} = \mathbf{a_t'}/\chi \tag{8.3}$$

Making the derivative of (8.1) with respect to s, substituting the result into (8.3), we have $\mathbf{a_n} = \mathbf{R}''/\chi$, then $\mathbf{a_n} \cdot \mathbf{a_n} = 1$ give us

$$\chi^2 = \mathbf{R}'' \cdot \mathbf{R}'' \tag{8.4}$$

Making a rotation from the unit vector $\mathbf{a_t}$ to the principle normal vector $\mathbf{a_n}$ in right-hand rule, we will have a definition of the sub-normal vector $\mathbf{a_b}$ as $\mathbf{a_b} = \mathbf{a_t} \times \mathbf{a_n}$. And then we have, from (8.1) and (8.3),

$$\mathbf{a_b} = \mathbf{R}' \times \mathbf{R}''/\chi \tag{8.5}$$

When the point moves from s to $s + ds$, the principle normal vector $\mathbf{a_n}$ becomes $\mathbf{a_n}(s + ds)$, and the difference, $\mathbf{a_n}(s + ds) - \mathbf{a_n}(s)$, is perpendicular to $\mathbf{a_n}$. As indicated in Figure 8-1, $\mathbf{a_n}(s+ds)$ and $\mathbf{a_n}(s)$ are the unit vectors, both of them form a sector in a unit circle and then the difference between $\mathbf{a_n}(s+ds)$ and $\mathbf{a_n}(s)$ is perpendicular to $\mathbf{a_n}$(s). Therefore, $\mathbf{a'_n}$, the derivative of $\mathbf{a_n}$ with respect to s, is the linear combination of $\mathbf{a_b}$ and $\mathbf{a_t}$ since both of them are perpendicular to $\mathbf{a_n}$ too. Following this point of view, it is able to consider that

$$\mathbf{a'_n} = A\mathbf{a_b} + B\mathbf{a_t} \tag{8.6}$$

In more detail, if s is a twist curve without any bending, then

$$\mathbf{a_n}(s+ds) - \mathbf{a_n}(s) = \mathbf{a_b}d\Psi \tag{8.7}$$

Where $d\Psi$ is the included angle between $\mathbf{a_n}(s+ds)$ and $\mathbf{a_n}(s)$. The direction of the difference, $\mathbf{a_n}(s+ds)$ - $\mathbf{a_n}(s)$, is coincident with $\mathbf{a_b}$, and then (8.7) is confirmed.

Definition: The definition of *the torsion of a twist curve* is

$$\tau = \frac{d\Psi}{ds} \tag{8.8}$$

Substituting (8.8) into (8.7), we have that A in (8.6) is equal to τ.
Substituting (8.1),(8.3),(8.5) into (8.6) we have

$$(\mathbf{R}'''/\chi) - (\mathbf{R}''\chi'/\chi^2) = A\mathbf{R}' \times \mathbf{R}''/\chi + B\mathbf{R}' \tag{8.9}$$

Then A may be obtained by means of the scale product $[(\mathbf{R}' \times \mathbf{R}'') \cdot (8.9)]$ as follows:

$$A = \tau = [(\mathbf{R}' \times \mathbf{R}'') \cdot \mathbf{R}''']/[(\mathbf{R}' \times \mathbf{R}'') \cdot (\mathbf{R}' \times \mathbf{R}'')] \tag{8.10}$$

Where the torsion τ is expressed by terms of the derivative of $\mathbf{R}$ with respect to s alone.

Now since $\mathbf{a_t} \cdot \mathbf{a_n} = 0$, we have $\mathbf{R}' \cdot \mathbf{R}'' = 0$. The derivative of $\mathbf{R}' \cdot \mathbf{R}'' = 0$ with respect to s gives $\mathbf{R}'' \cdot \mathbf{R}'' = -\mathbf{R}' \cdot \mathbf{R}'''$. And then the scale product $[(\mathbf{a_t} = \mathbf{R}') \cdot (8.6)]$ give us

$$B = \mathbf{R}' \cdot \mathbf{R}'''/\chi = -(\mathbf{R}'' \cdot \mathbf{R}'')/\chi = -\chi$$

Substituting A and B into (8.6) gives us

$$\mathbf{a'_n} = \tau\mathbf{a_b} - \chi\mathbf{a_t} \tag{8.11}$$

In addition, the derivative of $\mathbf{a_b} = \mathbf{a_t} \times \mathbf{a_n}$ with respect to s gives $\mathbf{a'_b} = \mathbf{a'_t} \times \mathbf{a_n} + \mathbf{a_t} \times \mathbf{a'_n}$. Where, from (8.3), $\mathbf{a'_t} = \chi\mathbf{a_n}$, and then $\mathbf{a'_t} \times \mathbf{a_n} = 0$. Then from (8.11), $\mathbf{a_t} \times \mathbf{a'_n} = -\tau\mathbf{a_n}$, we have

$$\mathbf{a'_b} = -\tau\mathbf{a_n}$$

All of those come to the conclusion: that the relationship between the three orthogonal vectors $\mathbf{a_t}$, $\mathbf{a_n}$, $\mathbf{a_b}$ in the twist coordinates system are as follows:

$$\begin{aligned} \mathbf{a_b} &= \mathbf{a_t} \times \mathbf{a_n} \\ \mathbf{a'_t} &= \chi\mathbf{a_n} \\ \mathbf{a'_n} &= -\chi\mathbf{a_t} + \tau\mathbf{a_b} \\ \mathbf{a'_b} &= -\tau\mathbf{a_n} \end{aligned} \tag{8.12}$$

8.2.2 Serret-Frenet Frame

$\mathbf{a_n}$, $\mathbf{a_b}$, $\mathbf{a_t}$ at any point along the curve s forms a right-hand orthogonal coordinates system, which is so called the well known *Serret-Frenet Frame*. The frame is rotating while the point in curve s is moving. If the curve is a plane curve, then the frame is rotating around $\mathbf{a_b}$ only and the rotating rate is described by χ. If the curve is rotating not only around $\mathbf{a_b}$ but also around $\mathbf{a_t}$, then the rotating rate is described by τ. From the geometry point of view, χ is the reciprocal of the curvature radius at this point. τ is the torsion at this point. Combination of these two rotations forms the rotating model of the frame for the case that the point s is moving around the twist curve. To measure the rotating rate for this frame, it is able to define a Darboux vector δ as

$$\delta = \tau \mathbf{a_t} + \chi \mathbf{a_b} \tag{8.13}$$

And then, from (8.12), we have

$$\begin{aligned} \mathbf{a}'_\mathbf{t} &= \chi \mathbf{a_n} = \delta \times \mathbf{a_t} \\ \mathbf{a}'_\mathbf{n} &= -\chi \mathbf{a_t} + \tau \mathbf{a_b} = \delta \times \mathbf{a_n} \\ \mathbf{a}'_\mathbf{b} &= -\tau \mathbf{a_n} = \delta \times \mathbf{a_b} \end{aligned} \tag{8.14}$$

where, the Darboux vector δ is able to be seen as an angular velocity vector, which is able to be used to describe the rotating rate as the three vector $(\mathbf{a_n}, \mathbf{a_b}, \mathbf{a_t})$ moving along the curve.

8.2.3 Non-orthogonal coordinates system

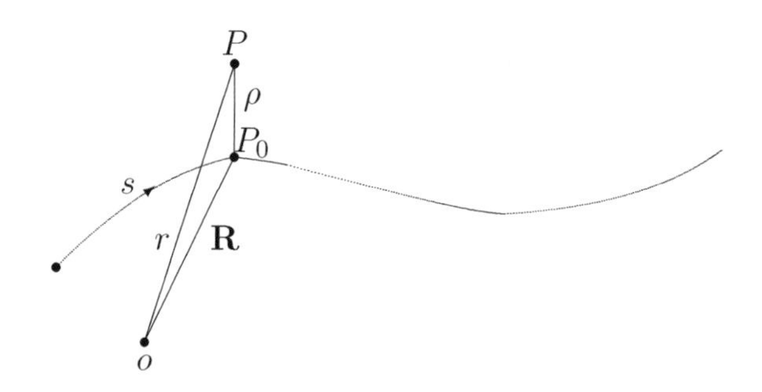

Figure 8-2 The structure of the local coordinates system

Note, that the Serret-Frenet frame is an orthogonal coordinates system only if the point is moving along the curve. If the point is moving away from the curve, the orthogonality will be destroyed by the torsion τ and then forms a non-orthogonal coordinates system. The farther away from the curve, the more serious the non-orthogonality.

The coordinates of P:

As shown in Figure 8-2, P is an arbitrary point away from the curve. $\mathbf{r}$ is the coordinate vector, i.e. $\mathbf{r} = \mathbf{R} + \rho$. ρ is a vector perpendicular to the curve. The end point of ρ is at $\mathbf{P}$, the start point of ρ is at P_0, P_0 is on the curve and $\mathbf{R}$ is a vector from o to P_0. Obviously, ρ is located at the plane contained $\mathbf{a_n}$ and $\mathbf{a_b}$, i.e. ρ is located at the transverse cross section of the curve as ρ is perpendicular to the curve. And then $\rho = n\mathbf{a_n} + b\mathbf{a_b}$. Where n, b are the coordinates value along $\mathbf{a_n}$ and $\mathbf{a_b}$, respectively. P_0 determines the s value corresponding to P. And then the coordinates of the P point can be expressed by means of the value of (n, b, s). Where n and b are the transverse coordinates and s is the longitudinal coordinate.

The line element $d\mathbf{r}$:

Now we will find the line element $d\mathbf{r} = d\mathbf{R} + d\rho$. From $\rho = n\mathbf{a_n} + b\mathbf{a_b}$, we have

$$d\rho = dn\mathbf{a_n} + db\mathbf{a_b} + (n\mathbf{a}'_\mathbf{n} + b\mathbf{a}'_\mathbf{b})ds = dn\mathbf{a_n} + db\mathbf{a_b} + \delta \times (n\mathbf{a_n} + b\mathbf{a_b})ds$$

Here use of (8.14) has been made. In addition, $d\mathbf{R} = \mathbf{a}_t ds$, and then $d\mathbf{r}$ in this coordinates system is able to be expressed as

$$d\mathbf{r} = \mathbf{a}_n dn + \mathbf{a}_b db + [\mathbf{a}_t + \delta \times (n\mathbf{a}_n + b\mathbf{a}_b)]ds \tag{8.15}$$

Considering that $(dr)^2 = d\mathbf{r} \cdot d\mathbf{r} = (8.15) \cdot (8.15)$, substituting the δ in (8.13) into (8.15), it is not difficult to obtain the line element as follows:

$$(dr)^2 = (dn)^2 + (db)^2 + [(1 - \chi n)^2 + \tau^2(n^2 + b^2)](ds)^2 + 2\tau(ndb - bdn)ds \tag{8.16}$$

Obviously, the last term in $(dr)^2$ is not available in orthogonal coordinates system. That means that (8.16) describes a non-orthogonal coordinates system. And the coordinates (n, b, s), the coordinates of the arbitrary point P are a non-orthogonal coordinates system. However, the point on the curve $(n = 0, b = 0)$ forms an orthogonal coordinates system. The non-orthogonal coordinates system at the point away from the curve is caused by the torsion τ, or in other words, is caused by the twist effect. If the curve is the one without twist, i.e. $\tau = 0$, then it is an orthogonal coordinates system even if the point is away from the curve.

Summary: In the case of non-plane curve, the Serret-Frenet frame is generally non-orthogonal coordinates system.

8.2.4 Tang's orthogonal coordinates system

As we mentioned above, the orthogonality of the Serret-Frenet frame is destroyed by the torsion τ. Tang proposed a new frame, which rotates around $\mathbf{a_t}$ with a rate of $-\tau$ related to the Serret-Frenet frame. And then the rotating rate of the new frame around $\mathbf{a_t}$ is zero, i.e. it doesn't rotate around $\mathbf{a_t}$.

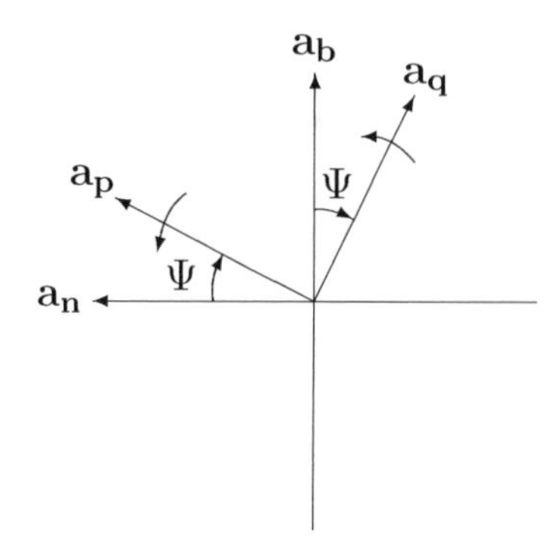

Figure 8-3 Tang's Frame

As shown in Figure 8-3, the new frame is denoted by the unit vectors $\mathbf{a_p}$, $\mathbf{a_q}$, $\mathbf{a_t}$. Where the included angle between $\mathbf{a_p}$ and $\mathbf{a_n}$, Ψ, is a function of s and satisfies [1]

$$\psi'(s) = -\tau(s) \tag{8.17}$$

since the net torsion of the Tang's frame is zero. In this case, the Darboux vector in the Tang's frame $\mathbf{a_p}, \mathbf{a_q}, \mathbf{a_t}$ becomes [2]

$$\delta_0 = \chi \mathbf{a_b} = \chi(\mathbf{a_p} sin\Psi + \mathbf{a_q} cos\Psi) \tag{8.18}$$

And then the line element becomes

$$(dr)^2 = (dn)^2 + (db)^2 + (1 - \chi n)^2 (ds)^2$$

Where $[(dn)^2 + (db)^2]^{1/2}$ is the projection of $d\mathbf{r}$ in the coordinates plane of nob. Considering that (n, b) and (p, q) are in the same plane, we have $(dn)^2 + (db)^2 = (dp)^2 + (dq)^2$, and $n = pcos\Psi - qsin\Psi$. And then the line element in the coordinates system (p, q, s) becomes

$$(dr)^2 = (dp)^2 + (dq)^2 + [1 - \chi(pcos\Psi - qsin\Psi)]^2 (ds)^2 \tag{8.19}$$

Obviously, Tang's coordinates system forms an orthogonal coordinates system, and the scale coefficients in the Tang's coordinates system are

$$h_p = h_q = 1, \qquad h_s = [1 - \chi(pcos\Psi - qsin\Psi)] \tag{8.20}$$

And then, the Maxwell equation in Tang's coordinates system may be written by means of an orthogonal coordinate system.

---

[1] According to (8.8) and the definition of Tang's frame.
[2] According to (8.13).

8.2.5 Application

1. For a rectangular waveguide with the axis along the end points of $\mathbf{R}(s)$, the transverse cross section of the waveguide can be defined by means of $p = \pm a/2$ and $q = \pm b/2$. Where a and b are the width and the high of the transverse cross section of the waveguide, respectively. In this case , it is easy to match the boundary condition.

2. For another shape waveguide with the axis along the end points of $\mathbf{R}(s)$, it is able to make use of a transformation

$$p = p(u, v), \qquad q = q(u, v) \tag{8.21}$$

to transform the coordinates system from $(p,\ q)$ into a new orthogonal coordinates system (u, v) so as to match the boundary condition. In which, the orthogonal condition for u and v is

$$\frac{\partial p}{\partial u}\frac{\partial p}{\partial v} + \frac{\partial q}{\partial u}\frac{\partial q}{\partial v} = 0$$

and the scale coefficients in the new coordinates system u, v, s are

$$\begin{aligned} h_u &= \left[\left(\frac{\partial p}{\partial u}\right)^2 + \left(\frac{\partial q}{\partial u}\right)^2\right]^{1/2} \\ h_v &= \left[\left(\frac{\partial p}{\partial v}\right)^2 + \left(\frac{\partial q}{\partial v}\right)^2\right]^{1/2} \\ h_s &= \left[1 - \chi\Big(p\cos\Psi - q\sin\Psi\Big)\right] \end{aligned} \tag{8.22}$$

For example, as shown in Figure 8-4, if we want to make the transformation from the coordinates system (p, q) into the polar coordinates system, we have

$$p = \rho cos\phi, \qquad q = \rho sin\phi \tag{8.23}$$

Then, from (8.22), the scale coefficients in the polar coordinates system are

$$h_\rho = 1, \qquad h_\phi = \rho, \qquad h_s = 1 - \chi\rho cos(\phi + \Psi) \tag{8.24}$$

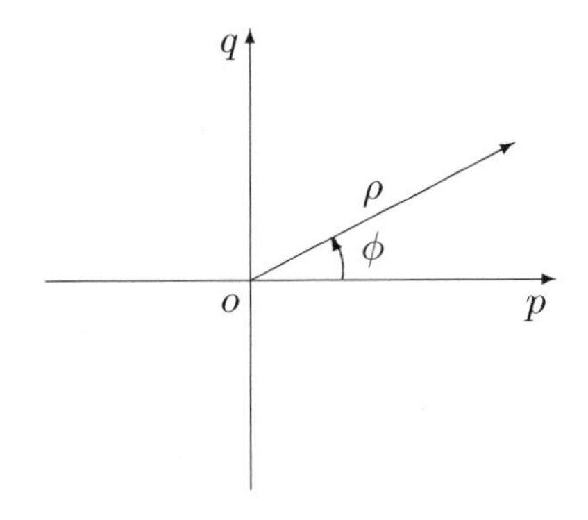

3. For a helical optical fiber as shown in Figure 8-5:

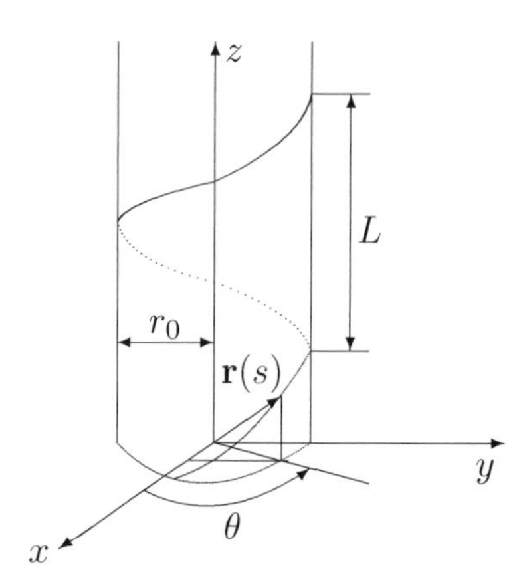

Figure 8-5 Helical optical fiber

The axis of the helical optical fiber forms a helical line. The radius of the helical line is r_0, the pitch of the helical line is L. Then the vector $\mathbf{r}(s) = \mathbf{a}_x r_0 cos\theta + \mathbf{a}_y r_0 sin\theta + \mathbf{a}_z z$. Where $s = z[1 + (2\pi r_0/L)^2]$, $\theta = 2\pi z/L$. Substituting all of those into (8.2) and (8.10) yields

$$\chi = \frac{r_0}{r_0^2 + (L/2\pi)^2}$$

$$\tau = \frac{L/2\pi}{r_0^2 + (L/2\pi)^2} \tag{8.25}$$

Substituting (8.25) into (8.17) yields

$$\Psi = -\frac{(L/2\pi)s}{r_0^2 + (L/2\pi)^2} \tag{8.26}$$

4. For a bending optical fiber as shown in Figure 8-6:

The axis of the optical fiber forms a half circle. Therefore, you can imagine that it is a helical line with zero pitch and zero torsion. Then, from (8.24), the scale coefficients for the bending optical fiber are

$$h_p = 1, \qquad h_\phi = \rho, \qquad h_s = 1 + (\rho/R_0)cos\phi \tag{8.27}$$

Where (ρ, ϕ, s) is the local coordinates system for the optical fiber. In this case, from (8.25), the curvature is

$$\chi = \frac{1}{R_0}$$

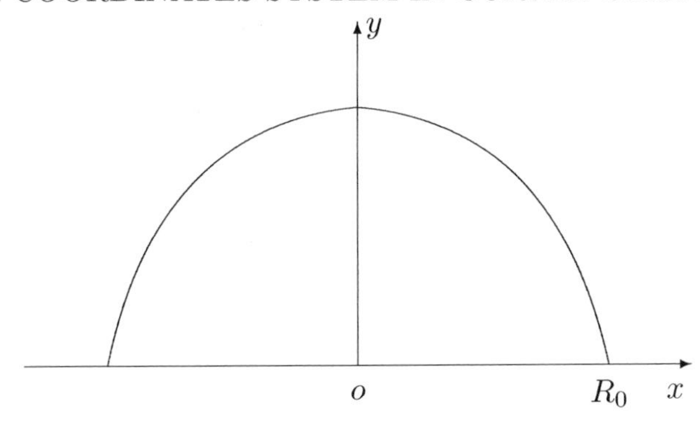

Figure 8-6 Bending optical fiber

8.2.6 Maxwell equations in Curvature structure

As we mentioned before that the Serret-Frenet frame (n, b, t) in the case of non-plane curve is generally a non-orthogonal coordinates system. However, it is able to be transformed into an orthogonal coordinates system such as the Tang's coordinates system (ν, β, t). And then the Maxwell equations may be expanded in the Tang's coordinates system.

However, the Tang's coordinates system is not always coincident with the boundary condition. For example, if we study the twist rectangular waveguide, then $\nu = constant$, $\beta = constant$ is always coincident with the wall of the waveguide. In this case, the problem is able to be solved by means of the Maxwell equations in Tang's coordinates system. However, if we study the bending optical fiber, the boundary condition of the optical fiber should be expressed in the polar coordinates system since the cross section of the fiber is a circle. And then the coordinates plane of the Tang's coordinates system is not coincident with the boundary condition for the optical fiber. In this case, we have to transform the Tang's orthogonal coordinates system from the rectangular Cartesian coordinates system into the polar coordinates system.

The transformation from the Tang's orthogonal coordinates system (ν, β, t) to a new orthogonal coordinates system (u, v, s) is involved in a scale coefficients h_ν, h_β, h_s given by (8.22). In this case, the Maxwell equations in the Tang's coordinates system are

$$\frac{\partial(h_s E_s)}{\partial v} - \frac{\partial(h_v E_v)}{\partial s} = -i\omega\mu h_v h_s H_u$$

$$\frac{\partial(h_u E_u)}{\partial s} - \frac{\partial(h_s E_s)}{\partial u} = -i\omega\mu h_s h_u H_v$$

$$\frac{\partial(h_v E_v)}{\partial u} - \frac{\partial(h_u E_u)}{\partial v} = -i\omega\mu h_u h_v H_s \qquad (8.28)$$

$$\frac{\partial(h_s H_s)}{\partial v} - \frac{\partial(h_v H_v)}{\partial s} = i\omega\varepsilon h_v h_s E_u$$

$$\frac{\partial(h_u H_u)}{\partial s} - \frac{\partial(h_s H_s)}{\partial u} = i\omega\varepsilon h_s h_u E_v$$

$$\frac{\partial(h_v H_v)}{\partial u} - \frac{\partial(h_u H_u)}{\partial v} = i\omega\varepsilon h_u h_v E_s \tag{8.29}$$

And then we can fine the solution from these equations in the new coordinates system.

For example, in the local coordinates system (ρ, θ, s) of the bending optical fiber, from (8.27), the scalar coefficients are $h_\rho = 1, h_\theta = \rho, h_s = 1 + (\rho/R_0)cos\theta$. Substituting the scala coefficients into (8.28) and (8.29), yields the Maxwell equations in ρ, θ, s coordinates system as follows:

$$\left(\frac{1}{\rho h_s}\right)\left[\frac{\partial(h_s E_s)}{\partial\theta} - \frac{\partial(\rho E_\theta)}{\partial s}\right] = -i\omega\mu_0 H_\rho$$

$$\left(\frac{1}{h_s}\right)\left[\frac{\partial E_\rho}{\partial s} - \frac{\partial(h_s E_s)}{\partial\rho}\right] = -i\omega\mu_0 H_\theta$$

$$\left(\frac{h_s}{\rho}\right)\left[\frac{\partial(\rho E_\theta)}{\partial\rho} - \frac{\partial E_\rho}{\partial\theta}\right] = -i\omega\mu_0 h_s H_s \tag{8.30}$$

$$\left(\frac{1}{\rho h_s}\right)\left[\frac{\partial(h_s H_s)}{\partial\rho} - \frac{\partial(\rho H_\theta)}{\partial s}\right] = i\omega\varepsilon_0 n^2 E_\rho$$

$$\left(\frac{1}{h_s}\right)\left[\frac{\partial H_\rho}{\partial s} - \frac{\partial(h_s H_s)}{\partial\rho}\right] = i\omega\varepsilon_0 n^2 E_\theta$$

$$\left(\frac{h_s}{\rho}\right)\left[\frac{\partial(\rho H_\theta)}{\partial\rho} - \frac{\partial H_\rho}{\partial\theta}\right] = i\omega\varepsilon_0 n^2 h_s E_s \tag{8.31}$$

8.3 The field in bending optical fiber

Figure 8-7 presents a bending optical fiber with the radius $2a$ of the core, refractive index n_1 inside the core and n_2 outside the core and n_1 and n_2 satisfy $\triangle = (n_1^2 - n_2^2)/2n_1^2 << 1$.

This system is involved with two coordinates systems: one is the circular cylinder coordinates system $R\Psi Z$ with the origin at o, which is an orthogonal coordinates system and is suitable to set up the Maxwell equations, another one is the local coordinates system $r\theta s$ with the origin at o', which is a bending coordinates system and the wave equation in this coordinates system is not orthonormal, however, it can be solved by the perturbation method under the slight bending condition, $R_0 >> a$. To this end, we have to set up the wave equation in the circular cylinder coordinates system $R\Psi Z$ first, and then making a coordinates transformation to obtain the wave equation in the local coordinates system $r\theta s$, and then find the solution by means of the perturbation method.

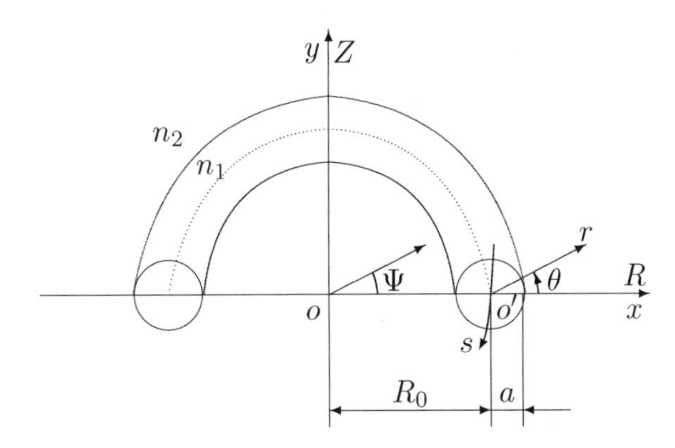

Figure 8-7 The coordinates system of the bending optical fiber

The coordinates transformation from $R\Psi Z$ coordinates system to $r\theta s$ coordinates system is

$$R = R_0 + r cos\theta, \qquad Z = r sin\theta, \qquad \Psi = -s/R_0 \tag{8.32}$$

For the field components, the relationship between the two coordinates systems are

$$E_R = E_x, \qquad E_Z = E_y, \qquad E_\Psi = -E_s \tag{8.33}$$

Where E_x, E_y are the field components in the local coordinates system $r\theta s$, they represent the fields with the perpendicular polarization and the horizontal polarization, respective, on the transverse cross section of the optical fiber with the reference at $\Psi = 0$. The relationship between E_x, E_y and E_r, E_θ are

$$E_x = E_r cos\theta - E_\theta sin\theta, \qquad E_y = E_r sin\theta + E_\theta cos\theta \tag{8.34}$$

And the scale coefficients for the $r\theta s$ coordinates system is available in (8.27), which are

$$h_r = 1, \qquad h_\theta = r, \qquad h_s = 1 + (r/R_0)cos\theta$$

8.3.1 The wave equation in bending optical fiber

First, from Maxwell equations

$$\nabla \times \mathbf{E} = -i\omega\mu_0\mathbf{H}$$

$$\nabla \times \mathbf{H} = i\omega\varepsilon_0 n^2 \mathbf{E}$$

and the vector identity

$$\nabla \times \nabla \times \mathbf{E} = \nabla(\nabla \cdot \mathbf{E}) - \nabla^2 \mathbf{E}$$

we have the wave equation

$$[\nabla^2 + n^2 k_0^2]\mathbf{E} = -\nabla[\mathbf{E} \cdot \nabla ln(n^2)] \tag{8.35}$$

Here use has been made of

$$\nabla \cdot \mathbf{D} = \nabla \cdot (\varepsilon_0 n^2 \mathbf{E}) = \varepsilon_0 n^2 \Big(\nabla \cdot \mathbf{E} + \mathbf{E} \cdot \frac{\nabla n^2}{n^2}\Big) = \varepsilon_0 n^2 (\nabla \cdot \mathbf{E} + \mathbf{E} \cdot \nabla ln(n^2)] = 0$$

Or

$$\nabla \cdot \mathbf{E} = -\mathbf{E} \cdot \nabla ln(n^2)$$

And then in $R\Psi Z$ coordinates system, the wave equation of the field components E_R, E_Z are able to be obtained by expanding (8.35) into

$$[\nabla^2 + k_0^2 n^2]E_R = -\frac{\partial}{\partial R}[\mathbf{E} \cdot \nabla ln(n^2)] + \Big(2\frac{\partial E_\Psi}{\partial \Psi} + E_R\Big)\frac{1}{R} \tag{8.36}$$

$$[\nabla^2 + k_0^2 n^2]E_Z = -\frac{\partial}{\partial Z}[\mathbf{E} \cdot \nabla ln(n^2)] \tag{8.37}$$

To make the transformation of the wave equations (8.36) and (8.37) from the $R\Psi Z$ coordinates system to the local coordinates system $r\theta s$, we have to make use of the coordinates transformation formulae (8.32) and (8.33), from which we have

$$\frac{\partial}{\partial R} = \frac{\partial}{\partial x}, \qquad \frac{\partial}{\partial Z} = \frac{\partial}{\partial y}, \qquad \frac{\partial}{\partial \Psi} = -R_0 \frac{\partial}{\partial s}$$

$$\nabla^2 = \frac{1}{R}\frac{\partial}{\partial R}\Big(R\frac{\partial}{\partial R}\Big) + \frac{\partial^2}{\partial Z^2} + \frac{1}{R^2}\frac{\partial^2}{\partial \Psi^2} = \frac{\partial^2}{\partial x^2} + \frac{\partial^2}{\partial y^2} + \frac{1}{R}\frac{\partial}{\partial R} - \frac{\beta^2}{(R/R_0)^2}$$

or

$$\nabla^2 = \nabla_t^2 + \frac{1}{R}\frac{\partial}{\partial R} - \frac{\beta^2}{(R/R_0)^2}$$

and

$$\nabla_t^2 = \Big[\frac{1}{\rho}\frac{\partial}{\partial \rho}\Big(\rho\frac{\partial}{\partial \rho}\Big) + \frac{1}{\rho^2}\frac{\partial^2}{\partial \theta^2}\Big]a^2$$

with

$$\rho = r/a$$

Substituting all of those transformation formulae into (8.36) and (8.37) yields the wave equation in the local coordinates system as follows

$$\Big[\nabla_t^2 + n^2 k_0^2 - \beta^2/(R/R_0)^2 + \frac{\partial}{R\partial R}\Big]E_x = -\frac{\partial[\mathbf{E} \cdot \nabla ln(n^2)]}{\partial x}$$

$$-\frac{(2i\beta R_0 E_s - E_x)}{R^2} \tag{8.38}$$

$$\Big[\nabla_t^2 + n^2k_0^2 - \beta^2/(R/R_0)^2 + \frac{\partial}{R\partial R}\Big]E_y = -\frac{\partial[\mathbf{E}\cdot\nabla ln(n^2)]}{\partial y} \tag{8.39}$$

Considering that the refractive index is n_1 inside the core and n_2 outside the core of the optical fiber, we have

$$n^2 = n_1^2[1 - 2\triangle H(\rho)], \qquad \triangle = (n_1^2 - n_2^2)/2n_1^2 \qquad (\triangle << 1)$$

$$H(\rho) = \begin{cases} 0 & (\rho < 1) \\ 1 & (\rho > 1) \end{cases}$$

and $\nabla ln[1 - 2\triangle H(\rho)] \approx \mathbf{a}_r\Big[-2\triangle\frac{\partial H(\rho)}{\partial\rho}\Big] = \mathbf{a}_r[-2\triangle\delta(\rho-1)].$

Substituting all of those into (8.38) and (8.39) yields the wave equation in the local coordinates system $r\theta s$ as follows

$$\Big[\nabla_t^2 + n^2k_0^2 - \beta^2/(R/R_0)^2 + \frac{\partial}{R\partial R}\Big]E_x = 2\triangle\frac{\partial[E_r\delta(\rho-1)]}{\partial x}$$

$$-\frac{(2i\beta R_0 E_s - E_x)}{R^2} \tag{8.40}$$

$$\Big[\nabla_t^2 + n^2k_0^2 - \beta^2/(R/R_0)^2 + \frac{\partial}{R\partial R}\Big]E_y = 2\triangle\frac{\partial[E_r\delta(\rho-1)]}{\partial y} \tag{8.41}$$

8.3.2 Perturbation method

For a slight bending optical fiber, it is able to find the solution of the wave equations (8.40) and (8.41) by means of the perturbation method. In which the dominate mode of the perturbation field E_q $(q = x, y, s)$ is the dominate mode of the straight optical fiber E_{0q} $(q = x, y, s)$, and the parameter of the perturbation field is

$$\varepsilon = V^2/R^* = 2\beta_0^2a^3/R_0 \tag{8.42}$$

(in which $V^2 = 2\triangle\beta_0^2a^2$, $R^* = R_0\triangle/a$), so that the perturbation solution is

$$E_x = (E_{0x} + \varepsilon E_{1x} + \varepsilon^2E_{2x} + \cdots)\exp(-i\beta s)$$

$$E_y = (E_{0y} + \varepsilon E_{1y} + \varepsilon^2E_{2y} + \cdots)\exp(-i\beta s)$$

$$E_s = (E_{0s} + \varepsilon E_{1s} + \varepsilon^2E_{2s} + \cdots)\exp(-i\beta s) \tag{8.43}$$

with

$$\beta^2 = \beta_0^2[1 + \beta_2^2\varepsilon^2 + \cdots] \tag{8.44}$$

Obviously, for a straight optical fiber, $R_0 \rightarrow \infty$, $\varepsilon \rightarrow 0$, and then E_{0x}, E_{0y}, E_{0s} are the field components in the straight optical fiber, β_0 is the propagation constant of the straight optical fiber. That means that the fields in the bending optical fiber are the perturbation expanding around the fields of the straight optical fiber.

In the straight optical fiber, there are two degeneration modes, one is the horizontal polarization mode HE^x_{**}, another one is the vertical polarization mode HE^y_{**}. In normal situation, the propagation constant β_0 of the HE^x_{**} mode is the same as that of the HE^y_{**} mode. However, in bending optical fiber, the propagation constant β_x of the HE^x_{**} mode is different from the propagation constant β_y of the HE^y_{**} mode.

Substituting (8.43) and (8.44) into (8.40) and (8.41), considering that the coefficient of the same order of ε in both sides should be equal to each other, we obtain a set of equations for HE^x_{**} mode, E^x_{nx}, and a set of equations for HE^y_{**} mode, E^y_{ny}. In which:

The zero order solution satisfies

$$a^2[\nabla_t^2 + n^2k_0^2 - \beta_0^2]E^2_{0x} = 0 \tag{8.45}$$

$$a^2[\nabla_t^2 + n^2k_0^2 - \beta_0^2]E^2_{0y} = 0 \tag{8.46}$$

The first order solution satisfies

$$a^2[\nabla_t^2 + n^2k_0^2 - \beta_0^2]E^x_{1x} = \\ -\rho E^x_{0x}\cos\theta - \left(\cos\theta\frac{\partial E^x_{0x}}{\partial\rho} - sin\theta\frac{\partial E^x_{0x}}{\rho\partial\theta} + 2i\beta_0 a E^s_{0x}\right)/(2\beta_0^2a^2) \tag{8.47}$$

$$a^2[\nabla_t^2 + n^2k_0^2 - \beta_0^2]E^y_{1y} = \\ -\rho E^y_{0y}\cos\theta - \left(\cos\theta\frac{\partial E^y_{0y}}{\partial\rho} - sin\theta\frac{\partial E^y_{0s}}{\rho\partial\theta}\right)/(2\beta_0^2a^2) \tag{8.48}$$

Then, from (8.45) and (8.46), the zero order solution is

$$E^x_{0x} = \begin{cases} J_0(u\rho)/J_0(u) & (\rho < 1) \\ K_0(w\rho)/K_0(w) & (\rho > 1) \end{cases} \tag{8.49}$$

$$\begin{aligned} E^y_{0y} &= E^x_{0x} \\ E^x_{0y} &= E^y_{0x} \end{aligned} \tag{8.50}$$

Where

$$u = (n_1^2k_0^2 - \beta_0^2)^{1/2}a, \qquad w = (\beta_0^2 - n_2^2k_0^2)^{1/2}a, \qquad n_2 = n_1(1 - \triangle)$$

The propagation constant β_0 may be found by an eigenvalue equation, which may be obtained by substituting (8.49) into the boundary condition of the straight single mode fiber at $\rho = 1$:

$$E^x_{0x}|_{\rho=1_-} = E^x_{0x}|_{\rho=1_+}, \qquad \frac{\partial E^x_{0x}}{\partial \rho}|_{\rho=1_+} = \frac{\partial E^x_{0x}}{\partial \rho}|_{\rho=1_-}$$

to obtain

$$uJ_1(u)/J_0(u) = wK_1(w)/K_0(w)$$

To find the first order solution from (8.47) and (8.48), it is necessary to find E^x_{0s} and E^y_{0s} first. Which can be obtain from the Maxwell equations (8.30), (8.31) in the local coordinates system $\rho\theta s$, and then via the transformation

$$E_x = E_\rho cos\theta - E_\theta sin\theta, \qquad E_y = E_r sin\theta + E_\theta cos\theta$$

to obtain

$$E_s = \left(1 - \frac{\beta_2 a^2}{2R_0^2}\right)\left\{\left[\left(-\frac{\partial E_x}{\rho \partial \rho} + \frac{\partial E_y}{\partial \rho}\right)\sin\theta + \left(\frac{\partial E_x}{\partial \rho} + \frac{\partial E_y}{\rho \partial \theta}\right)\cos\theta\right]\right.$$
$$\left.\left[1 + \cos\theta\left(\frac{\rho a}{R_0}\right)\right] + \frac{E_x a}{R_0}\right\}/i\beta_0 a \tag{8.51}$$

Substituting (8.49), (8.50) into (8.51) yields

$$E^x_{0s} = \begin{cases} -uJ_1(u\rho)\cos\theta/J_0(u)i\beta_0 a & (\rho < 1) \\ -wK_1(w\rho)\cos\theta/K_0(w)i\beta_0 a & (\rho > 1) \end{cases}$$

$$E^y_{0s} = \begin{cases} uJ_1(u\rho)\sin\theta/J_0(u)i\beta_0 a & (\rho < 1) \\ wK_1(w\rho)\sin\theta/K_0(w)i\beta_0 a & (\rho > 1) \end{cases}$$

Finally, substituting E^x_{0x}, E^x_{0s} and E^y_{0s} into (8.47) and (8.48) yields the first order solution as (see Appendix)

$$E^x_{1x} = \begin{cases} \{(-\rho^2/4u + C_1)J_1 - 3\rho J_0/4\beta_0^2 a^2\}\cos\theta/J_0(u) & (\rho < 1) \\ \{(\rho^2/4w^2 + C_2)K_1 - 3\rho K_0/4\beta_0^2 a^2\}\cos\theta/K_0(w) & (\rho > 1) \end{cases} \tag{8.52}$$

$$E^y_{1y} = \begin{cases} \{(-\rho^2/4u + C_1)J_1 - \rho J_0/4\beta_0^2 a^2\}\cos\theta/J_0(u) & (\rho < 1) \\ \{(\rho^2/4w^2 + C_2)K_1 - \rho K_0/4\beta_0^2 a^2\}\cos\theta/K_0(w) & (\rho > 1) \end{cases} \tag{8.53}$$

And

$$E_{1y} = E^y_{1x} = 0$$

since $E_{0y}^x = E_{0x}^y$. Where, C_1 and C_2 are determined by the boundary condition and are

$$C_1 = \frac{K_2(w)}{4uK_0(w)}, \qquad C_2 = \frac{J_2(u)}{4wJ_0(u)}$$

Substituting (8.52) and (8.53) into (8.51) yields E_{1s}^x and E_{1s}^x.

Following the same process, we may find E_{nx}^x, E_{ny}^y, E_{ns}^x, E_{ns}^y step by step so as to approximate to E_x^x, E_y^y, E_s^x, E_s^y step by step.

8.3.3 The propagation constant of the bending optical fiber

The HE_{**}^x mode and the HE_{**}^y mode are the degenerate modes in a perfect straight optical fiber, namely, the propagation constant is the same for both of them. However, the propagation constant of the HE_{**}^x mode is slight different from the propagation constant of the HE_{**}^y mode in a bending optical fiber. This is because that the variation of the propagation constant of the HE_{**}^x mode caused by the bending is different from that of the HE_{**}^x mode. Therefore, we have to find the propagation constant one by one.

For HE_{**}^x mode:

From (8.45) and (8.40), we have

$$a^2[\nabla_t^2 + n^2k_0^2 - \beta_0^2]E_{0x}^2 = 0 \tag{8.54}$$

$$a^2\left[\nabla_t^2 + n^2k_0^2 - \beta_x^2/\left(1 + \frac{a\rho}{R_0}\cos\theta\right)^2\right]E_x^x = -\frac{a^2}{R}\frac{\partial E_x^x}{\partial R} - \frac{(2i\beta_x R_0 a^2 E_s^x - a^2 E_x^x)}{R^2} \tag{8.55}$$

In which β_0 is the propagation constant of the straight optical fiber and β_x is the propagation constant of the HE_{**}^x mode in a bending optical fiber. The perturbation formula of β_x is, from (8.44),

$$\beta_x^2 = \beta_0^2[1 + \beta_{2x}\varepsilon^2 + \cdots] \tag{8.56}$$

Making comparison between (8.54) and (8.55) we find that the integral

$$\int_{A_\infty}[E_x^x \cdot (8.54) - E_{0x}^x \cdot (8.55)]dA$$

will results in some cancellation so that this formula contains β_{2x} term and $\int_{A_\infty}[E_x^x\nabla_t^2 E_{0x}^x - E_{0x}^x\nabla_t^2 E_x^x]dA$ term only. Where A_∞ is the big circle area centre in the central of the transverse cross section of the optical fiber. Considering that the field outside the core of the optical fiber is decaying in exponential law, and then the integral contribution is going to zero as the radius of the A_∞ circle is larger enough. Therefore

$$\varepsilon^2\,\beta_{2x} = \Big\{\int_{A_\infty}\Big[\frac{a^2\beta_0^2\Big(2\dfrac{a\rho}{R_0}\cos\theta + \dfrac{a^2R^2}{R_0^2}\cos^2\theta\Big)}{\Big(1+\dfrac{a\rho}{R_0}\cos\theta\Big)^2}E_{0x}^xE_x^x + \frac{a^2E_{0x}^x}{R}\frac{\partial E_x^x}{\partial R} + \frac{(2i\beta_xR_0a^2E_s^x - a^2E_x^x)E_{0x}^x}{R^2}\Big]dA\Big\} \Big\{\int_{A_\infty}\frac{a^2E_{0x}^xE_x^x}{\Big(1+\dfrac{a\rho}{R_0}\cos\theta\Big)^2}dA\Big\}^{-1} \tag{8.57}$$

Similarly, for HE_{**}^y, we have

$$\beta_y^2 = \beta_0^2[1+\beta_{2y}\varepsilon^2+\cdots]$$

And

$$\varepsilon^2\,\beta_{2y} = \Big\{\int_{A_\infty}\Big[\frac{a^2\beta_0^2\Big(2\dfrac{a\rho}{R_0}\cos\theta + \dfrac{a^2R^2}{R_0^2}\cos^2\theta\Big)}{\Big(1+\dfrac{a\rho}{R_0}\cos\theta\Big)^2}E_{0y}^yE_y^y + \frac{a^2E_{0y}^y}{R}\frac{\partial E_y^y}{\partial R}\Big]dA\Big\} \Big\{\int_{A_\infty}\frac{a^2E_{0y}^yE_y^y}{\Big(1+\dfrac{a\rho}{R_0}\cos\theta\Big)^2}dA\Big\}^{-1} \tag{8.58}$$

where ε has given in (8.42).

Summery

1. For a bending optical fiber, we need two coordinates systems: one is the circular cylinder coordinates system $R\Psi Z$ with the origin at o, which is an orthogonal coordinates system and can be used to expand the Maxwell equations, another one is the local coordinates system $r\theta s$ with the origin at o', which is a bending coordinates system and can be used to match the boundary conditions, however, it is a non-orthogonal coordinates system and then the wave equation in this coordinates system is not orthonormal.

2. Till now we can not find the accuracy solution for the fields in a bending optical fiber. However, we may get the approximation solution by means of the perturbation method for the slight bending optical fiber with $R_0 >> a$. To this end, we

have to set up the wave equation in the circular cylinder coordinates system $R\Psi Z$ first, and then transform to the local coordinates system $r\theta s$ via transformation (8.32) and (8.33), and then find the solution by means of the perturbation method.

8.4 The field in helical optical fiber

The field in helical optical fiber has been studied by Xi-sheng Fang, Zong-Qi Lin [71] and Jing-ren Qian [72]. The following discussion is based on their study.

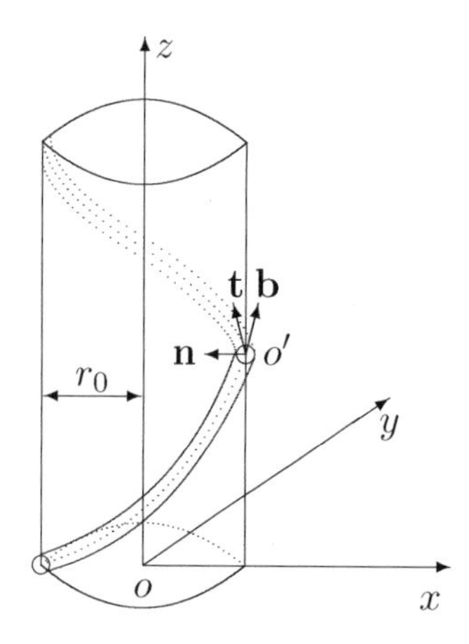

Figure 8-8 The coordinates system in helical optical fiber

Figure 8-8 presents a helical optical fiber with a radius of $2a$ and a pitch of $L = 2\pi B$. The radius of the cylinder is r_0. And then, from (8.25), the curvature χ and the torsion τ of the helical line are

$$\chi = r_0/(r_0^2 + B^2), \qquad \tau = B/(r_0^2 + B^2) \tag{8.59}$$

respectively. Here, the unit vectors $\mathbf{a}_n, \mathbf{a}_b, \mathbf{a}_t$ of the local coordinates system in the helical optical fiber $o'\mathbf{nbt}$ forms a Serret-Frenet frame.

As we mentioned before, since the effect of the torsion , the coordinates system $\mathbf{nbt}$ is a non-orthogonal coordinates system except at the central of the core of the optical fiber. Therefore, it is impossible to find the field solution by means of the Maxwell equation in this coordinates system.

How to set up an orthogonal coordinates system?

And then Tang proposed a new frame named *the Tang's frame*, which is rotating with a rate of $-\tau$ around $\mathbf{a}_t$ with respect to the Serret-Frenet frame, then, in Tang's frame, the net rate of the rotating around the $\mathbf{a}_t$ is zero. And then Tang's frame becomes an orthogonal coordinates system.

Assume that the unit vectors of the Tang's frame are $\mathbf{a}_p, \mathbf{a}_q, \mathbf{a}_s$, then the included angle between $\mathbf{a}_p$ and $\mathbf{a}_n$, Ψ, satisfies

$$\frac{d\Psi(s)}{ds} = -\tau \tag{8.60}$$

Then, from (8.59) and (8.60),

$$\Psi = -Bs/(r_0^2 + B^2) \tag{8.61}$$

where s is the axial coordinates along the helical optical fiber.

As shown in Figure 8-9, the relationship between the Tang's Frame $(\mathbf{a}_p, \mathbf{a}_q, \mathbf{a}_s)$ and the Serret-Frenet Frame $(\mathbf{a}_n, \mathbf{a}_b, \mathbf{a}_t)$ are

$$\begin{aligned} \mathbf{a}_s &= \mathbf{a}_t \\ \mathbf{a}_p &= \mathbf{a}_n \cos\Psi + \mathbf{a}_b \cos\Psi \\ \mathbf{a}_q &= -\mathbf{a}_n \sin\Psi + \mathbf{a}_b \cos\Psi \end{aligned}$$

And the scale coefficients of the Tang's Frame are available in (8.20), namely

$$h_p = h_q = 1, \qquad h_s = [1 - \chi(p\cos\Psi - q\sin\Psi)] \tag{8.62}$$

In this case, we can find the field solution in helical optical fiber by means of the Maxwell equations (8.28) and (8.29) in the Tang's coordinates system.

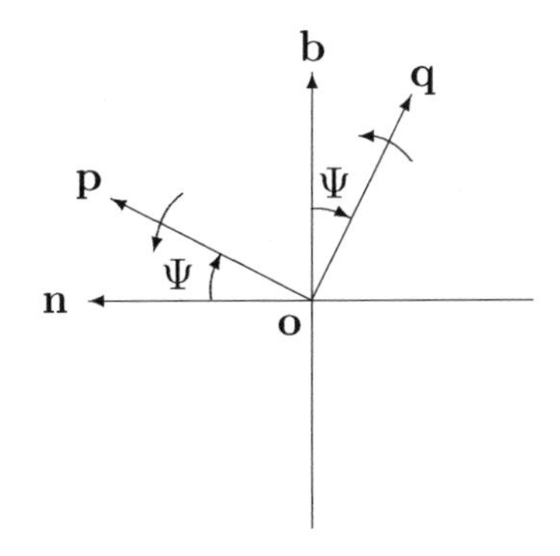

Figure 8-9 Serret-Frenet Frame and Tang's Frame

8.4.1 The wave equation in a helical optical fiber

The field in a single-mode helically-wound Optical fiber has been studied by Xi-sheng Fang, Zong-Qi Lin [71] and Jing-ren Qian [72], respectively. The following discussion is based on the paper by Xi-sheng Fang, Zong-Qi Lin [71] and the paper by Jing-ren Qian [72].

First, in the Tang's coordinates system, the electrical fields and the magnetic fields in a helical optical fiber are able to be expressed as

$$\mathbf{E} = \mathbf{a}_p E_p + \mathbf{a}_q E_q + \mathbf{a}_s E_s$$
$$\mathbf{H} = \mathbf{a}_p H_p + \mathbf{a}_q H_q + \mathbf{a}_s H_s$$

Secondary, to expand the wave equation

$$(\nabla^2 + n^2 k_0^2)\mathbf{E} = 0 \tag{8.63}$$

in the Tang's coordinates system , we need to make use of the vector identification

$$\nabla^2 \mathbf{E} = \nabla(\nabla \cdot \mathbf{E}) - \nabla \times \nabla \times \mathbf{E} \tag{8.64}$$

Where

$$\begin{aligned} \nabla \times \mathbf{E} &= \mathbf{a}_p \Big[\frac{\partial(h_s E_s)}{\partial q} - \frac{\partial E_q}{\partial s}\Big]/h_s + \mathbf{a}_q \Big[\frac{\partial E_p}{\partial s} - \frac{\partial(h_s E_s)}{\partial p}\Big]/h_s \\ &\quad + \mathbf{a}_s \Big[\frac{\partial E_q}{\partial p} - \frac{\partial E_p}{\partial q}\Big]/h_s \end{aligned} \tag{8.65}$$

$$\nabla \cdot \mathbf{E} = \Big[\frac{\partial(h_s E_p)}{\partial p} + \frac{\partial(h_s E_q)}{\partial q} + \frac{\partial E_s}{\partial s}\Big]/h_s \tag{8.66}$$

$$\nabla(\nabla \cdot \mathbf{E}) = \nabla\phi = \mathbf{a}_p \frac{\partial \phi}{\partial p} + \mathbf{a}_q \frac{\partial \phi}{\partial q} + \mathbf{a}_s \frac{\partial \phi}{h_s \partial s} \tag{8.67}$$

Substituting the vector identifications (8.64), (8.65), (8.66), (8.67) and scale coefficients equation (8.62) into the wave equation (8.63) yields the wave equation in the Tang's coordinates system. In which, the transverse part (along the unit vectors a_p and a_q)

$$[(\nabla^2 + n^2 k_0^2)\mathbf{E}]_p = 0, \qquad [(\nabla^2 + n^2 k_0^2)\mathbf{E}]_q = 0$$

become

$$\begin{aligned} &\Big[\Big(\frac{\partial^2}{\partial p^2} + \frac{\partial^2}{\partial q^2} + h_s^{-2}\frac{\partial^2}{\partial s^2} + n^2 k_0^2\Big) + \chi\Big(-\cos\Psi\,\frac{\partial}{\partial p} + \sin\Psi\,\frac{\partial}{\partial q}\Big)h_s^{-1} \\ &\quad + \chi^2\Big(-\cos^2\Psi h_s^{-2} + \Big(\tau/B\Big)\Big(p\sin\Psi + q\cos\Psi\Big)h_s^{-3}\frac{\partial}{\partial s}\Big)\Big]E_p \\ &+ \Big(\chi^2/2\Big)\sin 2\Psi h_s^{-2} E_q \end{aligned}$$

$$+\Big[2\chi h_s^{-2}\cos\Psi\frac{\partial}{\partial s}+\chi^2\Big(\frac{\tau}{B}\Big)h_s^{-2}\sin\Psi$$

$$-\chi^3\Big(\frac{\tau}{B}\Big)h_s^{-3}\Big(p\sin\Psi+q\cos\Psi\Big)\sin\Psi\Big]E_s=0 \tag{8.68}$$

and

$$\Big[\Big(\frac{\partial^2}{\partial p^2}+\frac{\partial^2}{\partial q^2}+h_s^{-2}\frac{\partial^2}{\partial s^2}+n^2k_0^2\Big)+\chi\Big(-\cos\Psi\frac{\partial}{\partial p}+\sin\Psi\frac{\partial}{\partial q}\Big)h_s^{-1}$$
$$+\chi^2\Big(-\sin^2\Psi h_s^{-2}+\Big(\tau/B\Big)\Big(p\sin\Psi+q\cos\Psi\Big)h_s^{-3}\frac{\partial}{\partial s}\Big)\Big]E_q$$

$$+\Big(\chi^2/2\Big)\sin 2\Psi h_s^{-2}E_p$$

$$+\Big[-2\chi h_s^{-2}\sin\Psi\frac{\partial}{\partial s}+\chi^2\Big(\frac{\tau}{B}\Big)h_s^{-2}\cos\Psi$$

$$+\chi^3\Big(\frac{\tau}{B}\Big)h_s^{-3}\Big(p\sin\Psi+q\cos\Psi\Big)\cos\Psi\Big]E_s=0 \tag{8.69}$$

respectively. Where, (8.68) and (8.69) are the wave equations in the Tang's coordinates system .

How to match the boundary condition?

To match the boundary condition of the helical optical fiber, the transversal partial differential $\left(\frac{\partial^2}{\partial p^2}+\frac{\partial^2}{\partial q}\right)$ should be written in a polar coordinates system as

$$\nabla_t^2=\frac{\partial^2}{\partial p^2}+\frac{\partial^2}{\partial q^2}=\Big[\frac{\partial^2}{\partial p^2}+\frac{\partial}{\rho\partial\rho}+\frac{\partial^2}{\rho^2\partial\theta^2}\Big]/a^2 \tag{8.70}$$

Where

$$\rho=r/a$$

then to match the boundary condition by the perturbation method.

8.4.2 The perturbation method

In the Tang's coordinates system, the wave equations in the helical optical fiber, (8.68) and (8.69), are the coupling wave equations and then it is difficult to find the accuracy solution. However, under the condition of

$$r_0>>a$$

it is able to be solved by the perturbation method, this is because that, under this condition, we also have, from (8.59),

$$\chi a<<1,\qquad \tau a<<1 \tag{8.71}$$

And then if we read (8.68) and (8.69) carefully, we will find that the coupling coefficient is in the order of $(\chi a)^2$ between E_p and E_q and is in the order of χa between E_p and E_s. And then

The zero order wave equations of (8.68) and (8.69) are

$$[\nabla_t^2 + n^2k_0^2 - \beta_0^2]E_{0p} = 0 \tag{8.72}$$

$$[\nabla_t^2 + n^2k_0^2 - \beta_0^2]E_{0q} = 0 \tag{8.73}$$

Therefore, in the Tang's coordinates system, the fields E_p, E_q in a helical optical fiber are able to be expanded to a power series around E_{0p} and E_{0q}, respectively. Where E_{0p} and E_{0q} are the zero order solution of the helical optical fiber and are the solution of the straight optical fiber. This is similar to that in the bending optical fiber except there is a torsion in the helical optical fiber.

In addition, we need to consider another two points:

First, from (8.44), in a bending optical fiber, the first order distortion of the field will results in the second order correction for the propagation constant. And then, it is reasonable to assume that the first order distortion of the field will results in the second order correction for the propagation constant in a helical optical fiber as well.

Secondary, there are two fundamental modes in a bending optical fiber, they are mutually orthogonal polarization waves, in which, the polarization directions are in parallel to $\mathbf{a}_n$ and $\mathbf{a}_b$, respectively. In a helical optical fiber, the torsion τ is not zero, and then the polarization directions of the field, $\mathbf{a}_p$ and $\mathbf{a}_q$, are rotating with a rate of $-\tau$ with respect to $\mathbf{a}_n$ and $\mathbf{a}_b$, respectively.

How to find the propagation constant in a helical optical fiber?

In a helical optical fiber, we can only find the propagation constants of the linear polarization waves along $\mathbf{a}_p$ and the $\mathbf{a}_q$, respectively, at the moment when the $\mathbf{a}_p$ and the $\mathbf{a}_q$ rotated to a position, where the $\mathbf{a}_p$ and the $\mathbf{a}_q$ are coincident with the $\mathbf{a}_n$ and the $\mathbf{a}_b$, respectively. In this case, the propagation constants are denoted by β_n and β_b, respectively, and the fields are denoted by E_p^n and E_q^b, respectively. At the other moment, the $\mathbf{a}_p$ and the $\mathbf{a}_q$ are not coincident with the $\mathbf{a}_n$ and the $\mathbf{a}_b$, respectively. Therefore, the fields at each s have to be decomposed into two polarization states, one is in parallel to the a_n at the coincident-point[3], another one is in parallel to the a_b at the coincident-point, and the propagation constant of the first one is β_n and the second one is β_b.

[3] The coincident-point is such a point where the $\mathbf{a}_p$ and the $\mathbf{a}_q$ are coincident with the $\mathbf{a}_n$ and the $\mathbf{a}_b$, respectively, at the moment when the $\mathbf{a}_p$ and the $\mathbf{a}_q$ rotated to this point.

Select a parameter related to the torsion

Finally, select a parameter related to the torsion, this could be $\varepsilon = 2\beta_0^2 a^2 \chi a$, so that the fields in the helical optical fiber are able to be expanded to

$$E_p^n \exp(-i\beta_n s) = (E_{0p}^n + \varepsilon E_{1p}^n + \cdots) \exp(-i\beta_n s)$$

$$E_q^b \exp(-i\beta_b s) = (E_{0q}^b + \varepsilon E_{1q}^b + \cdots) \exp(-i\beta_b s)$$

with

$$\beta_n^2 = \beta_0^2(1 + \beta_{2n}\varepsilon^2 + \cdots)$$

$$\beta_b^2 = \beta_0^2(1 + \beta_{2b}^2\varepsilon^2 + \cdots)$$

$$\varepsilon = 2\beta_0^2 a^2 \chi a \tag{8.74}$$

Where β_0 is the propagation constant of the straight optical fiber.

The longitudinal components of the fields will be $E_s^n \exp(-i\beta_n s)$ and $E_s^b \exp(-i\beta_b s)$, respectively.

In this case, the transverse electric field in the helical optical fiber is able to be expressed as

$$\mathbf{E}_t = A_p E_p \mathbf{a}_p + A_q E_q \mathbf{a}_q = A_n E_p^n \mathbf{a}_n + A_b E_q^b \mathbf{a}_b \tag{8.75}$$

In which

$$\begin{bmatrix} A_n \\ A_b \end{bmatrix} = \begin{bmatrix} \cos\Psi & -\sin\Psi \\ \sin\Psi & \cos\Psi \end{bmatrix} \begin{bmatrix} A_p \\ A_q \end{bmatrix} \tag{8.76}$$

Where $\mathbf{a}_n$ and $\mathbf{a}_b$ should be realized as the unit vectors at the coincident-point. And, from (8.68) and (8.69), E_p^n and E_q^b satisfy the following wave equations:

$$\begin{aligned} &a^2[\nabla_t^2 + n^2k_0^2 - \beta_n^2 h_s^{-2}]E_p^n = \\ &\quad \chi a\Big[\Big(\cos\theta\frac{\partial}{\partial p} - \sin\theta\frac{\partial}{\rho\partial\rho}\Big)E_p^n h_s^{-1} + 2i\beta_0 a E_s^n h_s^{-2}\Big] + O[(\chi a)^2] \end{aligned} \tag{8.77}$$

$$\begin{aligned} &a^2[\nabla_t^2 + n^2k_0^2 - \beta_b^2 h_s^{-2}]E_q^b = \\ &\quad \chi a\Big[\Big(\cos\theta\frac{\partial}{\partial p} - \sin\theta\frac{\partial}{\rho\partial\rho}\Big)E_q^b h_s^{-1}\Big] + O[(\chi a)^2] \end{aligned} \tag{8.78}$$

Substituting (8.74) into (8.77) and (8.78), considering that the coefficient of each power of ε in both sides of the equality should be equal to each other, yields two set (i.e. E_p^n and E_q^b) equations. In which, the zero order equations have been given in (8.72) and (8.73) already, and

The first order equations are

$$a^2[\nabla_t + n^2k_0^2 - \beta_0^2]E_{1p}^n =$$
$$\rho E_{0p}\cos\theta + \Big(\cos\theta\frac{\partial E_{0p}}{\partial\rho} - \sin\theta\frac{\partial E_{0p}}{\rho\partial\theta} + 2i\beta_0 aE_{0s}\Big)/2\beta_0^2a^2 \tag{8.79}$$

$$a^2[\nabla_t + n^2k_0^2 - \beta_0^2]E_{1q}^b =$$
$$\rho E_{0q}\cos\theta + \Big(\cos\theta\frac{\partial E_{0q}}{\partial\rho} - \sin\theta\frac{\partial E_{0q}}{\rho\partial\theta}\Big)/2\beta_0^2a^2 \tag{8.80}$$

Where E_{0s} is able to be obtained from the Maxwell equations in the Tang's coordinates system.

Now it is the time to find the zero order solution from (8.72) and (8.73) and the first order solution from (8.79) and (8.80), respectively. It is interested to point out that, in Tang's coordinates system, the zero order equations (8.72) and (8.73) in the helical optical fiber is the same as the zero order equations (8.45) and (8.46) in a bending optical fiber in form and the first order equations (8.79) and (8.80) in the helical optical fiber is the same as the first order equations (8.47) and (8.48) in a bending optical fiber in form. The only difference is that the ε in the helical optical fiber is different from the ε in the bending optical fiber.

From the discussion above it is known that, for a helical optical fiber with the refraction index distribution

$$n = \begin{cases} n_1 & (\rho < 1) \\ n_2 & (\rho > 1) \end{cases}$$

the zero solution of (8.72) and (8.73) are similar to (8.49), i.e.

$$E_{0p}^{(p)} = E_{0q}^{(q)} = \begin{cases} J_0(u\rho)/J_0(u) & (\rho < 1) \\ K_0(w\rho)/K_0(w) & (\rho > 1) \end{cases}$$
$$E_{0p}^{(q)} = E_{0q}^{(p)} = 0$$

Where

$$u = a(n_1^2k_0^2 - \beta_0^2)^{1/2}, \qquad w = a(\beta_0^2 - n_2^2k_0^2)^{1/2}$$

And $E_{0p}^{(p)}$ and $E_{0q}^{(q)}$ are the polarization waves in the direction of $\mathbf{a}_p$ and $\mathbf{a}_q$, respectively.

Substituting $E_{0p}^{(p)}$ and $E_{0q}^{(q)}$ into the Maxwell equations in the Tang's coordinates system yields $E_{0s}^{(p)}$ and $E_{0s}^{(q)}$, and then substituting all of $E_{0s}^{(p)}, E_{0s}^{(q)}, E_{0p}^{(p)}$ and $E_{0q}^{(q)}$ into (8.79) and (8.80) yields the first order solution as follows

$$E_{1p}^{n(p)} = \begin{cases} \Big[-\Big(-\dfrac{\rho^2}{4u} + C_1\Big)J_1 + \dfrac{3\rho J_0}{4\beta_0^2a^2}\Big]\cos\theta/J_0(u) & (\rho < 1) \\ \Big[-\Big(-\dfrac{\rho^2}{4w} + C_2\Big)K_1 + \dfrac{3\rho K_0}{4\beta_0^2a^2}\Big]\cos\theta/K_0(w) & (\rho > 1) \end{cases} \tag{8.81}$$

$$E_{1q}^{n(q)} = \begin{cases} \Big[-\Big(-\dfrac{\rho^2}{4u}+C_1\Big)J_1+\dfrac{\rho J_0}{4\beta_0^2 a^2}\Big]\cos\theta/J_0(u) & (\rho<1) \\ \\ \Big[-\Big(-\dfrac{\rho^2}{4w}+C_2\Big)K_1+\dfrac{\rho K_0}{4\beta_0^2 a^2}\Big]\cos\theta/K_0(w) & (\rho>1) \end{cases} \tag{8.82}$$

$$E_{1p}^{n(q)} = E_{1q}^{(p)} = 0$$

In which C_1 and C_2 are able to be determined by the boundary conditions, they are

$$C_1 = \frac{K_2(w)}{4uK_0(w)}, \qquad C_2 = \frac{J_2(u)}{4wJ_0(u)} \tag{8.83}$$

Obviously, the process of finding the first order solution of a helical optical fiber is the same as that of a bending optical fiber since both of them provides the same equations and the same boundary conditions in form. Where the superscripts (p) and (q) denote the polarization waves $HE_{**}^{(p)}$ and $HE_{**}^{(q)}$, respectively.

8.4.3 The bi-refraction in helical optical fiber

As we mentioned before, there are two linear polarization waves E_p^n (along a_p) and E_q^b (along a_q) in a helical optical fiber at the coincident-point where the $\mathbf{a}_p$ and the $\mathbf{a}_q$ are coincident with the $\mathbf{a}_n$ and the $\mathbf{a}_b$, respectively, at the moment when the $\mathbf{a}_p$ and the $\mathbf{a}_q$ rotated to this point. And then the propagation constant β_n of E_p^n is different from β_b of E_q^b. Which is so called the bi-refraction in the helical optical fiber

$$\triangle\beta = \beta_b - \beta_n \tag{8.84}$$

Substituting β_n and β_b from (8.74) into (8.84) reveals that the bi-refraction is mainly determined by the second order correction quantities $\beta_{2n}\varepsilon^2$ and $\beta_{2b}\varepsilon^2$. These second order correction quantities are able to be obtained by means of the Green integral. Namely, pick up a big circle area A_∞ with the radius $r = 100a$ on the transverse cross section of the helical optical fiber in the local coordinates system $O'\rho\phi s$ such that the central of the circle is located at O'. Then taking the integral

$$\int_{A\infty} [E_{0p} \times (8.77) - E_p^n a^2 \times (8.72)]dA$$

and substituting β_n^2 from (8.74) into the integral above, we will find that the part of the integral $\int_{A\infty} [E_{0p}\nabla_t^2 E_p^n - E_p^n \nabla_t^2 E_{0p}]dA \to 0$ due to the fields outside the core of the optical fiber is decaying in exponential law. And then we have

$$a^2\beta_0^2\varepsilon^2\beta_{2n} = \int_{A_\infty}\Big\{a^2\beta_0^2(1-b^2)E_p^n E_{0p} b^{-2}$$

$$-\chi a\Big[\Big(\cos\theta\frac{\partial}{\partial\rho}-\sin\theta\frac{\partial}{\rho\partial\theta}\Big)E_p^nE_{0p}h_s^{-1}$$
$$+2i\beta_0aE_s^nE_{0p}h_s^{-2}\Big]\Big\}dA\Big\{\int_{A_\infty}(E_p^nE_{0p}h_s^{-2})dA\Big\}^{-1}$$

Similarly, we have

$$a^2\beta_0^2\varepsilon^2\beta_{2b}=\int_{A_\infty}\Big\{a^2\beta_0^2(1-h_s^2)E_q^bE_{0q}h_s^{-2}-\chi a\Big[\Big(\cos\theta\frac{\partial}{\partial\rho}$$
$$-\sin\theta\frac{\partial}{\rho\partial\theta}\Big)E_q^bE_{0q}h_s^{-1}\Big]\Big\}dA\Big\{\int_{A_\infty}(E_q^bE_{0p}h_s^{-2})dA\Big\}^{-1}$$

8.4.4 The Properties of the helical optical fiber

To discuss the properties of the helical optical fiber we have to make the transformation of the fields from the Tang's coordinates system pqs to the Serret-Frenet frame coordinates system nbs.

We start from (8.75) and (8.76), in which, the transverse electric field of the helical optical fiber is able to be expressed as

$$E_t=A_pE_p\mathbf{a}_p+A_qE_q\mathbf{a}_q=A_nE_p^n\mathbf{a}_n+A_bE_q^b\mathbf{a}_b$$

Where

$$\begin{bmatrix}A_n\\A_b\end{bmatrix}=\begin{bmatrix}\cos\Psi & -\sin\Psi\\ \sin\Psi & \cos\Psi\end{bmatrix}\begin{bmatrix}A_p\\A_q\end{bmatrix}\tag{8.85}$$

and A_p, A_q denote the field amplitudes in the Tang's coordinates system, A_n, A_b denote the field amplitudes in the Serret-Frenet frame.

Suppose that $\mathbf{a}_p$ and $\mathbf{a}_q$ are coincident with $\mathbf{a}_n$ and $\mathbf{a}_b$, respectively, at a moment, which means that $\Psi=0$ at this moment, and two polarization waves $A_pE_p\mathbf{a}_p$ and $A_qE_q\mathbf{a}_q$ are incident at $s=0$ at this moment. After propagation a small distance $\triangle s$ along s in the Tang's coordinates system, the fields are transformed into the fields in the nb coordinates system. After taking the derivative of the fields with respect to s, the fields are transformed into the fields in the pq coordinates system. Taking the first order approximation of the field and considering the second order correction of the propagation constant, we have

$$\frac{d}{ds}\begin{bmatrix}A_p\\A_q\end{bmatrix}=$$
$$-i\begin{bmatrix}\cos\Psi & \sin\Psi\\ -\sin\Psi & \cos\Psi\end{bmatrix}\begin{bmatrix}\beta_n & 0\\ 0 & \beta_b\end{bmatrix}\begin{bmatrix}\cos\Psi & -\sin\Psi\\ \sin\Psi & \cos\Psi\end{bmatrix}\begin{bmatrix}A_p\\A_q\end{bmatrix}\tag{8.86}$$

Where, the last two matrices in (8.86) has been transformed into $col[A_n,A_b]$ by means of (8.85), after taking the derivative of the fields with respect to s, the fields are transformed into $col[A_p,A_q]$ via the transformation

$$\begin{bmatrix}A_p\\A_q\end{bmatrix}=\begin{bmatrix}\cos\Psi & \sin\Psi\\ -\sin\Psi & \cos\Psi\end{bmatrix}\begin{bmatrix}A_n\\A_b\end{bmatrix}.\tag{8.87}$$

Now taking the derivative of (8.85) with respect to s, considering (8.87) and (8.81) yields [71]

$$\frac{dA_n}{ds} = -i\beta_n A_n + \tau A_b$$
$$\frac{dA_b}{ds} = -i\beta_b A_b - \tau A_n \tag{8.88}$$

or [72]

$$\frac{dA_n}{ds} = -i(\beta_0 - \triangle\beta/2)A_n + \tau A_b$$
$$\frac{dA_b}{ds} = -i(\beta_0 + \triangle\beta/2)A_b - \tau A_n$$

Where $\triangle\beta$ is the total intrinsic bi-refraction of the helical optical fiber and β_0 is the propagation constant of the straight optical fiber.

The solution of (8.88) is able to be obtained by means of the Laplace transformation as

$$A_n(s) = [A_{p0}\cos(\tau s F^{-1/2}) + F^{1/2}(iA_{p0}\triangle\beta/\tau$$
$$+A_{q0})\sin(\tau s F^{-1/2})]\exp(-i\beta_m s) \tag{8.89}$$

$$A_b(s) = [A_{p0}\cos(\tau s F^{-1/2}) - F^{1/2}(iA_{q0}\triangle\beta/\tau$$
$$+A_{p0})\sin(\tau s F^{-1/2})]\exp(-i\beta_m s) \tag{8.90}$$

Where $A_{p0} = A_p(s=0,\ \Psi=0)$, $A_{qo} = A_q(s=0,\ \Psi=0)$, $F = [1 + (\triangle\beta/2\tau)^2]^{-1}$, $\beta_m = (\beta_n + \beta_b)/2$.

If $\triangle\beta << \tau$, (8.89) and (8.90) degenerate to

$$A_n(s) = A_{p0}\cos(\tau s) + A_{q0}\sin(\tau s)$$
$$A_b(s) = A_{q0}\cos(\tau s) - A_{p0}\sin(\tau s) \tag{8.91}$$

That means, if the linear polarization waves A_{p0} and A_{q0} are incident at $s = 0$ (with $\Psi = 0$), they are the qui-polarization waves propagating in the helical optical fiber. Since A_{p0} and A_{q0} are in Tang's coordinates system, the direction of them are rotating with $-\tau$ respect to the Serret-Frenet Frame.

In fact, in $n\,b\,s$ coordinates system, the normal modes in a helical optical fiber are not two mutually orthogonal linear polarization waves but two mutually orthogonal circle polarization waves. In more detail, they are two mutually orthogonal elliptic polarization waves with different propagation constant. This can be explained by

$$\begin{bmatrix} A_n \\ A_b \end{bmatrix} = \begin{bmatrix} \cos\phi & i\sin\phi \\ i\sin\phi & \cos\phi \end{bmatrix} \begin{bmatrix} A_r \\ A_l \end{bmatrix} \tag{8.92}$$

Where A_r is the wave amplitudes of the right-hand rotating circle polarization waves, A_l is the wave amplitudes of the left-hand rotating circle polarization waves. And

$$\phi = \frac{1}{2}\arctan(2\tau/\triangle\beta)$$

Then, from (8.92) and (8.88), we have

$$\frac{d}{ds}A_r = -i(\beta_0 - g)A_r$$
$$\frac{d}{ds}A_l = -i(\beta_0 + g)A_l \tag{8.93}$$

with

$$g = [(\triangle\beta)^2 + (2\tau)^2]^{1/2}$$

(8.93) said that, in $n\,b\,s$ coordinates system, the intrinsic modes in a helical optical fiber are two polarization waves. The propagation constant is $(\beta_0 + g)$ for the right-hand rotating circle polarization wave and is $(\beta_0 - g)$ for the left-hand rotating circle polarization wave, respectively.

Summery

1. To discuss the nature properties of a helical optical fiber, we have to find the fields of the helical optical fiber in a local coordinates system $n\,b\,t$ or $n\,b\,s$. And the unit vector of the local coordinates system $\mathbf{a}_n, \mathbf{a}_b, \mathbf{a}_t$ forms a Serret-Frenet Frame . However, the Serret-Frenet Frame is rotating around $\mathbf{a}_t$ with a rotating rate of τ, where τ is the torsion, this torsion causes the Serret-Frenet Frame becomes a non-orthogonal coordinates system except at the central of the core of the helical optical fiber.

2. To find the fields of the helical optical fiber from the Maxwell equations and the boundary conditions, we need to find an orthogonal coordinates system to express the Maxwell equations and the boundary conditions. Tang proposed a new frame, which is rotating with a rate of $-\tau$ around $\mathbf{a}_t$ with respect to the Serret-Frenet Frame. And then, in the Tang's frame, the net rate of the rotating around the $\mathbf{a}_t$ is zero. Therefore, the Tang's frame becomes an orthogonal coordinates system p, q, s.

3. After we find out the fields of the helical optical fiber in the Tang's Frame, we can make a transformation (8.76) to transform the fields from the Tang's Frame into the Serret-Frenet Frame to obtain the fields in the local coordinates system n, b, t.

8.5 Appendix

Find the solution of (8.47) E_{1x}^x:.

Substituting (8.49) and (8.52) into (8.47), we have

$$a^2[\nabla_t^2 + n^2k_0^2 - \beta_0^2]E_{1x}^x = A\cos\theta[-\rho J_0 + BJ_1] \tag{8.94}$$

where

$$A = 1/J_0(u), \quad B = 3u/(2\beta_0^2a^2)$$

Let

$$E_{1x}^x = R(Z)\cos\theta, \quad Z = u\rho \tag{8.95}$$

Substituting (8.95) into (8.94), we have

$$\left[\frac{\partial^2}{\partial Z^2} + \frac{\partial}{Z\partial Z} + \left(1 - \frac{1}{Z^2}\right)\right]R(Z) = A_1[-\rho J_0 + BJ_1] \tag{8.96}$$

where

$$A_1 = 1/[u^2J_0(u)]$$

Obviously, the homogeneous solution of (8.96) is

$$R_1(Z) = J_1(Z) \tag{8.97}$$

Let the general solution be

$$R(Z) = R_2(Z) + CJ_1(Z) \tag{8.98}$$

Substituting (8.98) into (8.96), we have

$$\left[\frac{\partial^2}{\partial Z^2} + \frac{\partial}{Z\partial Z} + \left(1 - \frac{1}{Z^2}\right)\right]R_2(Z) = A_1(-\rho J_0 + BJ_1) \tag{8.99}$$

Obviously, the solution of $R_2(Z)$ should contain J_1 and J_0. And then we may consider

$$R_2(Z) = R_3J_1(Z) + DJ_0(Z) \tag{8.100}$$

Substituting (8.100) into (8.99), we have

$$R_3'' - \frac{R_3'}{Z} - 2D' = A_1B \tag{8.101}$$

$$2R_3' + D'' + \frac{D'}{Z} - \frac{D}{Z^2} = -A_1\rho \tag{8.102}$$

where, the superscribe (') denotes the derivative with respect to Z. Let

$$R_3 = PZ^2, \quad D = QZ \tag{8.103}$$

Substituting (8.103) into (8.101) and (8.102), we have

$$2P - 2P - 2Q = A_1 B$$

$$4PZ + \frac{Q}{Z} - \frac{Q}{Z} = -A_1 \rho$$

And then we have

$$P = -A_1\rho/(4Z), \quad Q = -A_1 B/2$$

Then, from (8.103), we have

$$\begin{aligned} R_3 &= -A_1\rho Z/4 = -\rho^2/[4uJ_0(u)] \\ D &= -A_1 BZ/2 = -3\rho/(4J_0(u)\,\beta_0^2 a^2) \end{aligned}$$

Substituting R_3 and D into (8.100) and (8.98), we have

$$\begin{aligned} R(Z) &= R_2(Z) + CJ_1(Z) = R_3 J_1(Z) + DJ_0(Z) + CJ_1(Z) \\ &= \Big[\Big(-\frac{\rho^2}{4u} + C\Big)J_1(Z) - \Big(\frac{3\rho}{4\beta_0^2 a^2}\Big)J_0(Z)\Big]\frac{1}{J_0(u)} \end{aligned}$$

Substituting $R(Z)$ into (8.95), we will obtained the first order solution (8.52).

Chapter 9

Mode coupling theory in optical devices

9.1 Introduction

The mode coupling theory still is a powerful tool in the analysis of the light wave technology. For instance, the mode coupling theory has been successfully used to analyze the geometry distortion and the refraction index distortion caused by the process of the manufacture of the optical fiber and the optical fiber cable as well as the installation of the optical fiber cable system [73]. Meanwhile it has been successfully used to analyze the optical multi-waveguides system and the optical anisotropic multi-waveguides system etc.. Those systems are all of the linear systems, therefore, the mode of the system is able to be expressed as the linear superposition of the individual normal mode in a regular waveguide. Substituting these modes into the Maxwell equations, considering the orthogonality of the normal modes, it is not so difficult to obtain the coefficient equations, which consist of the coefficient of each normal mode. These coefficient equations are normally a coupling wave equations and can be expressed as a matrix equation. Then the coefficient of each normal mode is able to be obtained by means of the eigenvalues and the eigenfunctions of this coupling wave equations. This process is able to be used to variant liner coupling equations.

9.2 The mode coupling theory in optical multi-waveguides

The following discussion is based on the study of the coupled modes solutions of multiwaveguide systems by A. Hardy and W. Streifer [74].

9.2.1 The mode coupling equation

Suppose there is a system, which consists of N optical fibers (or N waveguides). The distribution of the dielectric constant is $\varepsilon(x, y)$ for the system and is $\varepsilon^{(m)}(x, y)$ for an individual waveguide. Which means that a perturbation of the dielectric constant in each individual waveguide caused by the rest waveguides is as follows:

$$\Delta\varepsilon^{(m)}(x, y) = \varepsilon(x, y) - \varepsilon^{(m)}(x, y) \qquad (m = 1, 2, \cdots N) \qquad (9.1)$$

Suppose that each individual waveguide can only support the lowest transmission mode and the others are all of the radiation modes. In which, suppose that $E_{t\eta}^{(p)}$ denotes the η mode in p waveguide, and $E_{t\eta}^{(p)}$ obeys the following normalization condition

$$2\hat{z}\cdot\int\int_{-\infty}^{\infty}[\mathbf{E}_{t\mu}^{(p)}\times\mathbf{H}_{t\eta}^{(p)}]dxdy=\delta_{\mu\eta}\qquad(p=1,2,\cdots N)\tag{9.2}$$

where t represent the transverse field.

In this system, each waveguide contains not only the field of this individual waveguide but also the fields from the other individual waveguides. Therefore, the transverse components of any waveguide is able to be expressed by a complete system, which consists of the components from each individual waveguide, i.e.

$$\mathbf{E}_{t1}^{(p)}(x,y)=\sum_{\eta=1}^{\infty}C_{\eta}^{(m,p)}\mathbf{E}_{t\eta}^{(m)}(x,y)\qquad(m,p=1,2,\cdots,N)\tag{9.3}$$

Where

$$C_{\eta}^{(m,p)}=2\hat{z}\cdot\int_{-\infty}^{\infty}\int_{-\infty}^{\infty}[\mathbf{E}_{t1}^{(p)}\times\mathbf{H}_{t\eta}^{(m)}]dxdy\qquad(m,p=1,2,\cdots N)\tag{9.4}$$

is the mode coefficient of each waveguide, and from (9.2), we have

$$C_{\eta}^{(m,m)}=\begin{cases}0 & (\eta\neq 1)\\ 1 & (\eta=1)\end{cases}$$

Then the transverse components of the total electric field and the magnetic field, $\mathbf{E}_t(x,y,z)$ and $\mathbf{H}_t(x,y,z)$, in N waveguides array are able to be expanded to a complete system, which consists of the modes of each individual waveguide as follows

$$\mathbf{E}_t(x,y,z)=\sum_{\eta=1}^{\infty}a_{\eta}^{(p)}(z)\mathbf{E}_{t\eta}^{(p)}(x,y)\qquad(p=1,2,\cdots,N)$$

$$\mathbf{H}_t(x,y,z)=\sum_{\eta=1}^{\infty}a_{\eta}^{(p)}(z)\mathbf{H}_{t\eta}^{(p)}(x,y)\qquad(p=1,2,\cdots N)\tag{9.5}$$

Where the total fields $\mathbf{E},\mathbf{H}$ satisfies

$$\nabla\times\mathbf{E}=-i\omega\mu_0\mathbf{H}$$

$$\nabla\times\mathbf{H}=i\omega\varepsilon\mathbf{E}\tag{9.6}$$

And the dominant mode in each individual waveguide satisfies

$$\nabla\times[\mathbf{E}_{\eta}^{(p)}(x,y)\exp(-i\beta_{\eta}^{(p)}z)]=-i\omega\mu_0\mathbf{H}_{\eta}^{(p)}(x,y)\exp(-i\beta_{\eta}^{(p)}z)$$

$$\nabla \times [\mathbf{H}_\eta^{(p)}(x,y)\exp(-i\beta_\eta^{(p)}z)] = i\omega\varepsilon\mathbf{E}_\eta^{(p)}(x,y)\exp(-i\beta_\eta^{(p)}z)$$
$$(\eta = 1, 2, \cdots \infty;\ p = 1, 2, \cdots, N) \tag{9.7}$$

Substituting (9.5) into (9.6), considering (9.7) and (9.2) we have

$$\frac{da_\mu^{(p)}}{dz} = i\beta_\mu^{(p)} a_\mu^{(p)}(z) + i\sum_{\eta=1}^{\infty} a_\eta^{(p)}(z)\tilde{K}_{\eta\mu}^{(p)}$$
$$(p = 1, 2, \cdots, N;\ \mu = 1, 2, \cdots \infty) \tag{9.8}$$

Where $\beta_\mu^{(p)}$ is the propagation constant of the μ mode in p waveguide, and

$$\tilde{K}_{\eta\mu}^{(p)} = \omega \int_{-\infty}^{\infty}\int_{-\infty}^{\infty} \triangle\varepsilon^{(p)}[\mathbf{E}_{t\eta}^{(p)} \cdot \mathbf{E}_{t\mu}^{(p)} - \frac{\varepsilon^{(p)}}{\varepsilon_0 n^2}\mathbf{E}_{z\eta}^{(p)} \cdot \mathbf{E}_{z\mu}^{(p)}]dxdy$$
$$(p = 1, 2, \cdots N) \tag{9.9}$$

is the mode coupling coefficient between μ mode and η mode in p waveguide. Obviously,

$$\tilde{K}_{\eta\mu}^{(p)} = \tilde{K}_{\mu\eta}^{(p)}$$

Now the total fields of the system, $\mathbf{E}_t(x,y,z)$ and $\mathbf{H}_t(x,y,z)$, are able to be expressed as

$$\mathbf{E}_t(x,y,z) = \sum_{m=1}^{N} u_m(z)\mathbf{E}_{t1}^{(m)}(x,y) + \mathbf{Q}_t(x,y,z)$$
$$\tag{9.10}$$
$$\mathbf{H}_t(x,y,z) = \sum_{m=1}^{N} u_m(z)\mathbf{H}_{t1}^{(m)}(x,y) + \mathbf{R}_t(x,y,z)$$

Where $\mathbf{Q}_t$ and $\mathbf{R}_t$ represent the radiation modes, which are usually expressed by means of an integral form. Making a comparison between (9.5) and (9.10), considering (9.3), we have

$$\mathbf{Q}_t = \sum_{\eta=1}^{\infty}\left[a_\eta^{(p)}(z) - \sum_{m=1}^{N} C_\eta^{(p,m)} u_m(z)\right]\mathbf{E}_{t\eta}^{(p)}$$
$$(p = 1, 2, \cdots, N) \tag{9.11}$$
$$\mathbf{R}_t = \sum_{\eta=1}^{\infty}\left[a_\eta^{(p)}(z) - \sum_{m=1}^{N} C_\eta^{(p,m)} u_m(z)\right]\mathbf{H}_{t\eta}^{(p)}$$

Obviously, $\mathbf{Q}_t$ and $\mathbf{R}_t$ shouldn't contain the lowest mode ($\eta = 1$), namely, (9.11) shouldn't contain

$$a_1^{(p)}(z) = \sum_{m=1}^{\infty} C_1^{(p,m)} u_m(z) \quad (p = 1, 2, \cdots, N) \tag{9.12}$$

where (9.12) is from the term of $\eta = 1$ in $\mathbf{Q}_t$ and $\mathbf{R}_t$, and then (9.11) becomes

$$\mathbf{Q}_t = \sum_{\eta=2}^{\infty} \Big[a_\eta^{(p)}(z) - \sum_{m=1}^{N} C_\eta^{(p,m)} u_m(z) \Big] \mathbf{E}_{t\eta}^{(p)}$$

$$(p = 1, 2, \cdots, N) \qquad (9.13)$$

$$\mathbf{R}_t = \sum_{\eta=2}^{\infty} \Big[a_\eta^{(p)}(z) - \sum_{m=1}^{N} C_\eta^{(p,m)} u_m(z) \Big] \mathbf{H}_{t\eta}^{(p)}$$

Thus, $\mathbf{Q}_t$ and $\mathbf{R}_t$ are orthogonal to the lowest mode ($\eta = 1$) of any individual waveguide.

Substituting (9.12) into (9.8) with $\mu = 1$ we have

$$\begin{aligned} \sum_{m=1}^{N} C_1^{(m,p)} \frac{du_m}{dz} &= i(\beta_1^{(p)} + \tilde{K}_{11}^{(p)}) \sum_{m=1}^{N} C_1^{(p,m)} u_m \\ &+ i \sum_{\eta=2}^{\infty} a_\eta^{(p)} \tilde{K}_{\eta 1}^{(p)} \qquad (p = 1, 2, \cdots, N) \end{aligned} \qquad (9.14)$$

Based on (9.14), adding and then subtracting a term

$$i \sum_{m=1}^{N} \Big[\sum_{\eta=2}^{\infty} C_\eta^{(p,m)} \tilde{K}_{\eta 1}^{(p)} \Big] u_m$$

we have the coupling equation of the multi-waveguides system as follows:

$$\sum_{m=1}^{N} C_{mp} \frac{du_m}{dz} = i \Big\{ \beta^{(p)} \sum_{m=1}^{N} C_{mp} u_m + \sum_{m=1}^{N} \tilde{K}_{pm} u_m + \tilde{Q}_p \Big\}$$

$$(p = 1, 2, \cdots, N) \qquad (9.15)$$

In which

$$C_{pm} = C_1^{(p,m)} = 2\hat{z} \cdot \int_{-\infty}^{\infty} \int_{-\infty}^{\infty} [\mathbf{E}_{t1}^{(m)} \times \mathbf{H}_{t1}^{(p)}] dx dy \qquad (9.16)$$

$$\begin{aligned} \tilde{K}_{pm} &= \sum_{\eta=1}^{\infty} C_\eta^{(p.m)} \tilde{K}(p)_{\eta 1} = \omega \int_{-\infty}^{\infty} \int_{-\infty}^{\infty} \triangle\varepsilon^{(p)} \Big[\mathbf{E}_{t1}^{(p)} \cdot \mathbf{H}_{t1}^{(m)} \\ &- \frac{\varepsilon^{(m)}}{\varepsilon_0 n^2} E_{z1}^{(p)} E_{z1}^{(m)} \Big] dx dy \qquad (p, m = 1, 2, \cdots, N) \end{aligned} \qquad (9.17)$$

$$\tilde{Q}_p = \sum_{\eta=2}^{N} \Big[a_\eta^{(p)} - \sum_{m=1}^{N} C_\eta^{(p,m)} K_{\eta 1}^{(p)} u_m \Big] \tilde{K}_{\eta 1}^{(p)}$$

$$(p, m = 1, 2, \cdots, N) \qquad (9.18)$$

Where use of (9.4), (9.9) and (9.3) have been made to obtain (9.16) and (9.17).

Now come back to the coupling equation of the multi-waveguides system (9.15). In which, $\tilde{Q}_p$ is from Q_t and Q_t is from the radiation modes. Considering that the amplitude of the radiation mode is at least one order lower less that of the dominant mode, and then it can be ignored. In this case, the coupling equation is able to be solved by means of the matrix form.

9.2.2 The solution from the matrix coupling equation

Matrix equation:

Considering that the radiation modes may be ignored and then Q_t may be ignored, therefore, (9.15) is able to be written as the matrix form as

$$C\frac{\partial U}{\partial z} = i(BC + K)U \tag{9.19}$$

Where $U = col[u_1, u_2, \cdots, u_N]$ is an unknown matrix. C is a $N \times N$ matrix with the elements C_{mp}. $B = diag[\beta^{(1)}, \beta^{(2)}, \cdots, \beta^{(N)}]$ is a diagonal matrix with the elements $\beta^{(m)}$. K is a $N \times N$ matrix with the elements $\tilde{K}_{pm}$. $\tilde{K}_{pm}$ $(p \neq m)$ is the coupling coefficient between the dominant mode of the p waveguide and the dominant mode of the m waveguide. And $\tilde{K}_{mm}$ and $\beta^{(m)}$ are related to the correction caused by $\triangle\varepsilon^{(p)}$.

Obviously, to make (9.19) to be solvable, it is required that $det(C) \neq 0$. And then the coupling equation (9.19) becomes

$$\frac{dU}{dz} = i\,M\,U \tag{9.20}$$

Where

$$M = C^{-1}BC + C^{-1}K \tag{9.21}$$

Decoupling:

Now, (9.20) is a coupling equation and is not so easy to be solved. For the sake of simplification, we need to make it decoupling. The way to make it decoupling is to project U on a W space. Namely, for a given U, it is able to find a matrix A anyway such that

$$W = AU \tag{9.22}$$

and then (9.20) is able to be written as

$$\frac{dW}{dz} = iSW \tag{9.23}$$

In which

$$S = AMA^{-1} \tag{9.24}$$

is a diagonal matrix of $N \times N$ with the elements $S_{mm} = \sigma_m$ $(m = 1, 2, \cdots, N)$. And σ_m is the propagation constant of the dominant mode in each individual waveguide.

Solution:

Now (9.23) is a decoupling first order of ordinary differential equation and then the solution is

$$W(z) = R(z)W(0) \tag{9.25}$$

In which $R(z)$ is a diagonal matrix of $N \times N$ with the elements $e^{i\sigma_m z}$ $(m = 1, 2, \cdots, N)$. And $W(0)$ is able to be obtained from (9.22), i.e.

$$W(0) = AU(0) \tag{9.26}$$

Substituting (9.25) and (9.26) into (9.22) gives

$$U(z) = Q(z)U(0) \tag{9.27}$$

with

$$Q(z) = A^{-1}R(z)A \tag{9.28}$$

The eigenvalue and the eigenfunction of M:

From another point of view, the solution U of the matrix equation (9.20) could be find by means of the eigenvalue and the eigenvector of M. Namely, the eigenvector $U^{(\mu)}(0)$ corresponding to the eigenvalue σ_ν of M should be propagating along z direction in form of

$$U^{(\nu)}(z) = \exp(i\sigma_\nu z)U^{(\nu)}(0) \tag{9.29}$$

Then, from (9.22), we have

$$W^{(\nu)} = AU^{(\nu)}(z) = \exp(i\sigma_\nu z)W^{(\nu)}(0) \tag{9.30}$$

Making a comparison between (9.30) and (9.25), we have that

$$\begin{aligned} W^{(1)}(0) &= col[1, 0, 0, \cdots] \\ W^{(2)}(0) &= col[0, 1, 0, \cdots] \\ \vdots & \quad\quad \vdots \end{aligned}$$

Then from the inverse of (9.22)

$$U^{(\nu)}(0) = A^{-1}W^{(\nu)}(0)$$

we have

$$A^{-1} = [U^{(1)}(0), U^{(2)}(0), \cdots, U^{(N)}(0)] \tag{9.31}$$

Where $U^{(\nu)}(0)$ $(\nu = 1, 2, \cdots, N)$ is the νth eigenvector of M and then (9.31) is unique and determinate. Since $U^{(\nu)}(0)$ is independent of z, the matrix A is a constant matrix. And then A is able to be obtained from A^{-1} as

$$A = (A^{-1})^{-1} = \frac{(A^{-1})^*}{det(A^{-1})} \tag{9.32}$$

Where $(A^{-1})^*$ is the adjoint matrix of (A^{-1}).

Substituting A^{-1} and A into (9.27) we will find the solution $U(z)$ of (9.20).

Summery

Now, the process of finding the solution of (9.20) is

(1) Finding the eigenvalue σ_ν and the corresponding eigenvector $U^{(\nu)}(0)$ of M from (9.21), we have

$$R(z) = \begin{bmatrix} e^{i\sigma_1 z} & & & \\ & \ddots & & \\ & & \ddots & \\ & & & e^{i\sigma_N z} \end{bmatrix}$$

(2) Find $A^{-1} = [U^{(1)}(0), U^{(2)}(0), \cdots, U^{(N)}(0)]$ from (9.31).

(3) Find $A = (A^{-1})^*/det(A^{-1})$ from (9.32).

(4) Find $Q(z) = A^{-1}R(z)A$ from (9.28).

(5) Then the solution of (9.20) is $U(z) = Q(z)U(0)$ from (9.27).

9.2.3 Application

Suppose that there are two identical waveguides and the coupling equations between the dominate mode of each individual waveguide are

$$\frac{du_1}{dz} = i\gamma u_1 + iKu_2$$

$$\frac{du_2}{dz} = iKu_1 + i\gamma u_2$$

Where γ is the propagation constant of the individual waveguide, and K is the coupling coefficient between the two waveguides. The coupling equation may be written in matrix form

$$\frac{dU}{dz} \quad iMU$$

where

$$M = \begin{pmatrix} \gamma & K \\ K & \gamma \end{pmatrix}$$

and the eigenvalue σ and the corresponding eigenvector U of M satisfy

$$(\sigma I - M)U = 0$$

Then the eigenvalue equation is

$$det(\sigma I - M) = \begin{vmatrix} \sigma - \gamma & -K \\ -K & \sigma - \gamma \end{vmatrix} = 0$$

and the eigenvalues are

$$\sigma_{1,2} = \gamma \pm K$$

The eigenvector $U^{(1)}$ corresponding to σ_1 satisfies

$$(\sigma_1 I - M)U^{(1)} = 0$$

and then

$$U^{(1)} = \begin{pmatrix} 1 \\ 1 \end{pmatrix}$$

The eigenvector $U^{(2)}$ corresponding to σ_2 satisfies

$$(\sigma_2 I - M)U^{(2)} = 0$$

and then

$$U^{(2)} = \begin{pmatrix} 1 \\ -1 \end{pmatrix}$$

Therefore

$$A^{-1} = (U^{(1)}\, U^{(2)}) = \begin{pmatrix} 1 & 1 \\ 1 & -1 \end{pmatrix}$$

And

$$(A^{-1})^* = \begin{pmatrix} A^{-1}_{11} & A^{-1}_{21} \\ \\ A^{-1}_{12} & A^{-1}_{22} \end{pmatrix} = \begin{pmatrix} -1 & -1 \\ -1 & 1 \end{pmatrix}$$

$$det(A^{-1}) = -1 - 1 = -2$$

$$A = (A^{-1})^{-1} = \frac{(A^{-1})^*}{\det(A^{-1})} = \frac{1}{2}\begin{pmatrix} 1 & 1 \\ 1 & -1 \end{pmatrix} = \frac{1}{2}A^{-1}$$

Additionally,

$$R(z) = \begin{pmatrix} e^{i\sigma_1 z} & 0 \\ 0 & e^{i\sigma_2 z} \end{pmatrix}$$

and

$$Q(z) = AR(z)A^{-1} = e^{i\gamma z}\begin{pmatrix} \cos(Kz) & i\sin(Kz) \\ i\sin(Kz) & \cos(Kz) \end{pmatrix}$$

Now, if the wave with amplitude V_0 is incident to the first waveguide alone, we have

$$U(z) = Q(z)U(0) = e^{i\gamma z}\begin{pmatrix} \cos(Kz) & i\sin(Kz) \\ i\sin(Kz) & \cos(Kz) \end{pmatrix}\begin{pmatrix} V_0 \\ 0 \end{pmatrix}$$

And the algebra form is

$$u_1(z) = V_0 e^{i\gamma z}\cos(Kz)$$
$$u_2(z) = iV_0 e^{i\gamma z}\sin(Kz)$$

This is a well known solution.

Exercise Suppose, there are three symmetry waveguides ray in parallel arrangement with a same distance between two adjacent waveguides. Please write down the mode coupling equations and find the solution by means of the matrix method.

9.3 Optical anisotropic multi-waveguides system

Since the problem to find the analytical solution for the anisotropic multi-waveguides system is rather complicated, we have to make some simplification as follows:

(1) Assume that the radiation mode of each individual waveguide is ignored.

(2) Considering that the optical waveguide is in weak-guiding condition (i.e. the longitudinal component of the field is much smaller than the transverse component of the field in waveguide), and then the variables of the longitudinal components caused by the system are able to be ignored.

In this case, it is much simpler than before and the accuracy is still rather good.

Suppose, there are N-anisotropic optical waveguides in parallel arrangement along z direction. The dielectric constant tensor is $\overline{\overline{\varepsilon}}(x, y)$ for the system and is $\overline{\overline{\varepsilon}}_n(x, y)$ $(n = 1, 2, \cdots, N)$ for each individual waveguide. Then the perturbation of the dielectric constant of each individual waveguide caused by the rest waveguides is

$$\triangle\overline{\overline{\varepsilon}}_n(x, y) = \overline{\overline{\varepsilon}}(x, y) - \overline{\overline{\varepsilon}}_n(x, y) \tag{9.33}$$

Suppose that each individual waveguide can only support one guided mode, $[\mathbf{E}_n(x, y)e^{-i\beta_n z},\ \mathbf{H}_n(x, y)e^{-i\beta_n z},\ n = 1, 2, \cdots, N]$, and the radiation modes can be ignored. In this case, the total field in the multi-waveguides system are

$$\mathbf{E}(x, y, z) = \sum_{n=1}^{N} u_n(z)\mathbf{E}_n(x, y)$$

$$\mathbf{H}(x,y,z) = \sum_{n=1}^{N} u_n(z)\mathbf{H}_n(x,y) \tag{9.34}$$

Where $u_n(z)$ represents the effect of the rest waveguides on nth waveguide.

In general, the transverse components of the total field should be expressed as the sum of the transverse components from each individual waveguide. And then the longitudinal components of the total field should be obtained by substituting the transverse components into the Maxwell equation. However, considering that the longitudinal components are very small under the weak-guiding condition, therefore, (9.34)[1] is able to be used to find the solution in reasonable accuracy.

Now the total fields, $\mathbf{E}(x,y,z)$ and $\mathbf{H}(x,y,z)$, in (9.34) satisfy the Maxwell equations

$$\begin{aligned} \nabla \times \mathbf{E} &= -i\omega\mu\mathbf{H} \\ \nabla \times \mathbf{H} &= i\omega\bar{\bar{\varepsilon}} \cdot \mathbf{E} \end{aligned} \tag{9.35}$$

And the guided mode of each individual waveguide $\mathbf{E}_n$, $\mathbf{H}_n$ satisfies the Maxwell equations

$$\begin{aligned} \nabla \times [\mathbf{E}_n(x,y)e^{-i\beta_n z}] &= -i\omega\mu\mathbf{H}_n(x,y)e^{-i\beta_n z} \\ \nabla \times [\mathbf{H}_n(x,y)e^{-i\beta_n z}] &= i\omega\bar{\bar{\varepsilon}}_n \cdot \mathbf{E}_n(x,y)e^{-i\beta_n z} \end{aligned} \tag{9.36}$$

Substituting (9.34) into (9.35), considering the vector operation of

$$\begin{aligned} \nabla \times \begin{pmatrix} \mathbf{E} \\ \mathbf{H} \end{pmatrix} &= \sum_{n=1}^{N} \nabla \times \begin{pmatrix} u_n e^{i\beta z} & \mathbf{E}_n e^{-i\beta_n z} \\ u_n e^{i\beta z} & \mathbf{H}_n e^{-i\beta_n z} \end{pmatrix} \\ &= \sum_{n=1}^{N} \left\{ u_n e^{i\beta_n z} \nabla \times \begin{pmatrix} \mathbf{E}_n e^{-i\beta_n z} \\ \mathbf{H}_n e^{-i\beta_n z} \end{pmatrix} \right. \\ &\quad \left. + \nabla \left(u_n e^{i\beta_n z} \right) \times \begin{pmatrix} \mathbf{E}_n e^{-i\beta_n z} \\ \mathbf{H}_n e^{-i\beta_n z} \end{pmatrix} \right\} \end{aligned}$$

we have

$$\sum_{n=1}^{N} \nabla u_n \times \mathbf{E}_n = \sum_{n=1}^{N} -i\beta_n u_n \mathbf{z}^0 \times \mathbf{E}_n \tag{9.37}$$

$$\sum_{n=1}^{N} \nabla u_n \times \mathbf{H}_n = \sum_{n=1}^{N} i\omega u_n \triangle\bar{\bar{\varepsilon}}_n \cdot \mathbf{E}_n - \sum_{i=1}^{N} i\,\beta_n u_n \mathbf{z}^0 \times \mathbf{H}_n \tag{9.38}$$

[1] $\mathbf{E}_n(x,y)$ and $\mathbf{H}_n(x,y)$ contain not only the the transverse components but also the longitudinal components in (9.34).

Where $\nabla u_n = \frac{du_n}{dz}\mathbf{z}^0$.

The integral over the transverse cross section s of the system

$$\int\int_s [\mathbf{H}_m \cdot (9.37) - \mathbf{E}_m \cdot (9.38)]dxdy$$

will gives the coupling wave equation as follows:

$$C\frac{dU}{dz} = -i(BC + K)U \tag{9.39}$$

Where $U = col[u_1, u_2, \cdots, u_N]$ is a column matrix, $B = diag[\beta_1, \beta_2, \cdots,$ $\beta_N]$ is a diagonal matrix. And C and K are the $N \times N$ matrixes. In which the elements C_{mn} and K_{mn} are

$$\begin{aligned} C_{mn} &= \int\int \mathbf{z}^0 \cdot (\mathbf{E}_m \times \mathbf{H}_n + \mathbf{E}_n \times H_m)dxdy, \; (m, n = 1, 2, \cdots, N) \\ K_{mn} &= \int\int \omega\mathbf{E}_m \cdot \nabla\bar{\bar{\varepsilon}}_n \cdot \mathbf{E}_n dxdy, \qquad (m, n = 1, 2, \cdots, N) \end{aligned} \tag{9.40}$$

respectively.

We note, that the coupling equation (9.39) is similar to (9.19) except that the element K_{mn} in (9.39) is different from $\tilde{K}_{mp}$ in (9.19) and the difference between them is caused by the simplification condition in (9.34).

Since the matrix equation (9.39) is similar to (9.19), the process to find the solution of (9.39) is the same as before. And then we don't need to repeat.

9.4 Optical modulation multi-waveguides system

The following discussion is based on the study of the anisotropic waveguide modulations by Tian Feng, Wu Yizun and Ye Peida [75].

9.4.1 The mode coupling equation

The electro-optical light intensity modulator and the optical switch are operation based on the mutual coupling of two intrinsic guided modes. For instance, the two intrinsic guided modes could be the TE mode and the TM mode in a single waveguide modulator or two intrinsic guided modes in a double-waveguide modulator, and the later is made of a double-waveguide directional coupler etc.. Therefore, we will discuss an optical waveguide system with two intrinsic guided modes as follows.

Without modulation:

In an optical waveguide system without modulation, the distribution of the dielectric constant tensor is

$$\overline{\overline{\varepsilon}} = \overline{\overline{\varepsilon}}(x, y) \tag{9.41}$$

Under the external modulation:

In an optical waveguide system with modulation, the dielectric constant tensor will be a slight variable function of the coordinates x, y, z and a slow variable function of the time t. And then, the distribution of the dielectric constant tensor for this optical waveguide system becomes

$$\overline{\overline{\varepsilon}}\,(x, y, z; t) = \varepsilon_0 \left[\overline{\overline{\varepsilon}}_r\,(x, y) + \delta\,\overline{\overline{\varepsilon}}\,(x, y, z; t)\right] \tag{9.42}$$

In which, $\varepsilon_0\,\overline{\overline{\varepsilon}}_r = \overline{\overline{\varepsilon}}$ represents the dielectric constant tensor for the optical waveguide system without modulation, and $\varepsilon_0\,\delta\,\overline{\overline{\varepsilon}}$ represents the infinitesimal disturbance of the dielectric constant tensor caused by the external modulation. Note, that (9.42) is obtained under the condition that the z axis is coincident with the principal axis of the crystal in modulator. In this case, the dielectric constant without modulation $\varepsilon_0\,\overline{\overline{\varepsilon}}_r\,(x, y)$ is uniform distribution along the propagation direction (z axis) of the optical wave. That is the true situation.

For the sake of simplification, we don't consider the weak dissipation in the optical waveguide system for the time being. In this case, according to the energy conservation law, it is provable that the tensor $\overline{\overline{\varepsilon}}$ is a Hermitian tensor. Namely $\overline{\overline{\varepsilon}}$ is equal to its own conjugate transpose,

$$\overline{\overline{\varepsilon}} = \overline{\overline{\varepsilon}}^{*} \tag{9.43}$$

Without modulation ($\delta\,\overline{\overline{\varepsilon}} = 0$), the two individual intrinsic guided modes in modulator (or switch) are

$$\left.\begin{aligned} \mathbf{E}_n &= \mathbf{e}_n(x, y)\exp[i(\omega t - \beta_n z)] \\ \mathbf{H}_n &= \mathbf{h}_n(x, y)\exp[i(\omega t - \beta_n z)] \end{aligned}\right\} \quad (n = 1, 2) \tag{9.44}$$

Suppose that the backscattering and the weak loss caused by the radiation mode are ignored. Then under the external modulation, the electric and magnetic fields, **E** and **H**, are able to be expressed by means of the linear superposition of the two individual intrinsic guided modes, i.e.

$$\begin{aligned} \mathbf{E} &= \sum_{n=1}^{2} u_n(z, t)\mathbf{E}_n \\ \mathbf{H} &= \sum_{n=1}^{2} u_n(z, t)\mathbf{H}_n \end{aligned} \tag{9.45}$$

Where the total fields **E** and **H** satisfies the Maxwell equations

$$
\begin{aligned}
\nabla \times \mathbf{E} &= -\mu \frac{\partial \mathbf{H}}{\partial t} \\
\nabla \times \mathbf{H} &= \frac{\partial}{\partial t}(\overline{\overline{\varepsilon}} \cdot \mathbf{E})
\end{aligned}
\tag{9.46}
$$

And the individual intrinsic guided modes satisfies the Maxwell equations

$$
\begin{aligned}
\nabla \times \mathbf{E}_n &= -\mu \frac{\partial \mathbf{H}_n}{\partial t} \\
\nabla \times \mathbf{H}_n &= \frac{\partial}{\partial t}(\varepsilon_0 \overline{\overline{\varepsilon}}_r \cdot \mathbf{E}_n)
\end{aligned}
\tag{9.47}
$$

Substituting (9.45) into (9.46), considering (9.47) and the operator operation (9.67), (9.68) and (9.70) in Appendix, we have

$$
\sum_{n=1}^{2} \nabla u_n \times \mathbf{E}_n = \sum_{n=1}^{2} \Big(-\mu \mathbf{H}_n \frac{\partial u_n}{\partial t} \Big) \tag{9.48}
$$

$$
\sum_{n=1}^{2} \nabla u_n \times \mathbf{H}_n = \sum_{n=1}^{2} \Big(i\omega\varepsilon_0 u_n \, \delta\overline{\overline{\varepsilon}} \cdot \mathbf{E}_n + \overline{\overline{\varepsilon}} \cdot E_n \frac{\partial u_n}{\partial t} \Big) \tag{9.49}
$$

Now considering the transverse cross-section integral

$$
\int\int [\mathbf{H}_m^* \cdot (9.48) - \mathbf{E}_m^* \cdot (9.49)] dxdy
$$

and making use of the general orthogonal normalization condition

$$
\int\int z^0 \cdot (\mathbf{E}_n \times \mathbf{H}_m^* + \mathbf{E}_m^* \times \mathbf{H}_n) dxdy = \begin{cases} 0 & m \neq n \\ 1 & m = n \end{cases} \tag{9.50}
$$

we have

$$
\begin{aligned}
\frac{\partial u_1}{\partial z} + a_{11}\frac{\partial u_1}{\partial t} + a_{12}\frac{\partial u_2}{\partial t} &= -i(K_{11}u_1 + K_{12}u_2) \\
\frac{\partial u_2}{\partial z} + a_{21}\frac{\partial u_1}{\partial t} + a_{22}\frac{\partial u_2}{\partial t} &= -i(K_{12}u_1 + K_{22}u_2)
\end{aligned}
\tag{9.51}
$$

Where

$$
a_{mn} = \int\int (\mathbf{E}_m^* \cdot \overline{\overline{\varepsilon}} \cdot \mathbf{E}_n + \mu \mathbf{H}_m^* \cdot \mathbf{H}_n) dxdy \quad (m, n = 1, 2) \tag{9.52}
$$

$$
K_{mn} = \omega\varepsilon_0 \int\int \mathbf{E}_m^* \cdot \delta\overline{\overline{\varepsilon}} \cdot \mathbf{E}_n dxdy \quad (m, n = 1, 2) \tag{9.53}
$$

Mode coupling equation:

(9.51) is the mode coupling equation for the modulator under the external modulation. This equation is able to be expressed in matrix form as

$$\left(\frac{\partial}{\partial z} + A\frac{\partial}{\partial t}\right)U = -iKU \tag{9.54}$$

Where

$$A(z,t) = \begin{pmatrix} a_{11} & a_{12} \\ a_{21} & a_{22} \end{pmatrix} \tag{9.55}$$

$$K(z,t) = \begin{pmatrix} K_{11} & K_{12} \\ K_{21} & K_{22} \end{pmatrix} \tag{9.56}$$

$$U(z,t) = \begin{pmatrix} u_1(z,t) \\ u_2(z,t) \end{pmatrix} \tag{9.57}$$

9.4.2 The solution of the matrix coupling equation

(9.54) is a two dimensional first order differential coupling equation. Obviously, it is more complicated than the matrix equation (9.19). Therefore, to obtain the analytical solution, we have to make some simplification.

First, from (9.52), a_{mn} depends on the overlap level of the field distribution of two modes. When $m = n$, both $\mathbf{E}_m$ and $\mathbf{E}_n$ represent the same mode, and then $\mathbf{E}_m$ and $\mathbf{E}_n$ are total overlap, and $\mathbf{H}_m$ and $\mathbf{H}_n$ are total overlap as well. Therefore $a_{mm} > a_{mn}$ $(m \neq n)$.

It is provable [75] that

$$a_{mn} \approx \begin{cases} 0, & m \neq n \\ \beta_n/\omega, & m = n \end{cases} \qquad (m,n = 1,2) \tag{9.58}$$

Meanwhile, the propagation constants for the two modes, β_1 and β_2, satisfy $|\beta_1 - \beta_2| << \beta_1$ or β_2. And then it is reasonable to consider that

$$a_{mm} \approx (\beta_1 + \beta_2)/2\omega = a \qquad (m = 1,2) \tag{9.59}$$

Substituting (9.58) and (9.59) into (9.51) yields

$$\begin{aligned} \frac{\partial u_1}{\partial z} + a\frac{\partial u_1}{\partial t} &= -i(K_{11}u_1 + K_{12}u_2) \\ \frac{\partial u_2}{\partial z} + a\frac{\partial u_2}{\partial t} &= -i(K_{21}u_1 + K_{22}u_2) \end{aligned} \tag{9.60}$$

Next, (9.60) are two dimensional first order differential equations. Which may be transformed into an one dimensional differential equations by means of the characteristic curve method. In which, the characteristic curve equation of (9.60) is

$$\frac{dt}{dz} = a \tag{9.61}$$

The characteristic curve family Γ may be obtained by the integral of (9.61) as follows

$$t = az + b$$

Where b is a constant. On the characteristic curve Γ

$$\frac{d}{dz} = \frac{\partial}{\partial z} + \frac{dt}{dz}\frac{\partial}{\partial t}$$

And then, on the characteristic curve, equation (9.60) can be expressed as

$$\begin{aligned} \frac{du_1}{dz} &= -i(K_{11}u_1 + K_{12}u_2) \\ \frac{du_2}{dz} &= -i(K_{21}u_1 + K_{22}U_2) \end{aligned} \tag{9.62}$$

Where all of K_{mn} $(m = 1, 2, n = 1, 2)$ are caused by $\delta\,\overline{\overline{\varepsilon}}$, or in more detail, caused by the field of the external modulation. Therefore, the time and space variation law for all of K_{mn} is the same. Meanwhile, considering that $\overline{\overline{\varepsilon}}$ is a Hermitian tensor, and then the matrix K is also the Hermitian matrix, namely, $K_{12} = K_{12}^*$, and then K_{11}, K_{22} are real functions. Therefore, K can be expressed as

$$K = K(z,t)\begin{pmatrix} d_1 & d_3e^{i(\gamma+\triangle\beta z)} \\ d_3e^{-i(\gamma+\triangle\beta z)} & d_2 \end{pmatrix} = K(z,t)K_0(z) \tag{9.63}$$

Where $K(z,t)$ is the modulation wave function, d_1, d_2, d_3, γ are real numbers, $\triangle\beta = \beta_1 - \beta_2$.

Substituting (9.63) into (9.62), we have, on the characteristic curve,

$$\frac{dU}{dz} = -iK(z,t)K_0(z)U \tag{9.64}$$

This is a first order linear differential equation and is similar to (9.20). Therefore the process of finding the solution is the same as that of (9.20). The analytical solutions are

$$u_1(z,t) = (s_+ - s_-)^{-1}\exp\Big\{S10\Big\}\Big\{S11e^{if(z,t)} - S12e^{-if(z,t)}\Big\} \tag{9.65}$$

$$u_2(z,t) = (s_+ - s_-)^{-1}\exp\Big\{S20\Big\}\Big\{S21e^{-if(z,t)} - S22e^{if(z,t)}\Big\} \tag{9.66}$$

with

$$\begin{aligned} S10 &= 0.5\,i\Big[\triangle\beta z - (d_1 + d_2)\int_0^z K(x, ax + t - az)dx\Big] \\ S11 &= (s_+ - s_-)u_{10} - d_3e^{i(\gamma+\triangle\beta\, z)}\,u_{20} \\ S12 &= (s_- - d_1)u_{10} - d_3e^{i(\gamma+\triangle\beta\, z)}\,u_{20} \\ S20 &= 0.5\,i\Big[\triangle\beta z - (d_1 + d_2)\int_0^z K(x, ax + t - az)dx\Big] \end{aligned}$$

$$S21 = d_3 e^{-i(\gamma+\Delta\beta z)}\, u_{10} + (s_+ - d_1)u_{20}$$
$$S22 = d_3 e^{-i(\gamma+\Delta\beta z)}\, u_{10} + (s_- - d_1)u_{20}$$

Where

$$s_\pm = \frac{1}{2}\left[d_1 + d_2 \pm \sqrt{(d_1-d_2)^2 + 4d_3^2}\right]$$

$$f(z,t) = \frac{1}{2}\int_0^z \Big\{\Big[\Big(s_+ - s_-\Big)^{-1}\Big(d_1 - d_2\Big)\Delta\beta + \Big(s_+ - s_-\Big)K\Big(x, ax + t - az\Big)\Big]^2 + \Big[2d_3\Delta\beta/(s_+ - s_-)\Big]^2\Big\}^{1/2} dx$$

The initial condition is

$$U(0,0) = col[u_{10}, u_{20}]$$

Now

(1) If the initial condition $U(0,0)$ and the structure parameters of the modulator are given, then it is able to find $K_{mn}(z,t)$ in (9.63).

(2) And then it is able to find the modulation properties equations (9.65) and (9.66).

This analytical approach is able to be used to analyze and design an optical fiber modulator, an optical waveguide modulator, a M-Z intensive modulator, an optical modulator by directional coupler and so on.

Summary:

1. The start point is that, under the external modulation, the dielectric constant tensor of the optical modulator becomes

$$\overline{\overline{\varepsilon}}(x,y,z:t) = \varepsilon_0[\overline{\overline{\varepsilon}}_r(x,y) + \delta\,\overline{\overline{\varepsilon}}(x,y,z:t)]$$

(9.42). Where

$\varepsilon_0\overline{\overline{\varepsilon}}_r(x,y)$ — is the dielectric constant tensor of the modulator without modulation.

$\varepsilon_0\delta\,\overline{\overline{\varepsilon}}(x,y,z:t)$ — is the infinitesimal disturbance of the dielectric constant tensor caused by the external modulation, which is related to the structure parameters of the modulator and the modulation signal.

2. $\varepsilon_0\overline{\overline{\varepsilon}}$ satisfies

$$\varepsilon_0\overline{\overline{\varepsilon}} = \varepsilon_0\overline{\overline{\varepsilon}}^*$$

(9.43) if the dissipation in the optical waveguide system is ignored.

3. The total electromagnetic field **E**, **H** are able to be expressed by the linear

superposition of the two individual intrinsic guided modes $\mathbf{E}_n$, $\mathbf{H}_n$, namely

$$\mathbf{E} = \sum_{n=1}^{2} u_n(z,t)\mathbf{E}_n$$
$$\mathbf{H} = \sum_{n=1}^{2} u_n(z,t)\mathbf{H}_n$$

(9.45) if the backscattering and the radiation mode are ignored. Where $\mathbf{E}_n$, $\mathbf{H}_n$ $(n = 1,2)$ are the individual intrinsic guided modes of the modulator without modulation. And $u_n(z,t)$ $(n = 1,2)$ are the external modulation signal.

4. Now, the total field $\mathbf{E}$ and $\mathbf{H}$ have to satisfy the Maxwell equations (9.46), and the individual intrinsic guided modes $\mathbf{E}_n$ and $\mathbf{H}_n$ have to satisfy the Maxwell equations (9.47) and the general normalization condition (9.50). All of those lead to the mode coupling equations for the modulator with external modulation as follows,

$$\frac{\partial u_1}{\partial z} + a_{11}\frac{\partial u_1}{\partial z} + a_{12}\frac{\partial u_2}{\partial t} = -i(K_{11}u_1 + K_{12}u_2)$$

$$\frac{\partial u_2}{\partial z} + a_{21}\frac{\partial u_1}{\partial z} + a_{22}\frac{\partial u_2}{\partial t} = -i(K_{21}u_1 + K_{22}u_2)$$

(9.51). These mode coupling equations can be written in matrix form

$$\left(\frac{\partial}{\partial z} + A\frac{\partial}{\partial t}\right) = -iKU$$

(9.54)

5. It is provable that $a_{mm} >> a_{mn}$ $(m \neq n)$ and the propagation constants of two modes in modulator satisfies $|\beta_1 - \beta_2| << \beta_1$ or β_2 and then (9.51) becomes

$$\frac{\partial u_1}{\partial z} + a\frac{\partial u_1}{\partial t} = -i(K_{11}u_1 + K_{12}u_2)$$
$$\frac{\partial u_2}{\partial z} + a\frac{\partial u_2}{\partial t} = -i(K_{21}u_1 + K_{22}u_2$$

(9.60). Where $a = a_{mm} = (\beta_1 + \beta_2)/2\omega$.

6. The coupling equations (9.60) may be transformed into

$$\frac{du_1}{dz} = -i(K_{11}u_1 + K_{12}u_2)$$
$$\frac{du_2}{dz} = -i(K_{21}u_1 + K_{22}u_2)$$

(9.62) by means of the characteristic curve method and the matrix form of (9.62)is

$$\frac{dU}{dz} = -iK(z,t)K_0(z)U$$

(9.64). This is similar to (9.20) and the process of finding the solution of (9.64) is

9.5 Appendix

[Operator operation]

From (9.45), (9.46) and (9.47)

$$\begin{aligned}
\nabla\times\begin{pmatrix}\mathbf{E}\\ \mathbf{H}\end{pmatrix} &= \sum_{n=1}^{2}\nabla\times\begin{pmatrix}u_n\mathbf{E}_n\\ u_n\mathbf{H}_n\end{pmatrix}\\
&= \sum_{n=1}^{2}\left\{u_n\nabla\times\begin{pmatrix}\mathbf{E}_n\\ \mathbf{H}_n\end{pmatrix}+\nabla u_n\times\begin{pmatrix}\mathbf{E}_n\\ \mathbf{H}_n\end{pmatrix}\right\} \qquad (9.67)\\
\frac{\partial\mathbf{H}}{\partial t} &= \sum_{n=1}^{2}\frac{\partial}{\partial t}(u_n\mathbf{H}_n)=\sum_{n=1}^{2}\left(u_n\frac{\partial\mathbf{H}_n}{\partial t}+\frac{\partial u_n}{\partial t}\mathbf{H}_n\right) \qquad (9.68)
\end{aligned}$$

$$\begin{aligned}
\frac{\partial}{\partial t}(\overline{\overline{\varepsilon}}\cdot\mathbf{E}) &= \sum_{n=1}^{2}\frac{\partial}{\partial t}(\overline{\overline{\varepsilon}}\cdot u_n\mathbf{E}_n)\\
&= \sum_{n=1}^{2}\left[u_n\frac{\partial}{\partial t}(\overline{\overline{\varepsilon}}\cdot\mathbf{E}_n)+\frac{\partial u_n}{\partial t}\overline{\overline{\varepsilon}}\cdot\mathbf{E}_n\right]\\
&= \sum_{n=1}^{2}\left[u_n\frac{\partial}{\partial t}(\varepsilon_0\overline{\overline{\varepsilon}}_r\cdot\mathbf{E}_n)+u_n\varepsilon_0\,\delta\overline{\overline{\varepsilon}}\cdot\frac{\partial\mathbf{E}_n}{\partial t}\right.\\
&\quad\left.+u_n\varepsilon_0\frac{\partial(\delta\overline{\overline{\varepsilon}})}{\partial t}\cdot\mathbf{E}_n+\frac{\partial u_n}{\partial t}\overline{\overline{\varepsilon}}\cdot\mathbf{E}_n\right]\\
&= \sum_{n=1}^{2}\left[u_n\frac{\partial}{\partial t}(\varepsilon_0\overline{\overline{\varepsilon}}_r\cdot\mathbf{E}_n)+u_n\varepsilon_0\left(i\omega\,\delta\overline{\overline{\varepsilon}}+\frac{\partial(\delta\overline{\overline{\varepsilon}})}{\partial t}\right)\cdot\mathbf{E}_n\right.\\
&\quad\left.+\frac{\partial u_n}{\partial t}\overline{\overline{\varepsilon}}\cdot\mathbf{E}_n\right] \qquad (9.69)
\end{aligned}$$

or

$$\frac{\partial}{\partial t}(\overline{\overline{\varepsilon}}\cdot\mathbf{E})=\sum_{n=1}^{2}\left[u_n\frac{\partial}{\partial t}(\varepsilon_0\overline{\overline{\varepsilon}}_r\cdot\mathbf{E}_n)+i\omega\varepsilon_0u_n\,\delta\overline{\overline{\varepsilon}}\cdot\mathbf{E}_n+\frac{\partial u_n}{\partial t}\overline{\overline{\varepsilon}}\cdot\mathbf{E}_n\right] \qquad (9.70)$$

Where, the second term $\dfrac{\partial(\delta\overline{\overline{\varepsilon}})}{\partial t}$ in the second brackets of (9.69) has been ignored, this is because that $\delta\overline{\overline{\varepsilon}}$ is caused by the external modulation and the variation rate of $\delta\overline{\overline{\varepsilon}}$ is proportional to the modulation frequency, considering that the modulation frequency is much lower that the optical frequency ω, this term could be ignored.

Index

Bibliography

[1] Lain Han Xiong and David C. Chang, "Scattering of a surface-wave mode incident obliquely onto a small dielectric step". National Radio Science Meeting, Commission B Session 2, B2-5, 1510, 5-7 January 1983, Sponsored by USNC/URSI, University of Colorado, Bouder, Colorado, U.S.A.

[2] Han-xiong Lian and David C. Cheng, "Theory of open dielectric waveguide and optical directional coupler with very flat response" Presented on the SINO-JAPANES Joint Meeting on Optical fiber Science and electromagnetic theory, 16-19 May 1985, Proceding pp.343-438, Begin, China.

[3] Han-xiong Lian and David C. Cheng, "The theory and conputer adied design of an optical directional coupler". Presented on the SINO-BRITISH Joint Meeting on optical fiber communications, 9-11 May 1986, Proceeding pp.307-312, Bejing, China.

[4] S.A. Boothroyd, J. Chrostowski, Han-xiong Lian, AS. Rawicz and J. Deen, "Second Harmonic Modulation in Potasium Niobate." Microwave and Optical Tehcnology Letters, Nov. 1991, pp.521-524.

[5] Han-xiong Lian and S. Stapleton, "The TM Mode in the Self-focusing Dielectric Plannar Waveguide." Microwave and Optical Technology Letters,Sept. 1992, pp.496-503.

[6] Han-xiong Lian, "Mathematic Method in Theory of Electromagnetic Field" Beijing University Publishing House, Beijing, 1990,2.

[7] Dun-ren Guo, "Shu Xue Wu Li Fan Fa",Renman Education Publishing House, 1978,9.

[8] Milton Abramowitz and Irene A. Stegun,"Hand Book of Mathematical Functions", Dover Publications, INC, New York, 1970, pp.256-257.

[9] J. Meixner, "The Behavior of Electromagnetic Fields at Edges" Inst. Math. Sci. Res. Rept. EM-72, New York University, New York, N.Y. Dec. 1954.

[10] Kun-miao Liang,"Mathematical and Physical Method",Renmin Education Publishing House, 1978,7. pp.379.

[11] B.A. Fukesi and B.B.Shabate, "Fu Bian Han Shu Ji Qi Ying Yong", Translated from Russian into Chinese by Gen-rong Zhao, Shangwu Publishing House 1955.

[12] R. Mittra and S. W. Lee, "Analytical Techniques in the Theory of Guided Wave", The Macmillan company, New York, 1971.

[13] Han-xiong Lian, Jia-min Chang and David C. Chang "A rigorous solution of the propagation constant for the optical modulation dielectric waveguide obtained by Winer-Hopf Techniques", Proceeding of Sino-Japanse Joint Meeting on Optical Fiber Science and Electromagnetic Theory, May 12-14, 1987, pp.71-76.

[14] C. Titchmarch "Theory of Functions", 2nd. New York, Oxford University Press. 1939, pp.99.

[15] David C. Chang and Edward F. Kuester, "Radio Science", Vol.16, No.1, 1981,pp.1-13.

[16] Kazuo Aoki, T. Miyazaki, K. Uchida and Y. Shimada, "On junction of two semi-infinty dielectric slabs", ≪ Radio Science ≫, Vol. 17, No. 1, 1982, pp.11-19.

[17] Kazunori Uchida and K. Aoki, "Scattering of Surface waves transverse discontinuities in symmetriical three-layer dielectric waveguide", IEEE Trans. of MTT, MTT-32, No. 1, 1984, pp.11-19.

[18] Kuniaki Yoshidomi, Kuzuo Aoki, and Kazuori Uchida, "Diffraction of an electromagnetic plane wave by a slit ina lossy dielectric slab", Electronis and communications in Japan, Vol. 66-B, No.8,1983.

[19] K. Uchida, T. Matsunaga and K. Aoki, "Scattering of a plane wave by a lossy dielectric rectangular cylinder", Dian Zi Tong Xin Xue Hui Lun Wen Zi J65-B, No.11,1982,pp.1417-1424.

[20] Eduard Prugovechi Quantum mechanics in Hilbert Space 2nd Edit, ISBN13:9780486453279, ISBN10:0486453278.

[21] Gustafson, Stephen J., Sigal and Issreal Michael, "Mathematics Concepts of Quantum Mechanics", 1st ed. 2003, Enlarged 2nd printing, 2006, XI, 286p., ISDN: 978-3-540-44160-1.

[22] From Wikipedia, the free encyclopedia.

[23] O. A. Treyakov, "New approch to electromagnetic field theory: electrodynamics without complex amplitudes", Proceeding of Sino-British Joint Meeting on Optical Fiber Communications, May 9-11, 1986, pp.333-338, Begin, China.

[24] M. S. Antyufeeva and O. A. Tretyakov, "Electromagnetic Field in a Resonant Cavity Filled with a Non-Stationary Dielectric Material". Telecommunications and Radio Engineering, 2002, Volume57, Issue 4, 101 pages, ISSN for PRINT: 0040-2508.

[25] Tamir, T., "Radio loss of lateral waves in forest enviromenrs", Radio Sci., Vol. 4, pp. 307-318, April, 1969.

[26] Tamir, T., "On radio wave propagation in Forest enviroment", IEEE Trans. Ant. and Prop. Vol. AP-15, No. 6, pp.806-817, November.

[27] Tamir, T., " Radio wave propagation in forest enviroment", IEEE Trans. Ant. and Prop., Vol. AP-25, No. 4, pp. 471-477, July, 1977.

[28] R. H. Lang, A. Schneider, S. Seker, and F. J. Altman, "UHF radiowave propagation through forests", CyberCOM Technical Report CTR-108-01, September, 1982, (Distribution Statement: Aproved for Public Release; Distribution Unlimited), which can be found from Internet:[PDF] RESEARCH AND DEVELOPMENT TECHNICAL REPORT CECOM-81-0136-4.

[29] Han-xiong Lian and Leonard Lewin, "The HF effective permeability and permitivity of a forest", ACTA Electronics Sinica, Vol. 15, No.1. Jan. 1987, pp.10-15, China.

[30] Han-xiong Lian and Leonard Lewin, "UHF Lateral wave loss in forests modeled by four-layered media with two anasotropic slabs", ACTA Electronics Sinica, Vol. 14, No.5. Sept. 1986, pp.12-20, China.

[31] Lewin, Leonard, "Theory of Waveguide", 1975.

[32] J. R. Wait, "Scattering of a Plane Wave from Right Circular Dielectric Cylinder at Oblique Incidence" Can. J. Phys. Vol. 33, pp. 189-195 (1955).

[33] G. T. Ruck, D. E. Barrick, W. D. Stuart, and C. K. Krichbaum, "Radar Cross Section Handbook", Vols 1 and 2 , New Yory: Pleum (1970).

[34] Wait, James R., "Electromagnetic waves in Stratified Media", New York: Copyright @ 1962 and 1970, Pergamon Press Ltd.

[35] T. Tamir, "On radio wave Propagation in Forest Environment," IEEE Trans. Antanas and Propagation, Vol. AP-15, No. 6, pp. 806-817, November, 1967.

[36] Vogler, L. E., and J. L. Noble (1963), "Curves of ground proximity loss for dipole antenas," Tech. Note 175, National Bureau od Standards, Boulder, Colorado,

[37] Rogers, D. A., A. J. Giarola, et al., "The radio loss in forests using a model with four layer media," *Radio Science*, Vol. 18, No. 15, pp. 691-695, Sept,-Oct. 1983.

[38] Tosiya Taniuti and Katsunobu Nishihara, "Nonlinear Waves", Pitman Publishing Limited, 1977.

[39] Lax, Peter, "Integrals of nonlinear equations of evoluation and solitary waves", Comm. Pure Applied Math. 21: pp. 467-490, 1968, doi: 10.1002/cpa.3160210503.

[40] Peter Lax and R.S. Phillips, Scattering Theory for Automorphic Functions, (1976) Princeton University Press.

[41] A Levy Yeyati, A Martin-Rodero and J C Cuevas, "The phase-dependent linear conductance of a superconducting quantum point contact", J. Phys.: Condens. Matter 8 pp.449-456, 1996.

[42] Han-xiong Lian and Jia-min Zhang, "The nonlinearity in single mode fiber and the solution of thetransist wave equation" Communications in China,No.1 1989, pp.1-7.

[43] Han-xiong Lian and Shawn Stapleton, "The TM mode in the Self-focusing Dielectric Planar Waveguide", Microwave and Optical Technology Letters, Vol. 5, No. 10, September, 1992, pp.496-503.

[44] S. A. Boothroyd, J. Chrostowski, Han-xiong Lian, Andrew Rawicz, and Jamel Deen, "Second Harmonic Modulator in Potassium Niobate" Microwave and Optical Technology Letters, Vol. 4, No. 12, November 1991, pp.521-524.

[45] G. I. Stegeman, E. M. Wright, N. Finlayson, R. Zanoni, and C. T. Seaton, "Third Order Nonlinear Integrated Optics," J. Lightwave Technology, Vol. 6, June, 1988, pp.953-970.

[46] Marie Fontaine, "Scaling Rules for Nonlinear Thin Film Optical Waveguides," Appl. Opt., Vol. 29, No. 27, 1990, pp.3891-3899.

[47] Marie Fontaine, "Universal Dispersion and Power Curves for TM wave Propagating in Slab Waveguide with Nonlinear Self-Focusing Substrate," J. Appl., Phys., Vol. 69, No.6, 1991, pp.2826-2834.

[48] V. E. Wood, E. D. Evan, and R. P. Kenan, "Soluble Satrurable Refrative-Index Nonlinearity Model," Opt. Commun., Vol. 69, 1988, pp.156-160.

[49] K. Hayata, and M. Koshiba, "Full Vectorial Analysis of Nonlinear-OPtical Waveguides," J.oc. Am. B, Vol. 5, 1988, pp. 2494-2501.

[50] G. I. Stegeman, and C. Seaton, "Nonlinear Wave Guided by Thin Films," Appl. Phys. Letter, Vol. 44, 1884, pp.830-832.

[51] A. Boadman and P. Egan, "Optically Nonlinear Waves in Thin Films," IEEE J. Quantumn, Vol. QE.-22, 1986, pp.139-324.

[52] P. M. Lambkin and K. A. Shore, ""Defocusing Nonlinearity," IEEE J. Quantum Electron., Vol. QE-24, 1988, pp.2946-2051.

[53] G. I. Stegeman, J. Ariyasu, C. T. P. Shen, and J. V. Moloney, "Nonlinear Thin-Film Guided Waves in Non-Kerr Media," Appl. Phys. Lett., Vol. 47, 1985, pp. 1245-1256.

[54] U. Langbein, F. Lederer, and H.-E. Ponath, "Generalized Dispersion Relations for Nonlinear Slab-Guided Waves," Opt. Commun., Vol. 53, 1985, pp. 417-420.

[55] U. langbein, F. Lederer, T. Peschel, and H. E. Ponath, "Nonlinear Duided waves in Saturable Nonlinear Media, "Opt. Lett., Vol. 10, 1985, pp.571-573.

[56] G. I. Stegeman, E. M. Wright, C. T. Seaton,.J. V. Moloney, Tsea-Pyng Shen, A. A. Maradudin, and R. F. Wallis, "Nonlinear Slab-Guided Waves in Non-Kerr Media." IEEE J. Quantumn Eletron., Vol. QE-22, 1986, pp. 977-983.

[57] S. Chelkowski and J. Chrostowski, "Scaling Rules for Slab Waveguides with Nonlinear Substrate, "Appl. Opt., Vol. 26, 1987, pp. 3681-3686.

[58] Sang-Yung Shin, Ewan M. Wright, and George I. Stegeman, "Nonlinear TE Waves of Coupled Waveguids Bounded by Nolinear Media," J. Lightwave Technol., Vol. 6, June 1988, pp. 977-983.

[59] G. I. Stegeman, E. M. Wright, N. Finlayson, R. Zanoni, and C. T. Seaton, "Third Order Nonlinear Integrated Optics," J. Lightwave Technol., Vol. 6, No. 6, June 1988, pp. 953-970.

[60] G. I. tegeman, "Guided Waves Approches to Optical Bistablity," IEEE. J. Quantum Electron., Vol. QE-18, 1982, pp.119-1618.

[61] Michael Cada and Jone D. Begin, "An analysis of a planar Optical Directional Coupler With Lossless Kerr-Like Coupling Medium," IEEE J. Quantum Electron., Vol. QE-26, No. 2, 1990, pp. 361-371.

[62] S. Trillo, S. Wabnitz, and G. I. Stegeman, "Nonlinear Codirectional Guided wave Mode Conversion in Grating Structures," J. Lightwave Technol., Nol. 6, No. 6, 1988, pp. 971-976.

[63] Han-xiong Lian and Jia men Zhang, "A novel General Qunersi-steady-state solution of the simulated Raimann Scattering in the optical fiber", Sine-Japenese electromagnetic field theory and optical fiber proceding, pp.34-39,1987,5, Nan-Jing, China.

[64] Jia-min Zhang and Han-xiong Lian, "The theoretical analysis of the CW optical fiber oscillator", Communications in China, No.1, 1990.

[65] Han xiong Lian and Jia-min Zhang, " The nonlinearity in single mode fiber and the transient solution of the transient coupling wave equations".nal of China Institute of Communications, Vol. 10, N0. 1, Jan. 1989, pp. 1-7. China.

[66] Jia-min Zhang and Han-xiong Lian,"The analysis of transient process of the stimulated scattering in single mode fiber" Journal of Beijing University of posts and telecommunications, No.1, 1989, pp.35-41.

[67] Wei Jian Jiang and Pei Da Ye, "A novel method of solving SBS equations in coherent system". Sine-Japanese electromagnetic field theory and optical fiber science meetng proceding pp.297-302, 1987,5. Nan Jing.

[68] Lian-xiang Wang, De-zhi Fang, et al. "Mathematic Hand Book" People's Education Publicing House, Beijing. 1979, pp.763.

[69] Kato, Y., prog. "Theory phys", Suppl. No.55, 1974, pp.247.

[70] L. Lewin, David, C. Chang and E. F. Kuester, "Electromagnetic wave and curved structures", Peter Peregrinus LTD., 1977.

[71] Xi-sheng Fang and Zong-Qi Lin, "Field in Single-Mode Hellically-wound Optical Fibers", IEEE MTT, Vol.MTT-33, No.11, Nov.1985, pp.1150-1154.

[72] Jing-ren Qian, "Coupled-Mode Theory for Helical Fibers", IEE Proceeding, Vol.135, Pt.J. No.2, April 1998, pp.175-182.

[73] Ye, Paida, "Optical fiber theory" Zhi-Shi Publishing House, Beijing, 1985.6. pp.42-71.

[74] A. Hardy and W. Streifer, "Coupled modes solutions of multiwaveguide systems", IEEE J. Quantum Electron, Vol. QE-22, 1986 pp.528-534.

[75] Tian Feng, Wu Yizun and Ye Peida, "Improved Coupled Mode Theory for Anisotropic Waveguide modulations", IEEE J. Quantum Electronics, Vol. QE-24, pp.531-536, 1988.